ENABLING IP ROUTING

WITH CISCO ROUTERS

D1089388

LIMITED WARRANTY AND DISCLAIMER OF LIABILITY

THE CD-ROM THAT ACCOMPANIES THE BOOK MAY BE USED ON A SINGLE PC ONLY. THE LICENSE DOES NOT PERMIT THE USE ON A NETWORK (OF ANY KIND). YOU FURTHER AGREE THAT THIS LICENSE GRANTS PERMISSION TO USE THE PRODUCTS CONTAINED HEREIN, BUT DOES NOT GIVE YOU RIGHT OF OWNERSHIP TO ANY OF THE CONTENT OR PRODUCT CONTAINED ON THIS CD-ROM. USE OF THIRD-PARTY SOFTWARE CONTAINED ON THIS CD-ROM IS LIMITED TO AND SUBJECT TO LICENSING TERMS FOR THE RESPECTIVE PRODUCTS.

CHARLES RIVER MEDIA, INC. ("CRM") AND/OR ANYONE WHO HAS BEEN IN-VOLVED IN THE WRITING, CREATION, OR PRODUCTION OF THE ACCOMPANY-ING CODE ("THE SOFTWARE") OR THE THIRD-PARTY PRODUCTS CONTAINED ON THE CD-ROM OR TEXTUAL MATERIAL IN THE BOOK, CANNOT AND DO NOT WARRANT THE PERFORMANCE OR RESULTS THAT MAY BE OBTAINED BY USING THE SOFTWARE OR CONTENTS OF THE BOOK. THE AUTHOR AND PUBLISHER HAVE USED THEIR BEST EFFORTS TO ENSURE THE ACCURACY AND FUNCTION-ALITY OF THE TEXTUAL MATERIAL AND PROGRAMS CONTAINED HEREIN. WE HOWEVER, MAKE NO WARRANTY OF ANY KIND, EXPRESS OR IMPLIED, REGARDING THE PERFORMANCE OF THESE PROGRAMS OR CONTENTS. THE SOFTWARE IS SOLD "AS IS" WITHOUT WARRANTY (EXCEPT FOR DEFECTIVE MATERIALS USED IN MANUFACTURING THE DISK OR DUE TO FAULTY WORKMANSHIP).

THE AUTHOR, THE PUBLISHER, DEVELOPERS OF THIRD-PARTY SOFTWARE, AND ANYONE INVOLVED IN THE PRODUCTION AND MANUFACTURING OF THIS WORK SHALL NOT BE LIABLE FOR DAMAGES OF ANY KIND ARISING OUT OF THE USE OF (OR THE INABILITY TO USE) THE PROGRAMS, SOURCE CODE, OR TEXTUAL MATERIAL CONTAINED IN THIS PUBLICATION. THIS INCLUDES, BUT IS NOT LIMITED TO, LOSS OF REVENUE OR PROFIT, OR OTHER INCIDENTAL OR CONSEQUENTIAL DAMAGES ARISING OUT OF THE USE OF THE PRODUCT.

THE SOLE REMEDY IN THE EVENT OF A CLAIM OF ANY KIND IS EXPRESSLY LIMITED TO REPLACEMENT OF THE BOOK AND/OR CD-ROM, AND ONLY AT THE DISCRETION OF CRM.

THE USE OF "IMPLIED WARRANTY" AND CERTAIN "EXCLUSIONS" VARIES FROM STATE TO STATE, AND MAY NOT APPLY TO THE PURCHASER OF THIS PRODUCT.

ENABLING IP ROUTING WITH CISCO ROUTERS

R. DAS

K. CHAKRABARTY

CHARLES RIVER MEDIA, INC.
Hingham, Massachusetts

Acquisitions Editor: James Walsh
Cover Design: The Printed Image

CHARLES RIVER MEDIA, INC.
10 Downer Avenue
Hingham, Massachusetts 02043
781-740-0400
781-740-8816 (FAX)
info@charlesriver.com
www.charlesriver.com

This book is printed on acid-free paper.

Rajarshi Das and Koel Chakrabarty. *Enabling IP Routing with Cisco Routers.*
ISBN: 1-58450-335-1

Library of Congress Cataloging-in-Publication Data
Das, Rajarshi.
  Enabling IP routing with Cisco routers / Rajarshi Das and Koel Chakrabarty.
     p. cm.
  ISBN 1-58450-335-1 (pbk. with cd-rom : alk. paper)
  1. Routers (Computer networks) 2. TCP/IP (Computer network protocol) I. Chakrabarty, Koel. II. Title.
  TK5105.543.D35 2004
  004.6'2—dc22
                                2004003394
Printed in the United States of America
04 7 6 5 4 3 2 First Edition

CHARLES RIVER MEDIA titles are available for site license or bulk purchase by institutions, user groups, corporations, etc. For additional information, please contact the Special Sales Department at 781-740-0400.

Contents

v

Part

I

IP Routing

1   Introduction to Routing

EVOLUTION OF NETWORKING

A network is a group of hosts connected to each other, using different physical media. In the 1960s and 1970s, IBM™ introduced the concept of networks with the advent of mainframes. The earliest network consisted of a mainframe connected to a group of "dumb" terminals. These terminals accessed mainframes over slow serial link terminals and had no processing capability of their own. This model of networking was called the centralized computing environment, where the processing overhead was borne by the central mainframe server. However, the model did not address complex issues pertaining to interconnecting centralized servers, unique addresses for each node, synchronization of data transfer, and error correction.

In the 1980s, IBM introduced the personal computer (PC), which had independent processing capabilities and could run its own applications. Aptly called "desktop," PCs were being used in almost every office. While the stand-alone PC could execute all types of applications, it was not much use unless it was connected with other PCs. This led to the introduction of the Local Area Network (LAN), where a group of PCs were connected to share data and expensive network resources. As the importance of interconnecting computers within one's own organization grew, the initiative to build a LAN grew too.

With the introduction of PCs and subsequently LANs, organizations were increasingly dependent upon the concept of networking for executing day-to-day business functions. Connecting and communicating were of paramount importance, and this is where the LAN suffered a serious drawback. While the LAN allowed a group of hosts to connect and share a set of common resources, it was confined to a limited geographical area.

Enterprises were now looking at sharing data and resources among hosts spread across different geographical locations. Metropolitan Area Networks (MANs) were introduced to connect stand-alone LANs across the city. This led to the concept of Wide Area Networks (WANs), which connected an organization's stand-alone LANs, located in different cities across the country. The Internet was introduced as a type of Global Area Network (GAN), connecting users across the world via a single network. As a result, communication became simpler, and it was possible to manage network resources centrally, reducing costs with increased manageability and productivity.

As networks evolved, a standard was required for network communication. Communication methodologies and standards differed with respect to the vendor, and it had become virtually impossible for different systems to connect. In 1984, the International Organization for Standardization (IOS) introduced the Open Systems Interconnect (OSI) Model—a conceptual model designed to standardize internetwork communication.

OSI Reference Model

A model is a reference point for designing and planning a technology. Any technology, in the field of networking, is defined as a combination of networking devices, applications, and protocols. All of these components can fit into a model that provides a reference point for the industry.

The networking industry required a model for developing different components of a technology, such that maximum interoperability could be achieved. Interoperability could only be possible when all vendors followed a set of standards. For this purpose, ISO introduced the OSI Reference Model. The OSI Model is a set of protocol specifications, standardized to accomplish similar tasks and used by all applications.

The OSI Model has become an industry standard and is considered the key reference of networking. It is used as a point of reference when comparing and discussing different networking protocols. Let us look at this model in detail and understand the working of its architecture. This section is important for the day-to-day work of any networking professional.

The primary goal of the OSI Model is to provide a set of guidelines for developing and implementing network equipment, applications, and protocols. These rules and guidelines create a path for effective and fast transmission of data across the network.

The OSI Model presents a seven-layer approach, where a complex task is divided into smaller tasks and where each layer implements a separate set of tasks.

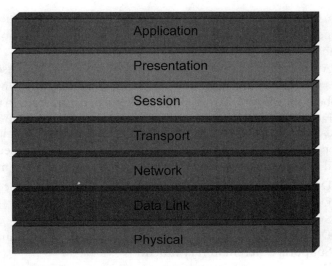

FIGURE 1.1   Seven layers of the OSI Model.

The OSI Model states that data flowing from source to destination must always pass through each layer of the model at both the ends. The seven layers of the OSI Model are shown in Figure 1.1.

You can memorize the sequence of the layers of the OSI Model using the sentence "**A**ll **P**eople **S**eem **T**o **N**eed **D**ata **P**rocessing."

The benefits of a layered protocol model are:

- Simple to understand.
- Interoperability of products from different vendors .
- Programming interfaces can be easily modified.
- Structured model because each layer introduces headers and trailers as the data move across the different layers.
- Each layer performs a specific function.
- Troubleshooting is simplified.
- Layered model enhances compatibility.
- Future upgrade of a single layer does not affect other layers.

Application Layer

The Application layer is the uppermost layer of the OSI Model. It is also known as Layer 7 of the OSI Model. The interaction of users with the network and desktop applications occurs at the Application layer. It serves as the user interface because users interact only with this layer of the OSI Model. The Application layer supports all the communication aspects of an application, such as the medium of communication, resources required for communication, and the identification of communicating partners. All host requests for network services and applications are communicated at the Application layer level.

As the name suggests, this layer is concerned with end use applications. It provides services to all applications, such as Microsoft Word™, Microsoft Excel™, and Adobe Photoshop™. The Application layer performs certain functions, such as the interaction between the user and the desktop application and other users with the network application.

The functions of the Application layer are:

- Determines the source and destination of the intended communication
- Determines resources required for communication
- Synchronizes the communicating applications
- Determines authorization and authentication of communicating users
- Supports different library functions
- Provides services for uniting the components of network applications, such as remote access, directory access, e-mail, and file transfer
- Provides measures required for detection of any error during the communication process
- Determines error recovery procedures
- Handles flow control during communication

The Application layer uses services of protocols, such as Hyper Text Transfer Protocol (HTTP), File Transfer Protocol (FTP), and Simple Mail Transfer Protocol (SMTP) to run an application. Telnet also resides on the Application layer. In addition, the Application layer also supports protocols, such as Domain Name Server (DNS), Trivial File Transfer Protocol (TFTP), Bootstrap Protocol (BOOTP), Simple Network Management Protocol (SNMP), RLOGIN, Multipurpose Internet Mail Extension (MIME), Network File System (NFS), FINGER, Network Control Protocol (NCP), Advanced Program-to-Program Communication (APPC), AppleTalk Filing Protocol (AFP), and Server Message Block (SMB). These protocols are used by the Application layer to communicate with the application on the destination system. This layer does not provide services to other OSI layers. Instead, it supports user applications.

With the increase in internetworking environments, the Application layer now provides services to internetworking applications, such as the World Wide Web, e-mail gateways, and Electronic Data Interchange (EDI).

Presentation Layer

This layer is also known as Layer 6 of the OSI Model. Different applications represent data in different formats. This has led to the need for a layer that translates data into a standard format during transmission and converts it back into a presentable format. The Presentation layer performs this process of translation. It formats the data received from the Application layer into a proper format as desired by the destination node.

For example, a source computer using ASCII-based character codes transmits data to a destination computer using Extended Binary-Coded Decimal Interchange Code (EBCDIC) character codes. It uses the Presentation layer to convert the ASCII data to EBCDIC data.

The Presentation layer performs the coding and decoding function by converting the user-identifiable language into machine language. The user-identifiable language is delivered to the Presentation layer from the Application layer. It ensures that the data from one machine are readable by the other machine.

The Presentation layer performs these functions:

- Translates different application formats into standard format and vice versa
- Transforms different data structure programs into standard format
- Formats data transfer syntax
- Performs encryption and decryption of data before transmitting across the network
- Performs data compression and decompression while transmitting data
- Performs expansion of graphic commands
- Uses Abstract Syntax Notation 1 (ASN.1) as standard data syntax

Tasks such as compression, decompression, encryption, and decryption are associated with the Presentation layer. For quick delivery, the Presentation layer compresses data at the source level before transmission. This helps reduce bandwidth consumption. When the data reach the destination level, the Presentation layer decompresses the data and sends them to the Application layer.

Session Layer

The connection between the source and the destination during data transmission is termed a session. A communication session is divided into three processes: connection-establishment, data transmission, and connection-termination. The Session layer gets its name from its functions and executes the three processes of a communication session. This layer establishes a connection, transfers data, and finally terminates the connection.

The Session layer identifies and authenticates the hosts for communication and then establishes the connection between two hosts. In the authentication process, this layer ensures the security of data during the transmission phase. After data transmission, the Session layer terminates the connection between the two hosts. This layer also enables some applications to maintain session-based connections between clients and servers.

The communication between systems or applications can be in any of the three modes: simplex, half-duplex, and full-duplex. The Session layer organizes communication in these three modes. Let us look at each of these modes:

Simplex

The simplex mode of communication refers to one-sided communication between the source and destination. This means that communication is always one way, with one device sending and the other device receiving. It is similar to a one-sided lane in which traffic can flow only from street A to street B, and no traffic can flow from street B to street A. Figure 1.2 depicts the simplex mode of communication.

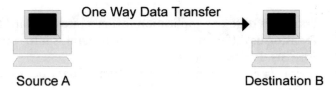

FIGURE 1.2   Simplex mode of communication between source and destination.

Half-Duplex

In the half-duplex mode of communication, both devices can communicate. However, both devices cannot talk simultaneously; one device has to wait until the other has finished. This concept is similar to that of a traffic signal. When the light turns green, traffic flows from street A to street B, and then the traffic flows from street B to street A. Figure 1.3 depicts the half-duplex mode of communication.

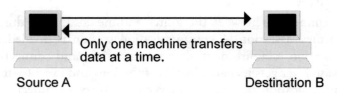

FIGURE 1.3   Half-duplex mode of communication between source and destination.

Full-Duplex

Full-duplex is the mode of communication where both devices can communicate simultaneously. This is similar to a two-way lane. The traffic can flow from street A

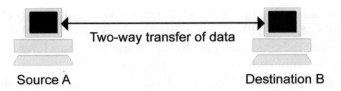

FIGURE 1.4 Full-duplex mode of communication between source and destination.

to street B as well as from street B to street A simultaneously. Figure 1.4 depicts the full-duplex mode of communication.

The functions of the Session layer are:

- Coordinates the interaction between two applications
- Establishes, maintains, and terminates the connection between two applications
- Manages more than one connection for one application
- Reconnects two hosts in case of an error
- Coordinates dialogs and data exchange
- Provides security to the system
- Determines the duration of a session
- Authenticates and identifies hosts
- Incorporates checkpoints in connection-oriented sessions
- Ensures measures for errors occurring during data transfer
- Performs log-on authentication
- Serves as an administrator for a session

The Session layer works with different protocols, such as NetBIOS, Network File System (NFS), Names Pipes, Digital Network Architecture Session Control Protocol (DNA SCP), Mail Slots, and Remote Procedure Call (RPC).

Transport Layer

The Transport layer is responsible for the quality of data transmission from source to destination. This layer provides reliability to the data delivery mechanism of the OSI Model. For effective and efficient transfer of large blocks of data, the Transport layer divides data into different segments. These segments are independent from each other and can be sent to the same segment or to multiple hosts, or different segments can be sent to the same hosts.

The data segments are arranged into a predefined sequence before the sequence is transmitted to the destination. During sequencing, the Transport layer numbers each segment and asks the destination to acknowledge receipt of each segment. If the destination does not give an acknowledgment, this layer re-transmits that particular segment. This is also known as the connection-oriented feature of the

Transport layer. The protocol that is responsible for this feature is Transmission Control Protocol (TCP). In addition, User Datagram Protocol (UDP) is a protocol that provides the connectionless feature to the Transport layer. At the destination, this layer recombines data segments into a correct sequence.

For example, node A sends a 2 MB file to node B. However, due to some error, node B fails to receive the entire file. Some parts of the file are lost. Node B sends a request to node A to resend the file. Node A resends the file to node B, which, in turn, sends an acknowledgment after receiving the file. Figure 1.5 depicts the Transport layer in the OSI Model.

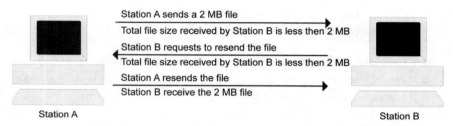

Station A sends a 2 MB file

Total file size received by Station B is less then 2 MB

Station B requests to resend the file

Total file size received by Station B is less then 2 MB

Station A resends the file

Station B receive the 2 MB file

Station A   Station B

FIGURE 1.5   Data transfer at the Transport layer.

The functions of the Transport layer are:

- Transfers data between hosts
- Performs error checking and recovery in case of loss of transmission
- Ensures data integrity by maintaining flow control
- Prevents buffer overflows for avoiding data loss
- Provides acknowledgment for successful transmissions
- Ensures complete data transfers, using techniques such as Cyclic Redundancy Checksum (CRC), windowing, and acknowledgments

Transmission of a large volume of data over the network results in congestion. The Transport layer effectively resolves this problem of congestion. When a destination receives too many data segments from multiple sources, it stores them in the memory. However, if the buffer overflows and the memory is unable to store more data segments, the Transport layer controls the flow. This layer of destination issues a "not ready" indicator so that the source does not send any further segments. When the destination machine processes stock segments, it sends a "ready" indicator for resuming data transfer.

Network Layer

Layer 3, or the Network layer, defines the network address. It is responsible for routing packets through an internetwork. Routers and other Layer 3 devices are placed in the Network layer. This layer provides a protocol-specific type of addressing known as Network layer addressing or Layer 3 addressing. It is also called "logical addressing."

For example, if you are visiting your friend's place for the first time, you may need a map to guide you. Similarly, path determination is important for communication of data from source to destination. The path for data communication is determined by different protocols featured in the Network layer. These protocols determine the best possible path for end-to-end delivery of packets. Figure 1.6 depicts how path determination occurs on the network layer.

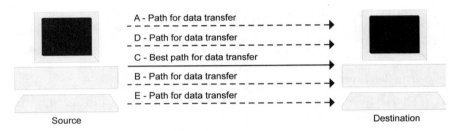

FIGURE 1.6 Path determination on the Network layer.

The Network layer logically demarcates the network. It defines a mechanism to deliver data packets from one network to another in an internetwork. This is delivered via the best possible path from the source to the destination. Path determination depends on different factors, such as network traffic, information priority, and the amount of information. No physical encoding takes place in Layer 3. Logical addressing only defines the path, with the help of the routing table, according to the configured network address. After the routers define the path, data packets are forwarded from source to destination. All paths connected to the routers are assigned some numbers, which are used as network addresses. These network addresses are used by protocols to pass packets from source to destination.

The Network layer of the source sends packets of information to the Network layer of the destination via a router. However, the router is not capable of simultaneously sending such large bits of information. To counter this problem, the data are divided into small fragments by the Network layer of the source destination, which is reassembled by the Network layer of destination. Figure 1.7 shows that the router is unable to process a large volume of data. As a result, the data are divided into smaller chunks, making it possible for the router to process the data sequentially.

The functions of the Network layer are:

- Transmits data between devices that are not locally connected
- Controls congestion and internetwork traffic
- Performs encapsulation of data packets before transmission

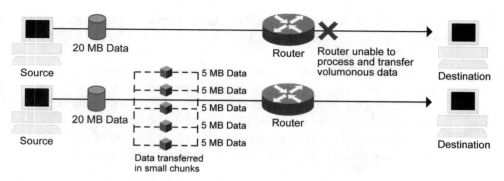

FIGURE 1.7   Breaking up data into smaller fragments.

Data-link Layer

The Data-link layer converts packets into frames, and its main function is to scramble and assemble the frames into bits for the use of the Physical layer at both the source and destination. The Data-link layer receives packets and converts these into frames, bits into bytes, and bytes into frames. Protocols at this layer convert the information received from the Network layer into packets and frames. These frames are transferred from a node to the next along the transmission path.

The Data-link layer has two sublayers:

■ Media Access Control (MAC)
■ Logical Link Control (LLC)

LLC is the upper sublayer and, in conjunction with the Network layer, provides flexibility to the protocols. LLC supports both connectionless and connection-oriented services. MAC is the bottom sublayer, adjacent to the Physical layer, and is responsible for framing. This sublayer provides the addressing for each device on the network, while the LLC establishes and maintains links between the devices communicating with each other. MAC addressing provides a unique address to each device present in any network. Figure 1.8 displays the Data-link layer of the OSI Model.

The Data-link layer formats the data received from the Network layer into frames. It adds a header to the original data, which contains information pertaining to the destination hardware and the source address.

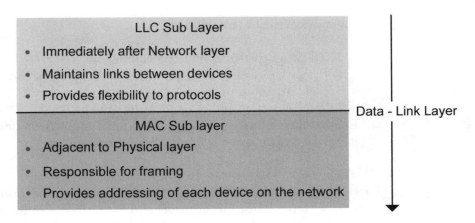

FIGURE 1.8   Data-link layer showing the two sublayers.

Figure 1.9 shows the general format of a frame.
Table 1.1 describes the different parts of a frame and the functions of each.

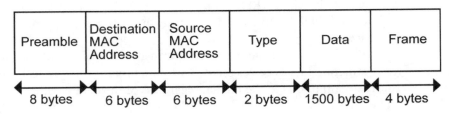

FIGURE 1.9   Data-link layer frame showing different fields.

TABLE 1.1   Parts of a Frame and Their Respective Functions

Parts of a Frame	Functions
Preamble	Known as Start Indicator Marks the beginning of a data frame
Destination MAC Address	Stores the address of the destination node
Source MAC Address	Stores the address of the source node
Type	Known as Length field Indicates the type of data
Data	Contains the actual message Contains information of above layers
Frame	Determines the status of frame such as any damage, corruption of data Performs function of Cyclic Redundancy Checksum (CRC)

The functions of the Data-link layer are:

- Sends data across a particular link or medium
- Defines physical addressing, sequencing of frames, and flow control
- Defines control methods to prevent data bottlenecks and traffic congestion
- Performs error detection and error correction methods
- Ensures that there is no loss of data packets by counting the frames
- Organizes packets in correct sequence before transmission

Physical Layer

The Physical layer is Layer 1 of the OSI Model. The primary function of this layer is to interact with the transmission media and forward data to the media in the form of bits. The Physical layer is the hardware layer where all the network devices are connected to this layer. This layer represents the physical characteristics of the transmission medium and defines the electrical and mechanical characteristics of the physical link. It defines the cables, cards, and other physical devices needed for transmission of bits from source to destination. Voltage, connectors, cables, data rates, Registered Jack 45 (RJ45), EIA/TIA-232, V.35, Ethernet, Fiber Distributed Data Interface (FDDI), NRZ, and transmission distance are examples of Physical layer devices.

The Physical layer of the source host transmits the bits, and the Physical layer of the destination host receives them. The main task for the Physical layer is to forward the stream of bits to the media, regardless of the data type.

The Physical layer also deals with the voltage level of transmissions through pins of a connector and physical interface of the Network Interface Card (NIC). This layer also performs a function known as signal encoding. An important feature of this layer is bit formation. The Physical layer considers data to be in the form of bits. It sends and receives bits, in the form of 0 and 1 between devices. It is often misunderstood that the NIC card functions at Layer 1. This is not true because the NIC card contains the MAC address, which works at Layer 2 of the OSI. However, the physical interface of the NIC card is originally a part of Layer 1.

The functions of the Physical layer are:

- Defines how electrical signals should be transferred across physical media
- Controls modulation, demodulation, and decoding of electrical signals
- Converts bits into electrical signals
- Deals with all physical aspects of a network

Table 1.2 displays details about the seven layers of the OSI Model for quick reference.

TABLE 1.2   Seven Layers of the OSI Reference Model

Layer	Functions	Protocols Used
Application	Communication partners Identification Host authorization Flow control	DNS FTP, TFTP BOOTP SNMP, SMTP MIME NFS FINGER TEL NET
Presentation	Translation Encryption Decryption Compression Decompression	PICT TIFF MIDI MPEG
Session	Connection establishment and termination Activity management Dialog coordination Synchronization	NetBIOS NFS RPC Mail Slots DNA SCP
Transport	Complete data transfer Connection control Flow control Error control and recovery	TCP SPX NetBIOS/NetBEUI ATP ARP, RARP
Network	Logical addressing Routing	IGMP IPX NetBEUI OSI DDP
Data Link	Frames of data Physical addressing Flow control Error detection and control Access control	HDLC SDLC LAPB PPP ISDN SLIP
Physical	Physical aspects of resources, interface, cards, and cables Bits encoding and decoding Modulation and demodulation of electrical signals Transmission of bits	IEEE 802 IEEE 802.2 EIA/TIA-232 EIA-530 ISDN

Data Communication in OSI Layers

The OSI Model has two effective features for open system connectivity. The first feature is that all the layers in the OSI network architecture communicate with each other with the help of a set of rules. Since certain rules are followed within the OSI Model, it becomes easy for the vendors to develop new products for any specific OSI layer. This assures that all the products seamlessly interact with each other. The second feature is layer independence, which means that all layers work independently of each other. All the layers perform unique functions, which enable communication between products of different layers.

Data in the OSI Model move from one layer to another. Each layer provides a service to the layer above and below it. After the data reach the Physical layer, the data are handed over by communication protocols to transmission protocols. The transmission protocols move data across physical channels, such as cable and radio frequencies. When data reach the destination, they move upwards, from the Physical to the Application layer by corresponding layer protocols.

Let us consider a scenario where node A is the source and node B is the destination. According to the OSI Model, all data start from the Application layer of source A and travel up to the Physical layer. After reaching the Physical layer of the destination—node B—they move to the Application layer of node B. At each layer, these data or messages are modified according to the functionality of that respective layer. During the entire process, each layer of node A will only communicate with the corresponding layer of node B. This means that the Presentation layer of node A will communicate with the Presentation layer of node B, and the Session layer of A will communicate with the Session layer of node B. This example is illustrated in Figure 1.10.

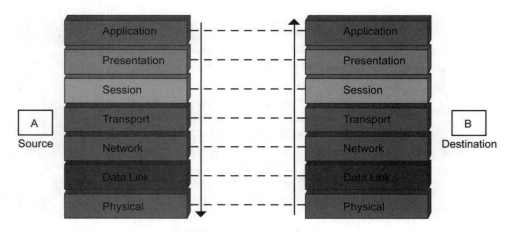

FIGURE 1.10   Movement of data from source to destination.

There are two important features pertaining to the movement of data:

■ Each OSI layer of the source node will only communicate with the corresponding OSI layer of the destination node. For example, the Network layer of source A will talk to the Network layer of destination B. The data sent by a layer of the source machine to the corresponding layer of the destination machine is called a Protocol Data Unit (PDU).

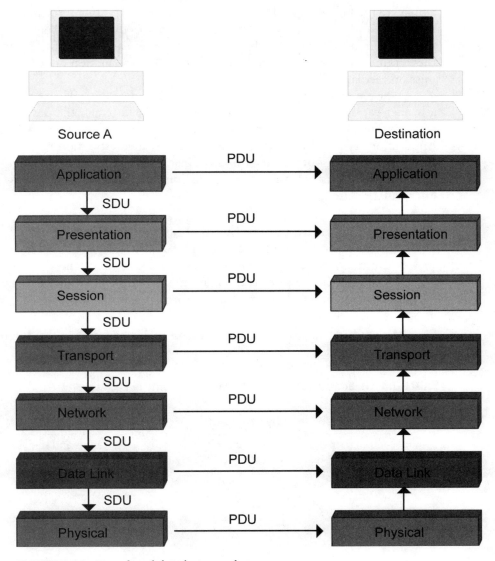

FIGURE 1.11 Transfer of data between layers.

■ Each layer of one machine transfers some data to the layer below it in the case of the source node and to the layer above it in the case of the destination node. This transfer of data from one layer to another of the same machine is called a Service Data Unit (SDU). Figure 1.11 depicts how data travel between layers of the source and destination hosts and upper and lower layers.

Besides transferring the data from one layer to another, each layer of the OSI Model performs certain functions on the data. Data are converted into segments, which are further converted into packets. Headers containing information such as the source and destination IP addresses are added to the packet. The packet is further converted to frames, to which some headers are added. At the Physical layer, these frames are converted into bits and sent to the destination node via the transmission media.

As the data move from the Application layer to the Physical layer of the source, a header capsule is added to the data. After the data reach the destination, the additional header is detached from data while moving from the Physical layer to the Application layer of the destination node. Thus, the destination receives original data as sent by the source. This process of inserting a header into the original message by each layer is called Data Encapsulation. The five conversion steps for Data Encapsulation are:

1. Data are created, and the user application is ready to send the data.
2. The Transport header is created, and data are placed behind it.
3. Destination Network layer address is added to the data.
4. Data-link address of the destination is added.
5. The data bits are transmitted.

Table 1.3 lists the headers added to a data packet or frame in the different layers of the OSI Model.

TABLE 1.3   Headers Added to a Data Packet or Frame at Each Layer

OSI Layer	Header Added	Information in Header
Application Layer	Application Header	Application and network processes information
Presentation Layer	Presentation Header	Data representation information
Session Layer	Session Header	Interhost communication information

TABLE 1.3   (continued)

OSI Layer	Header Added	Information in Header
Transport Layer	Segment Header	Sequencing information
Network Layer	Network Header	Source and destination logical address
Data Link Layer	Frame Header and Trailer	Source and destination physical address
Physical Layer	Physical Header	Bit formation information

TCP/IP Reference Model

The Transmission Control Protocol/Internet Protocol (TCP/IP) model evolved in the 1960s, when researchers of the Advanced Research Projects Agency (ARPA)™ realized the need for a communications network that would enable the interconnection of research laboratories and data centers. The objective of constructing an interconnecting communication network was to avoid communication breakdown in the event of a nuclear attack on any of the data centers.

TCP/IP is the most widely used networking protocol used on the Internet. Both TCP and IP work together to facilitate safe and fast delivery of data through a network. To increase the transmission speed of large blocks of data over the network, the data are split into smaller data fragments. These fragments are called data packets. TCP/IP plays a pivotal role in this context. TCP affixes a header to the data packet, which contains information pertaining to the destination address, source address, and length of the data. The role of IP is to ensure that when the data packets reach the destination address, they are reassembled into the original data block, and sent for use by the intended application.

TCP/IP is a condensed version of the OSI Model and has fewer layers. Let us look at the four layers of the TCP/IP model in detail.

Application Layer

The Application layer is the top layer of the TCP/IP reference model. It includes both the Presentation and Session layers of the OSI Model. The word "application" is used to define any process that takes place above the Transport layer. The Application layer encompasses all processes, which use the Transport layer protocols and transfer data to the Internet layer. There are different application level protocols. The most widely used protocols are shown in Table 1.4.

TABLE 1.4   Application Layer Protocols and Their Functions

Protocol	Function
Simple Mail Transfer Protocol (SMTP)	Delivers electronic mail messages over a network.
Post Office Protocol (POP)	Enables users to access electronic mails over a network from a central server
Hypertext Transfer Protocol (HTTP)	Enables transfer of text, music, pictures and other multimedia data through a Graphical User Interface (GUI)
File Transfer Protocol (FTP)	Enables the download and upload files over a network
Domain Name Service (DNS)	Maps the IP addresses to the Internet domain names
Network Terminal Protocol (Telnet)	Enables textual communication for remote log-in and communication via the network
Routing Information Protocol (RIP)	Enables the exchange routing-related information by network devices

Transport Layer

The Transport layer lies between the Application layer and the Internet layer of the TCP/IP model. The two primary protocols associated with this layer are the Transmission Control Protocol (TCP) and the User Datagram Protocol (UDP).

TCP is a connection-based protocol, which enables error detection and correction in the data packets. It also ensures reliable delivery of data packets from the source to the destination.

UDP is a connectionless and faster protocol as compared to TCP. It has low overhead costs and time associated with it, because it provides quick transfer of data. However, it does not provide the error detection and correction features that TCP provides.

The selection of protocols in the Transport layer depends on user needs. TCP is used if a reliable connection session with two-way communication of data is of paramount importance. UDP is used to develop applications that are low on overheads.

Internet Layer

The Internet layer lies immediately below the Transport layer and above the Network Access layer of the TCP/IP model. The operating protocol of this layer is the Internet Protocol (IP). The Internet Protocol builds the foundation for the packet delivery system. The entire concept of TCP/IP networking is based on this foundation. This protocol manages connections over networks when data packets are transferred from the source to the destination.

IP is a connectionless protocol, which means that it does not provide features such as a source-to-destination control of communications flow. IP relies on other layers and their associated protocols to provide this feature. Even functions such as error detection and correction in the data packets are executed by other layers. In this context, IP is sometimes referred to as an unreliable protocol. This does not imply that IP is not to be relied upon to deliver data via a network. It means that IP itself does not execute error checking and correcting functions. All information that flows through the TCP/IP networks uses the Internet Protocol.

Network Access Layer

The Network Access layer is the lowest level of the TCP/IP model hierarchy. The functions provided by the Network Access layer include encapsulation of the IP datagrams into frames and mapping of IP addresses with physical devices.

All the processes in the Network Access layer are carried out by software applications and drivers, which are customized to suit individual parts of hardware. Configuration often involves selecting the required driver for loading and selecting TCP/IP as the protocol.

However, some Network Access layer protocols might require extensive work when it comes to configuration. For example, in serial line communication, using ISDN lines, remote sites have to be accessed. This is done using different hardware and software, such as Serial Line Internet Protocol (SLIP) and Point-to-Point Protocol (PPP).

Remote authentication protocols, such as Terminal Access Controller Access Control System (TACACS) and Remote Authentication Dial-in User Service (RADIUS), are used for the remote access of sites that use Network Access Server equipment. Modem stacks and terminal servers are types of Network Access Server equipment.

The functions of the protocols used at each layer within the TCP/IP model are briefly explained in Table 1.5.

TABLE 1.5   Network Layer Protocols and Their Functions

Layer	Associated Protocols	Functions
Application	SMTP, POP, HTTP, FTP, Telnet, DNS, RIP	Focuses on application protocols and how the host programs interact with transport layer services to use the network.
Transport	TCP, UDP	Enables control and management over the connection established between source and destination computers.

(continued)

TABLE 1.5   *(continued)*

Layer	Associated Protocols	Functions
Internet	IP	Fragments data into IP datagrams, which include information pertaining to the source and destination addresses. These datagrams are then sent over the network.
Network Access	SLIP, PPP, TACAS, RADIUS	Encapsulates IP datagrams into frames that are transmitted via the network. Also maps IP addresses with the physical devices.

Mapping the TCP/IP and OSI Models

The TCP/IP suite of protocols works in Layers 3 and 4 of the OSI Model. TCP/IP is made up of two protocols, TCP and IP. TCP/IP is the protocol used for communication over the Internet. Besides working in Layers 3 and 4, TCP/IP also has specifications for applications such as mail and FTP in the upper layers.

Table 1.6 maps the TCP/IP and OSI Models.

TABLE 1.6   Mapping the TCP/IP and OSI Models

TCP/IP	OSI
The protocols were developed prior to the model, so the functions are well defined and can be replaced with relative ease.	The model was developed before the protocols. The functions of the protocols in the OSI layers are not well defined nor have they been increased over time.
The TCP/IP Model consists of four layers with the Transport and Application layers being the same as in the OSI Model.	The OSI Model is composed of seven layers.
The TCP/IP Model focuses on providing interconnectivity and is less rigid about the functions of the various layers.	The OSI Model is rigid and strictly adheres to the functions of the individual layers.
The Internet is based on the foundation of the TCP/IP Model.	The development of the TCP/IP Model was based on the OSI Model.

Mapping TCP and IP to the OSI Model

The Internet layer of the TCP/IP Model maps to the third layer of the OSI Model and enables the logical transmission of packets over the internetwork. The Internet layer handles the IP addressing of the network nodes.

TCP and UDP map onto the fourth layer of the OSI Model. In this layer, data are segmented for transport across the network. This layer ensures that the data are delivered error-free and in the correct sequence. This is achieved by a mechanism called flow-control. This layer also provides features such as data buffering, multiplexing, and virtual circuits.

Data buffering is used to store data temporarily during transmission. Multiplexing is a process of sending multiple virtual-paths over the same physical path. The primary differences between TCP and UDP are reliability and speed. TCP is more reliable, and UDP is faster.

THE NEED FOR ROUTING

Routing is the process of transferring data between hosts residing in different networks. The concept of routing evolved with the introduction of internetworking. As LANs evolved into WANs, data had to be transmitted across networks spanning different geographical locations. With the evolution of LANs into WANs, the devices used for data transmission also evolved. Let us first look at the different devices used to transmit data in routerless environments and how these devices evolved to create a router.

Routerless Networks: LANs

A LAN refers to a group of computers connected to each other, with the help of cables and wires, within a limited geographical area. Data transmission among hosts within a LAN does not require a router, because the data have to traverse a short distance and are intended for the local network. These devices enable data transmission between hosts situated on a LAN:

- Repeater
- Hub
- Bridge
- Switch

These devices have been designed to enable data transmission across small distances, making these suitable and cost efficient for LAN. Before looking at each of these devices in detail, let us understand some concepts pertaining to the methods of data transfer within a LAN.

Concepts Associated with Data Transmission over LANs

Some concepts associated with data transmission over local area networks are:

- Broadcast Domain
- Collision Domain
- Carrier Sense Multiple Access

Broadcast Domain

A broadcast domain is a network segment, where a residing node can view and process broadcast messages sent by other nodes in the same segment. For example, there are 50 nodes in a broadcast domain. If any node sends a broadcast message, the other 49 nodes can process that message. Let us understand the concept of broadcast domain with the help of Figure 1.12.

Figure 1.12 shows three interconnected LAN switches supporting multiple workstations. These workstations are on the Ethernet and use TCP/IP for communication. The source computer is A, and it needs to send a 5 MB file to D. To find the physical MAC address of D, A sends a broadcast message on the segment to all nodes—B, C, and D. Only D responds to the broadcast message, with its MAC address and the method of data transfer. All nodes that can process the broadcast message generated by A are said to be in a single broadcast domain.

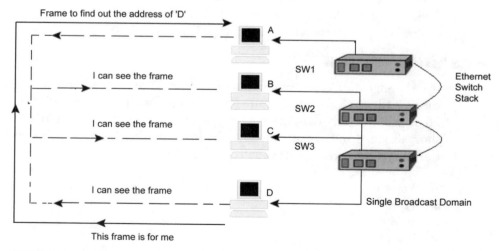

FIGURE 1.12   Communication in a broadcast domain.

Collision Domain

A collision domain is a limited domain where two or more nodes contend for the same media. A collision occurs when these nodes are unable to sense the traffic on the media and send data on the network. It is possible that two or more nodes may send out a frame on the network at the same time, leading to collision. A collision makes data transmission invalid, because it involves retransmission of the same frame over the network segment.

You can look at the collisions on a switch, using the show interface fastEthernet command.

Each time a collision occurs, the devices are required to retransmit the frames. As the number of devices accessing the same segment increases, the number of collisions increases, and subsequently, the rate of retransmissions also increases. As a result, the network is clogged due to transmission and retransmission of data, and this leads to poor network performance.

If the number of nodes on a collision domain is limited, there is reduced network traffic and increased network performance. Collision domains function on Layer 2 of the OSI Model, unlike broadcast domains, which operate on Layer 3. It is possible to divide a collision domain into smaller collision domains by introducing bridges or switches; this is discussed later in the chapter.

Carrier Sense Multiple Access

Carrier Sense Multiple Access (CSMA) is a technique that enables each node attached to Ethernets and other bus-oriented LANs to check carrier frequency and determine network availability. The nodes can access the shared devices and transmit data only after checking the network availability on that segment. In the event of a busy condition or collision, the node attempts data transfer or access to shared devices on the network after a random interval.

Carrier Sense Multiple Access with Collision Detection (CSMA/CD) is one of the most common media access control methods used in bus and cabled Ethernets. If any node on the bus detects a data collision, it sends a signal, which jams the network. This signal is sent over a lower frequency of the network and informs all the nodes on the network about the collision. Thereafter, all devices calculate a certain random interval before attempting retransmission.

Carrier Sense Multiple Access with Collision Avoidance (CSMA/CA) is a technique that gives priority to a few nodes on the network. This ensures transmission privileges to high-priority nodes. CSMA/CA is more expensive to implement because it requires additional programmed logic to be embedded in each device or Network Interface Card (NIC). CSMA/CA offers the advantage of improved access control in order to reduce collisions and improve overall network performance.

Repeaters

A repeater is a device that regenerates or amplifies a signal transmitted across a LAN. It was the earliest attempt at extending a cable such as a cat5 or a cat 5e to connect computers situated at remote ends of a LAN. After every 500 meters (100 meters for cat5), the signal transmitted across a cable usually deteriorates. The repeater regenerates and retransmits the signal to a connecting cable. It cannot segregate the collision or broadcast domain. There are two types of repeaters: analog and digital. An analog repeater amplifies the signal, and the digital repeater is capable of regenerating the signal to its original intensity.

A repeater is a Layer 1 device—it works on the Physical layer of the OSI Model. It is only concerned with transmitting bits of data, regardless of the data type. It contains two ports, each required to receive and transmit signals, as shown in Figure 1.13.

The use of repeaters follows the 5-4-3 rule; that is, there should be 5 segments with 4 repeaters and 3 populated segments.

NOTE

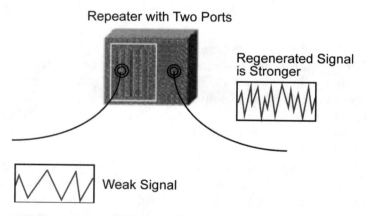

FIGURE 1.13 Repeater with two ports.

Hubs

As the number of nodes on a network increased, there was a need for a repeater with multiple ports. A hub is a multi-port repeater, which serves as a point of convergence for all nodes connected to a LAN, as shown in Figure 1.14.

All frames sent by the nodes pass through the hub, which forwards them to the server. A network cable is inserted into the ports situated on the computer and hub in order to connect the two. There are two types of hubs: passive and active hubs. A passive hub is a simple physical attachment for the multiple cable segments in a network. Active hubs are those that boost the signal from one node to the other and work with cables of greater length. Figure 1.14 shows a hub with multiple nodes connected to it.

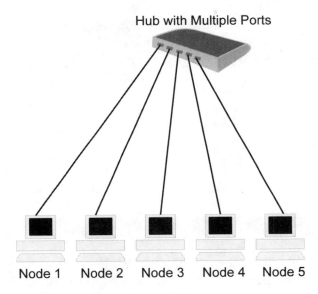

FIGURE 1.14   Hub with multiple ports.

A hub is not an intelligent device; that is, it does not understand the source and destination address of a frame. It merely broadcasts the received frame on the network. As a result, if node A needs to send a message to node B, the hub broadcasts the message to every node in the LAN segment, and only node B picks up the message.

Bridges

When a hub transmits a message between two nodes, it broadcasts the message to all nodes in the broadcast domain. This increases the amount of traffic on the network, increasing the number of collisions on the network and reducing network performance. Figure 1.15 shows a bridged network environment where four nodes—A, B, C, and D—are connected over an Ethernet.

The nodes are connected over the same Ethernet collision segment. Nodes B and C send data packets simultaneously, over the Ethernet. This results in a collision, and the Ethernet retransmits the frame. If A and D also simultaneously send data packets on the network, the number of collisions and retransmissions

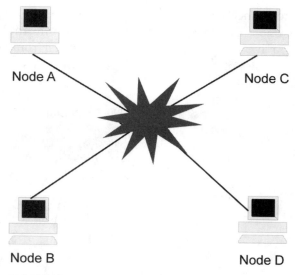

FIGURE 1.15   Data transmission in a hub environment.

increases. Figure 1.15 shows data transmission within devices connected via a hub.

A bridge is a device used to divide a large network segment into smaller segments, reducing the size of each collision domain. In Figure 1.16, a bridge has been introduced between nodes A, B, and C, D, dividing the Ethernet segment into two collision domains.

In this setup, if nodes A and B simultaneously transmit frames, there would be no collision. The introduction of a bridge has reduced the number of collisions and retransmissions, resulting in an improvement of the overall network performance.

Bridges are data communication devices that operate on Layer 2 of the OSI Model. As a result, these are called Data-link Layer devices. When bridges were first introduced, these permitted connection and transmission of data packets between computers connected to a homogeneous network. A bridge operates on MAC-Layer addresses. In addition, a bridge is protocol-independent and requires little or no configuration.

TIP

It was only recently that bridging between different networks became possible. A bridge connects two LAN segments into one larger continuous LAN.

A bridge builds a table of physical addresses of the network devices connected to both sides of it. When a bridge sees a packet, it checks the address of the packet

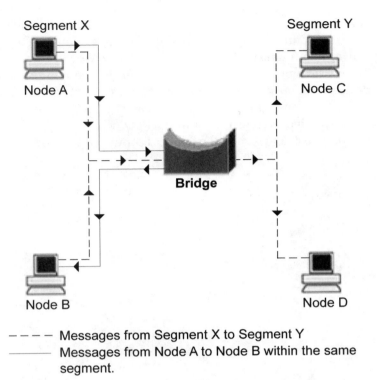

Segment X   Segment Y

Node A   Node C

Bridge

Node B   Node D

– – – – Messages from Segment X to Segment Y
———— Messages from Node A to Node B within the same
        segment.

FIGURE 1.16 Data transmission using a bridge.

against its table. If the destination physical address is on the opposite segment or if
the bridge does not have the address stored in the list, the bridge forwards the information. A bridge can distinguish between local and remote data. As a result, data
traveling from one workstation to another in the same segment are not broadcast
to all nodes on the network.

For example, nodes A and B lie in segment X, and node C lies in segment Y. If
node A sends a message to node B, the bridge sends the message directly to node B.
If node A or B need to communicate with node C, the bridge checks the table and
forwards the message to segment Y.

Switches

Switching emerged as an enhancement over bridging-based internetworking so-
lutions. A switch is a multi-port bridge, and as a result, it can create a larger
number of segments. It provides a higher port density and lower per-port cost.
Bridges and switches are used to increase the length of the LAN segment allowing

communication between nodes separated by large distances. In a switch, each port maintains the MAC address of the device connected to it and is used for forwarding frames.

A switch provides a dedicated, collision-free communication between network devices. The most important advantage of using a bridge or a switch is reduced size of the collision segment, as shown in Figure 1.17.

A switch provides superior throughput performance, higher port density, lower per-port cost, and greater flexibility as compared to bridges.

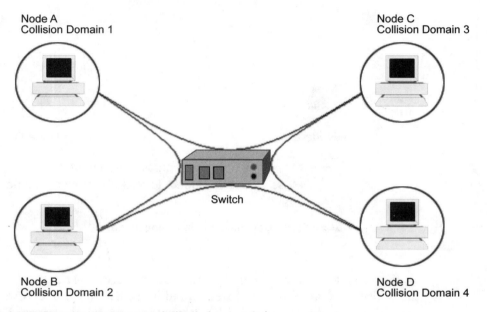

FIGURE 1.17   Data transmission using a switch.

Networks That Use Routers: WANs, MANs, and the Internet

Bridges and switches connect different segments of a LAN to ensure smooth flow of network traffic. A router is a Layer 3 device and is placed at the gateway of the LAN to streamline data transmission sent or received from external networks.

The primary difference between bridging, switching, and routing is that bridges filter traffic at Layer 2 by using the MAC address, routers filter traffic at Layer 3 by using both MAC and IP addresses, and switches cannot connect network segments with different network addresses. Since routers operate at layer 3, they can segregate traffic into different segments or broadcast domains. This prevents unnecessary traffic from being broadcast to segments, where it is not required.

In Figure 1.18, two network segments having different network addresses at Layer 3 of the OSI Model cannot be connected using a switch.

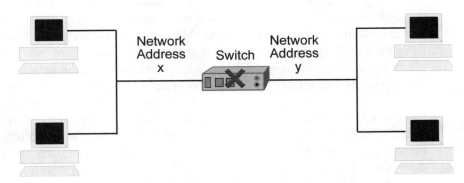

FIGURE 1.18   A switch cannot connect network segments with different network addresses.

A router operates in Layer 3 and possesses the intelligence to forward packets at Layer 3 to establish connectivity between two networks with different network addresses, as shown in Figure 1.19.

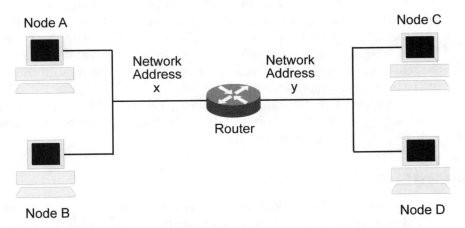

FIGURE 1.19   Router connecting networks with different network addresses.

Routers build routing tables for forwarding packets. The information in the routing table specifies the next-hop router address for reaching the destination for a particular network address. The content of the routing table and the way it is presented varies for different protocols. These will be discussed later in the book. A router maintains a separate routing table for different routed and routing protocols—the routing table for IP is different from that of Internetwork Packet Exchange (IPX).

THE CISCO HIERARCHICAL DESIGN

An ideal network is one that provides optimal bandwidth utilization, efficiently manages network traffic, and offers maximum uptime. Keeping these parameters in mind, Cisco introduced the Hierarchical Design model. This model enables you to design a scalable, reliable, and cost-effective hierarchical internetwork. It is a simple and efficient design topology that allows you to divide the network into three distinct layers. These layers are a logical and not a physical segmentation of the network. Each layer performs a specific set of tasks. However, the functions of each layer may overlap during a physical implementation of a hierarchical network.

The hierarchical model has been built on the premise that the volume and patterns of network traffic significantly impact network performance. As a result, the three layers have been designed to manage network traffic efficiently, reducing the network load and increasing scalability.

The three layers of hierarchy in Cisco are Core, Distribution, and Access. The Core layer is the topological center of the network and provides high-speed switching, redundancy, high availability, and bandwidth to the traffic passing through it. It does not process or route any data packet. The Distribution layer is responsible for packet filtering and routing. This layer provides high density and processing capacity per interface for packet manipulation. The Access layer is the outermost layer, where the nodes of the network are connected to each other.

When designing and implementing the network, it is important to ensure that the layers are placed such that the traffic is efficiently forwarded between layers. This is because the network rejects any unnecessary packet that a lower layer forwards to an upper layer, increasing the upload time. Figure 1.20 shows the three layers of the Cisco hierarchical model.

In Figure 1.20, the different workgroups in the access layers are connected to routers placed in the distribution layer. Each of these routers is connected to a high-end router in the core layer.

Core Layer

The Core (or backbone) layer is present in the middle of the network and at the top of the hierarchical layer of Cisco. It is a high-speed switching backbone that transports large volumes of traffic reliably and quickly. The traffic transported across the core is usually between nodes and enterprise services, such as Internet and e-mail.

The Core layer is only responsible for providing high-speed transport. As a result, there is no room for latency and complex routing decisions pertaining to filters, and access lists are not implemented at this layer. Instead, routing decisions are usually implemented at the Distribution layers. Quality of Service (QoS) may be implemented at this layer to ensure a higher priority to traffic that may be lost or delayed in congestion.

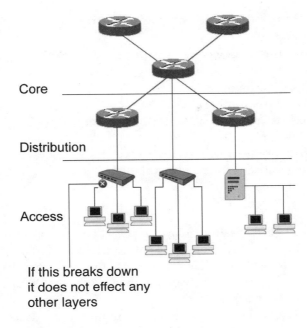

Core

Distribution

Access

If this breaks down
it does not effect any
other layers

FIGURE 1.20 The Cisco hierarchical model
showing the three layers.

The Core layer is also required to build a high degree of redundancy. As a result, if there is a failure in the active path, an alternate path is used to transport the packet. The following functions are implemented at the Core layer:

- Switching of large volumes of traffic at high speed
- Redundancy in the network
- High availability and fault tolerance

Distribution Layer

The Distribution (or policy) layer is placed between the Access and Core layers. This layer performs complex, CPU-intensive calculations pertaining to routing, filtering, inter-VLAN routing, access-lists, address or area-aggregation, security, and identifying alternate paths to access the core. It also determines the best possible path to forward host requests to the Core layer. The Core layer is then required to transport host requests and responses to and from enterprise services.

Members of this layer include most of the LAN servers, routers, and backbone switches. The Distribution layer implements network policies, making the network flexible.

The functions implemented at the Distribution layer are:

- Access lists and network security policy
- Packet filtering and queuing
- Redistribution of routing protocols
- Inter-VLAN routing
- Broadcast and multicast domain definition
- Address or area aggregation
- Departmental or workgroup access
- Media transitions

Access Layer

The Access (or workgroup) layer is the outermost layer of the Cisco hierarchical model. On this layer, all hosts are connected to the LAN. All user workstations and local resources, like printers, are placed at this layer. Routers serve as gatekeepers at the entry and exit to this layer, and they ensure that local server traffic is not forwarded to the wider network.

The Access layer controls user and workgroup access to internetwork resources. Other functions at this layer are shared and switched bandwidth, MAC-layer filtering, and microsegmentation.

The functions implemented at the Access layer are:

- Access control lists and network policies
- Segmentation
- Workgroup connectivity into the distribution layer
- MAC layer filtering
- Microsegmentation
- Switched and shared bandwidth

SUMMARY

In this chapter, you learned about the evolution of networks and the need for seamless communication across networks. We also reminded you about the need for routing and how, with the introduction of routers, it became possible to exchange data between networks. In the next chapter, we move on to TCP/IP network layer IP addressing and routing.

All code listings, figures, and tables presented in this book can be found on the book's companion CD-ROM.

ON THE CD

POINTS TO REMEMBER

- The seven layers of the OSI Model are Application, Presentation, Session, Transport, Network, Data-link, and Physical.
- The Application layer enables interaction of users and desktop applications, using protocols such as HTTP, FTP, SMTP, TFTP, and SNMP.
- The Presentation layer formats data received from the Application layer into the proper format as desired by the destination node.
- The Session layer establishes a connection, transfers data, and terminates a connection between two hosts.
- The Transport layer is responsible for quality data transmission from the source to the destination using protocols, such as TCP and UDP.
- The Network layer is responsible for routing packets through an internetwork using logical addressing.
- The Data-link layer converts packets into frames, and it scrambles and assembles the frames into bits for the use of the Physical layer at the source and destination. MAC and LLC are two sublayers of this layer.
- The Physical layer interacts with transmission media and forwards data to the media as bits.
- PDU enables communication among OSI layers of source and destination, and SDU is used among layers in the same model.
- The Application, Transport, Internet, and Network Access layers constitute the TCP/IP Model.
- The Internet layer of the TCP/IP model maps to the third layer of the OSI Model, and the Transport layer maps to the fourth layer.
- A broadcast domain is a network segment, where a residing node can view and process broadcast messages sent by other nodes in the same segment.
- A collision domain is a limited domain where two or more nodes contend for the same media.
- CSMA enables nodes attached to Ethernets and other bus-oriented LANs to check carrier frequency and determine network availability.
- A repeater regenerates a signal transmitted across a LAN.
- A hub is a multi-port repeater that is a point of convergence for all nodes connected to a LAN.
- A bridge is used to divide large network segments into smaller segments to reduce the size of each collision domain.
- A switch is a multi-port bridge to create a larger number of network segments.
- The three layers of hierarchy in Cisco are Core, Distribution, and Access.
- The Core layer is a high-speed switching backbone that transports large volumes of traffic.

- The Distribution layer performs routing calculations, filtering, inter-VLAN routing, access-lists, address or area-aggregation, and security, and it identifies alternate paths to access the core.
- The Access layer controls user and workgroup access to internetwork resources.

2 | The OSI Network Layer

IN THIS CHAPTER

- ■ Network Layer Addressing
- ■ Network Layer Routing

IP ADDRESSING

IP belongs to the Network layer of the OSI Model and plays a key role in routing packets from one network to another. As discussed earlier in this book, you can divide a network into several broadcast domains and route the packets among each of these domains. This is done with the help of an IP addressing scheme.

An IP address is a unique 32-bit Layer 3 address, which uniquely defines a host in the network and allows it to participate in a TCP/IP network. The IP address is a 32-bit series divided into four groups of eight bits (octets) each. This series is written in a decimal notation with numbers ranging from 0 to 255.

The IP address is assigned by the network administrator and differs from a MAC address, which is allocated by the hardware manufacturer.

An IP address is made up of 32 bits. This address is broken into four bytes or octets. Table 2.1 shows a 32-bit IP address in a structured addressing scheme.

TABLE 2.1 The 32-bit IP Addressing Scheme

IP Address Notation	Address
Dotted decimal	192.168.12.1
Binary	11000000.10101000.00001100.00000001
Subnet mask in dotted decimal	255.255.255.0
Subnet mask in binary	11111111.11111111.11111111.00000000

Each network can again be subdivided into a number of subnets, depending on the requirement. Finally, each host within the same network or subnet should have a unique IP address that identifies it in the entire network. This is also called the host address. In Table 2.1, the host address is 192.168.12.1. The network part of the IP address identifies the network to which the particular host belongs. All hosts should have the same network address in order to communicate. In Table 2.1, the network number is 192.168.12.0.

When a greater number of hosts and lesser number of networks are required, more bits are allocated to the hosts and less to the networks and vice-versa.

For example, 192.168.2.4 is an IP address with a subnet mask of 255.255.0.0. The first two bytes or octets represent a network address, and the last two octets are the host address, as shown in Figure 2.1.

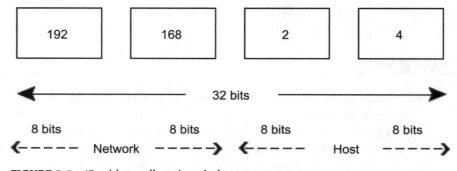

FIGURE 2.1 IP address allocations in bytes or octets.

The number of bits assigned to the network and host portions depends on the number of networks to be configured. In a public network, the Internet Network Information Center (InterNIC) assigns this network number. To identify the number of bits that determine the network portion, a subnet mask has to be used along with the IP address. The subnet mask determines the demarcation point between the network and the noted portion of the IP address. This is explained later in the chapter.

Classes of IP Addressing

The requirement to allocate IP addresses to networks of varying sizes was addressed by introducing the concept of Address Classes. The hierarchical model of the IP addressing scheme gave rise to different classes of IP addressing. The InterNIC assigns the classes of IP addresses to an internetwork with respect to the size of the network. This is to avoid any confusion during the allocation and distribution of IP address. There are five different address classes:

■ Class A
■ Class B

- Class C
- Class D
- Class E

Internetworks are divided into three sizes:

Big internetworks: Assigned a larger number of hosts and fewer networks. These organizations are allocated Class A addresses.

Small internetworks: Assigned a smaller number of hosts and larger number of subnetworks. These organizations are allocated Class B addresses.

Medium internetworks: Assigned a requirement of hosts and subnetworks in between the big and small internetworks. These organizations are allocated Class C addresses.

Table 2.2 lists the specifications and options associated with different classes.

TABLE 2.2   Class Requirement and Available Options

Class	Purpose	Maximum Networks	Maximum Hosts
A	Large organizations	127	16,777,214
B	Medium-size organizations	16,384	65,543
C	Small organizations	2,097,152	254

The Internet community defined a set of rules in the hierarchical IP addressing scheme. For addresses in Class A, the leading bits of the first octet should always start with 0. The leading bits should be 10 for Class B, 110 for Class C, 1110 for Class D, and 1111 for Class E. Table 2.3 shows address ranges of different classes.

TABLE 2.3   Leading Bits and Address Ranges of Classes

Class	Leading Bit	Address Range
A	0	1.0.0.0 - 127.255.255.255
B	10	128.1.0.0 - 191.254.0.0
C	110	192.0.0.0 - 223.255.255.255
D	1110	224.0.0.0 - 239.255.255.255
E	1111	240.0.0.0 - 254.255.255.255

Class A

Class A addresses range from 1.0.0.0 to 126.0.0.0, where the first octet represents the network portion, and the last three octets represent the host. The Class A address format is used for large organizations with networks supporting a large number of end users. The maximum number of networks possible with Class A addressing is 127, and the maximum number of hosts per network number is 16,777,214. The highest order of the network bits is always the most significant bits and defines the class of the network. In case of Class A networks, the highest order bit—the first bit of the first octet—is zero. Figure 2.2 depicts the Class A addressing format.

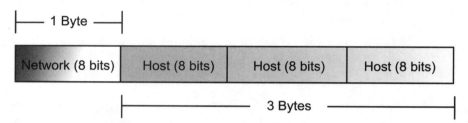

FIGURE 2.2   Class A addressing format showing networks and hosts.

Class B

Class B addresses range from 128.1.0.0 to 191.254.0.0, where the first two octets represent the network portion, and the other two octets represent the host. The Class B address format is used for networks of mid-sized organizations. The maximum number of networks possible with Class B addressing is 16,384, and the maximum number of hosts per network is 65,543. The highest order of the network bits is always 10. The first bit of the first octet is set to 1, and the second bit is set to 0. Figure 2.3 depicts the Class B addressing format.

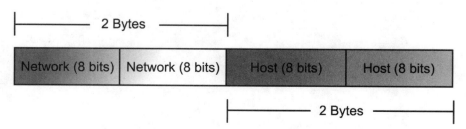

FIGURE 2.3   Class B addressing format showing networks and hosts.

Class C

Class C addresses range from 192.0.1.0 to 223.255.254.0, where the first three octets represent the network portion, and the last octet represents the host. The Class C addressing format is used for small organizations, with networks supporting a large number of users. The maximum number of networks possible with Class C addressing is 2,097,152, and the maximum number of hosts per network number is 254. The highest order of the network bits is always 110. Figure 2.4 depicts the Class C addressing format.

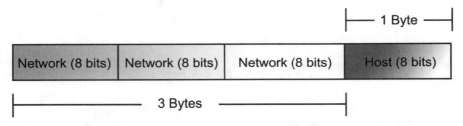

FIGURE 2.4   Class C addressing format showing networks and hosts.

Classes D and E

Unlike Classes A, B, and C, Classes D and E are not for commercial use. A Class D address is used for multicast groups and ranges from 224.0.0.0 to 239.255.255.255. Class E addresses are used for experimental purpose and range from 240.0.0.0 to 254.255.255.255.

Table 2.4 gives a breakdown of IP addresses of Class A, B, and C in binary format.

In Table 2.4, note that a Class A address has "0" in the first bit of the first octet, Class B has "10" in the first two bits of the first octet, and Class C has "110" as the first three bits of the first octet.

TABLE 2.4   IP Addresses in Binary Format

Network Number (Dotted Decimal)	Network Number (Binary)
10.1.1.0	00001010.00000001.00000001.00000000 (Class A)
150.5.5.0	10010110.00000101.00000101.00000000 (Class B)
192.1.1.0	11000000.00000001.00000001.00000000 (Class C)

Tables 2.5 and 2.6 list the characteristics of all the five classes of IP addresses.

TABLE 2.5 Characteristics of Classes A, B, C, D, and E

Class	Format	Purpose	Leading Bit
A	N.H.H.H	Large organizations	0
B	N.N.H.H	Medium-size organizations	10
C	N.N.N.H	Small organizations	110
D	N/A	Multicast groups	1110
E	N/A	Experimental	1111

TABLE 2.6 Characteristics of Classes A, B, C, D, and E

Class	Address Range	Maximum Networks	Maximum Hosts
A	1.0.0.0 -126.0.0.0	127	16,777,214
B	128.1.0.0 - 191.254.0.0	16,384	65,543
C	192.0.1.0 - 223.255.254.0	2,097,152	254
D	224.0.0.0 - 239.255.255.255	N/A	N/A
E	240.0.0.0 - 254.255.255.255	N/A	N/A

It is possible to connect networks with different classes of IP addresses. Figure 2.5 depicts a scenario where the networks 10.1.1.0 (Class A), 150.5.5.0 (Class B), and 192.1.1.0 (Class C) are connected in an internetwork.

In Figure 2.5, the networks belong to three different classes of address: 10.1.1.0/8 (Class A), 150.5.5.0/16 (Class B), and 192.1.1.0/26 (Class C).

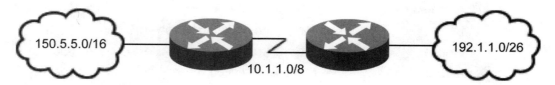

FIGURE 2.5 Internetworking different classes of an IP network.

Extended IP Addressing

The number of networks and hosts decides the class of IP address to be used by a particular organization. After the Class of the IP address is fixed, it is divided into the required number of networks by using subnet masking for each individual department. The subnet masking is not always placed exactly on the octet boundaries. Calculating the subnet masking may be a bit cumbersome at times because the number of hosts does not require a full subnet.

For example, you need to create four subnetworks of 30 hosts each, of one Class C address, 192.168.1.0/24. In such a case, the default subnet mask of 255.255.255.0 will not give the required solution. The fourth subnet needs to be subnetted to create the required number of four subnetworks. On using the first three bits of the fourth octet, you get $2^3 - 2 = 6$ networks. The remaining five bits of the last octet will give $2^5 - 2 = 30$ hosts. This suits the requirement of four subnetworks and 30 hosts out of this Class C network. Thus, the subnet masking to be used will be 255.255.255.224.

Subnetting

An organization may break down the range of registered public IP addresses obtained from the InterNIC into several smaller internal networks, depending on the requirement. This calculation is based on the number of networks and hosts required by the organization. The subnet mask has an important role in implementing this.

The IP networks are subdivided according to the network design and requirement by using subnet masking. A subnet mask is a 32-bit address that indicates the number of bits of the IP address that represent the network address. Each subdivided network is called a subnetwork or subnet. The different subnets in a network are connected using a router. Subnetting efficiently uses subnet masks to provide flexibility for creating the requisite number of networks and the required number of hosts in a network, without wasting IP addresses. This decreases the broadcast domains, which, in turn, increases the performance of the network.

Subnets are locally created and administered within an organization's network. This means that the Internet community or users from other networks are unable to see the individual subnets within a network.

The binary representation of an IP address and its subnet mask perform a logical AND operation to find the network portion of the network. A logical AND operation is explained in Table 2.7.

TABLE 2.7 Logical AND Operation between A and B

A	B	A AND B
0	0	0
0	1	0
1	0	0
1	1	1

For example, a host with the IP address 132.16.12.129 has a subnet mask, 255.255.255.128. You can determine the network and the host portions by using a logical AND operation of IP address and subnet mask, as shown in Table 2.8.

TABLE 2.8 And Operation of IP Address and Subnet Mask

IP Address Portion	Address
IP address in dotted decimal	132.16.12.129
IP address in binary format	10000100.00010000.00001100.10000001
Subnet mask in dotted decimal	255.255.255.128
Subnet mask in binary format	11111111.11111111.11111111.10000000
Logical AND between IP address and subnet mask gives the network portion	10000100.00010000.00001100.10000000
Derived network address in decimal format	132.16.12.128

In Table 2.8, the objective is to find out the network portion of the IP address, which is the network address to which a host having this IP address belongs. Both the IP address and the subnet address are broken down into their binary format. A logical AND between these two values derives the network address. Therefore, the network portion of the IP address 132.16.12.129 with the subnet mask 255.255.255.128 is 132.16.12.128.

As a part of subnetting, you borrow bits from the high-order bits of the host portion of the IP address and use them to define the various subnets. The value 1 in a subnet mask specifies the network, and 0 specifies the host.

Consider a major network 192.168.12.0/24 with a subnet mask 255.255.255.0. This network is divided into three different subnets of 192.168.12.128/28, 192.168.12.64/28, and 192.168.12.192/28. For an observer looking at the network from outside, say from the Internet cloud, the network 192.168.12.0/24 is the only one that is visible; each individual subnet is not visible. Figure 2.6 shows this case.

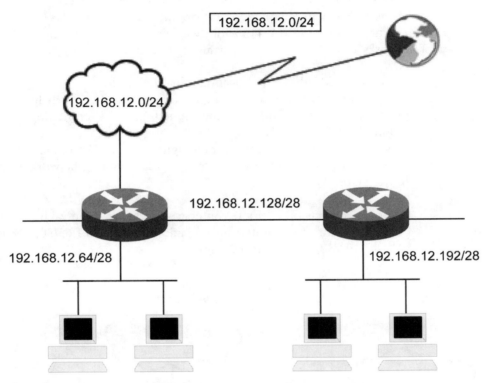

FIGURE 2.6   Subnetting 192.168.12.0/24 into three different subnets.

Subnetting is performed to create a smaller number of networks from one big network. A router interconnects these smaller networks. The router stops broadcasts, and the size of the broadcast domains becomes smaller. This reduces the network traffic on each subnetwork, thus increasing the performance. Small subnetworks are easy to manage, and the security policies can be fine-tuned to suit the requirement of each small subnet. The rules of subnetting are:

- Network bits cannot be all 0s or 1s.
- Network and host boundary can be anywhere in the 32-bit range.
- The subnet mask determines the boundary between the network and the host portion of the IP address.

Variable Length Subnet Mask

Variable Length Subnet Mask (VLSM) is a method of dividing a network into different subnets by assigning different subnet masks. VLSM is designed to conserve IP addresses and optimize the use of the available address space.

Using VLSM, you can divide a network into granular subnets based on the requirement. In addition, you can allocate a subnet mask according to the number of networks and hosts. The rules pertaining to subnetting using VLSM are:

- Use a network address by allocating it to different hosts.
- Enable VLSM only for supporting routing protocols. This helps the routing protocol understand and carry the subnetting and supernetting information required to route the network traffic.
- Have the same high-order bits for contiguous networks to be summarized.
- Take routing decisions based on the longest matching network entry in the routing table.

Figure 2.7 depicts four remote routers connected to a central hub router. The LAN interfaces of the four routers have IP addresses 172.16.1.65/30, 172.16.1.129/30, 172.16.1.193/30, and 172.16.1.225/30.

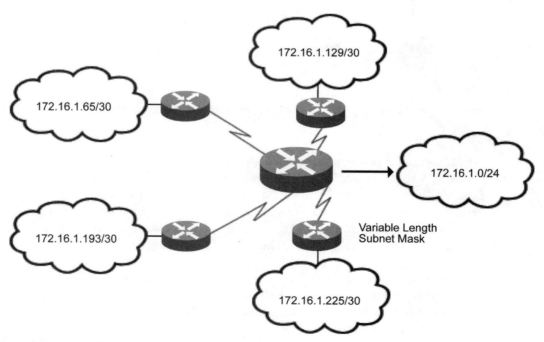

FIGURE 2.7   VLSM using four remote routers.

In Figure 2.7, after performing a logical AND operation between the IP address and the subnet mask, you will get the network address of each LAN interface. The derived network addresses are 172.16.1.64/30, 172.16.1.128/30, 172.16.1.192/30, and 172.16.1.224/30. This shows that a Class B network, which has a default subnet mask of 255.255.0.0, has a VLSM of 255.255.255.252. Using summarization, the network address 172.16.1.0/24 is sent as a routing update. This step reduces the length of the routing table and efficiently uses bandwidth.

The advantages of VLSM are:

- Ensures efficient use of the IP address range
- Allows summarization, which reduces the length of the routing tables and saves the network bandwidth

Protocols, such as Open Shortest Path First (OSPF), Enhanced Interior Gateway Routing Protocol (EIGRP), Routing Information Protocol version 2 (RIP v2), and Border Gateway Protocol (BGP), support VLSM. Other protocols, such as RIP v1 and Interior Gateway Routing Protocol (IGRP), do not support VLSM. VLSM is different from Classless Interdomain Routing (CIDR) because it is used within an organization. CIDR, which is discussed later in this chapter, is used within the Internet.

Classless Interdomain Routing

CIDR, or prefix routing, is a method to replace multiple routes by a single route entry in a top-level global routing table. Classless routing differs from the classful routing of A, B, and C. Classful routing did not provide the granularity required by the Internet. IP addressing was not completely used when allocated to organizations that required a small number of hosts. An example of this would be an instance in which an entire Class C address had to be allocated to an organization that required only 20 hosts. The remaining 234 (255–20) addresses were not used.

Prefix routing was introduced to use the IP addressing space efficiently. Prefix routing and summarization have an important role in the Internet today because they provide a legitimate solution to the problem of the exhaustion of IP addresses.

If the prefix is shorter, the network defined is more general and would represent a larger number of subnetworks. If the prefix is longer, the network defined is more specific and would represent a lesser number of subnetworks. At the top of the hierarchical design, the subnet masks get shorter because they aggregate the subnets that are lower down in the hierarchy. These summarized networks are also called supernets. The Internet uses supernets to aggregate different classes of addresses.

CIDR is defined in the Request For Comments (RFC) 1817 and defines the process of aggregating multiple network numbers in a single routing entry. Route aggregation allows routers, which are at the top of the hierarchy, to group various routes into few routes. This aggregation saves bandwidth when the routing information is transferred to other routers. With CIDR, several networks appear as a single entity to the outside network. Table 2.9 lists some examples of prefix masks:

TABLE 2.9 Examples of Prefix Masks

Prefix	Mask
/23	255.255.254.0
/27	255.255.255.224
/10	255.192.0.0
/6	252.0.0.0

CIDR allows flexibility in allocating networks and subnets. The optimization of routing table size and reduction of router resource overhead reduces the usage of network bandwidth.

For example, 192.168.48.0 is the network address of Abacus Inc. If it were the default subnet mask of a Class C network, it would have been 255.255.255.0. By using a subnet mask 255.255.248.0, the organization can use three additional right-most bits of the third octet to get $2^3 = 8$ "extra" networks. Table 2.10 shows sample prefix masks to provide "extra" networks to an organization.

TABLE 2.10 Prefix Masks to Create Extra Networks

Description	First Octet	Second Octet	Third Octet	Fourth Octet
IP address (decimal)	192	168	48	0
IP address (binary)	11000000	10101000	00110000	00000000
Prefix mask (decimal)	255	255	248	0
Prefix mask (binary)	11111111	11111111	11111000	00000000

Abacus Inc. is a large organization with a big setup. The network setup consists of two routers connected over the Internet. One of the routers has networks 192.16.12.160/27 and 192.16.12.192/27, summarized to 192.16.12.128/25. The other router has networks 192.16.11.64/27 and 192.16.11.32/27, summarized to 192.16.11.0/25. Instead of sending the routing updates of each of the individual subnets of the Internet, only the summarized address is sent, as shown in Figure 2.8. As a result, the use of prefix routing has helped preserve bandwidth.

Table 2.11 lists the prefix routing summarization for networks shown in Figure 2.7. For example, networks 200.16.1.0/24, 200.16.2.0/24, and 200.16.3.0/24 are networks that belong to different customers in an organization. These networks are summarized to 200.16.0.0/22 by Internet Service Provider 1 (ISP 1) and sent as routing updates to ISP 2. The ISPs conserve bandwidth by sending out prefixes during routing updates.

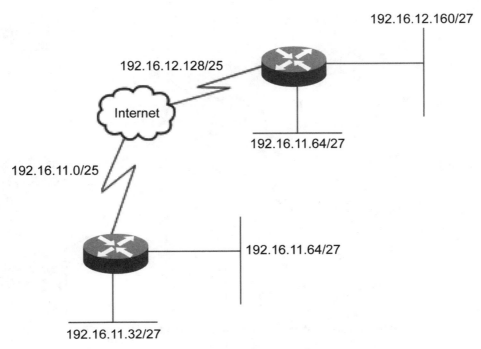

FIGURE 2.8 Prefix routing by sending only the summarized address.

TABLE 2.11 Prefix Routing Summarization

Networks (Dotted decimal)	Networks (Binary)
192.16.12.160/27	11000000.00010000.00001100.10100000
192.16.12.192/27	11000000.00010000.00001100.11000000
192.16.12.128/25 (summarized)	11000000.00010000.00001100.10000000
192.16.11.64/27	11000000.00010000.00001011.01000000
192.16.11.32/27	11000000.00010000.00001011.00100000
192.16.11.0/25 (summarized)	11000000.00010000.00001011.00000000

Figure 2.9 shows the use of prefix routing summarization by different ISPs while exchanging routing updates.

To create supernets, you use different commands for each routing protocol. This will be explained in detail when discussing routing protocols. For example, the supernet command in EIGRP is:

```
Router (conf)# ip summary-address eigrp 1 200.16.0.0 255.255.252.0
```

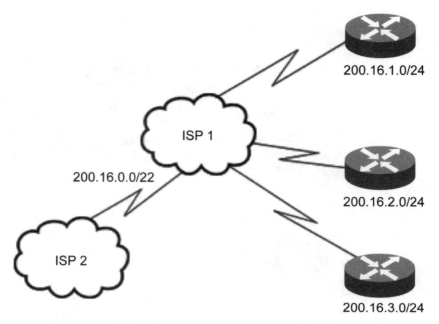

FIGURE 2.9   Prefix Routing Summarization by different ISPs.

Internet Protocol Version 6

Internet Protocol version 6 (IP v6) is an enhancement over IP v4 and was designed to overcome the problem of diminishing IP addresses. Unlike the 32-bit IP v4, IP v6 uses a 128-bit address space and hexadecimal numbers, generating multibillion IP addresses. Additional improvements have been incorporated in IP v6 in areas such as routing and network auto configuration. IP v6 is gradually replacing IP v4, with the two coexisting for a number of years during a transition period.

IP v6 is backward compatible with IP v4 and incorporates features such as extension headers, flow labeling, authentication, and security checks of the IP packet. IP v6 is the industry standard Layer 3 protocol. It is supported by major operating systems and applications.

Each IP v6 address has a length of 128 bits comprising 8 fields of 16 bits each. The 8 fields are separated by a colon (:) and are expressed in hexadecimal numbers. An example of IP v6 is FF03:0101:0202:FEAC:4536:8734:0A0B:0051.

The features of IP v6 are:

- Overcomes the problems faced with IP v4, which has a shortage of IP addresses
- Supports authentication

- Supports Quality of Service (QoS)
- Fragments the packets at the end points, making it granular and more efficient (IP v4 performs fragmentation at the routers.)
- Simplifies IP header packets
- Enables Internet connection for mobile hand-held devices, which support IP v6

Unlike IP v4, an IP v6 address can be classified into three distinct types. The type of IP v6 address that is assigned depends on the function to a network device. Different types of IP v6 addresses are:

Unicast Address: Uniquely identifies a host and resolves to one interface.

Anycast: Uniquely identifies a router. It resolves to one interface of a host in the domain, which is selected from a group of hosts based on the shortest distance routing protocol.

Multicast: Represents a group of nodes using a single address. It resolves all hosts in a domain.

ROUTING

Routing is a process of sending data from the source to the destination host, using different protocols.

Routing protocols will be discussed in detail in the next chapter.

Classful Routing

The routing protocols that do not transfer information about the prefix length are called classful routing protocols. In addition, protocols that do not send the subnet mask along with their routing updates are called classful routing protocols. They summarize routing information by using major network numbers. Examples of classful routing protocols are Routing Information Protocol (RIP) and IGRP (Interior Gateway Routing Protocol (IGRP).

The classful routing protocol uses the first octet rule. Per the rule, the bit pattern of the first four bits of the first octet in the IP address determines the class of the address. Table 2.12 shows the first octet rule applied to the classful routing protocols.

TABLE 2.12 First Octet Rule

Class of Address	Bit Pattern	Range of the First Octet
A	0	0 – 127
B	10	128 – 191
C	110	192 – 223
D	1110	224 – 239
E	1111	240 – 254

Classful routing protocols do not transfer information about the prefix length and do not support VLSM. As a result, classful routing protocols do not summarize classes of addresses that use VLSM within the Internet. This incapability to carry the prefix length or subnetting information leads to a design constraint.

Features of classful routing protocols are:

■ Perform summarization at the network boundary.
■ Do not provide support for VLSM.
■ Do not break up an address space into subnets.
■ Summarize routing updates between different networks into network boundaries.
■ Routing updates within the same network do not carry the subnet mask information, because this information is assumed to be consistent.

Classless Routing

Classless routing is the process of splitting a network into multiple subnets and routing the same, using a protocol that carries subnet information. Routing protocols that transfer information about the prefix length during routing updates are called classless routing protocols. A classless routing protocol makes it possible to represent contiguous blocks of networks or subnets. Examples of classless routing protocols are OSPF, EIGRP, RIP v2, IS_IS, and BGP. These protocols support VLSM.

Features of classless routing protocols are:

■ Support VLSM, and as a result, carry subnetting information
■ Summarize routes at major network boundaries
■ Support CIDR

Figure 2.10 depicts a scenario where Router B needs to send route data to Routers A and C, using classless routing protocols. If classful routing protocols

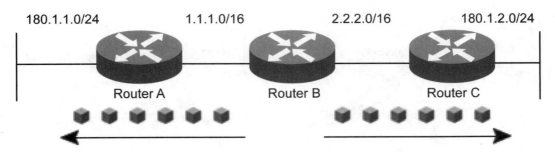

FIGURE 2.10 Using classless routing protocols to avoid routing loops.

were used in such a scenario, it would lead to routing loops. This is because class-ful routing protocols do not carry subnet mask information. Router B does not have any interface in the 180.1.x.0 subnet, and as a result, it sees a classful summary address of 180.1.0.0/16. This leads to routing loops, because Router B does not know that in order to route packets to 180.1.1.0/24, the next-hop router is Router A. For routing packets to 180.1.2.0/24, its next-hop router is Router C. Figure 2.10 shows routers A, B, and C, such that Router B has noncontiguous networks 180.1.1.0/24 and 180.1.2.0/24 on either side.

In the figure, if classless routing was not used as the routing protocol in each of the three routers, then it would lead to a routing loop.

Routing Design Consideration

Before implementing a network, you need a thorough study to analyze the requirement. During this study, gather information pertaining to:

- Number of subnets required immediately in a new network (commonly called "green-field network")
- Number of hosts on each subnet and the entire network
- Number of subnets required for the future
- Number of hosts on each subnet and the entire network in the future
- Rate of user growth and plans for future expansion
- Type of application that the network will support
- Type of model used—peer-to-peer or client-server model
- Volume and nature of traffic
- Security requirements with respect to the nature of business operations
- Accessibility matrix of each subnet
- Requirements for the Internet connectivity
- Requirements for NAT and DHCP
- Protocols that would be used in the network and their level of use

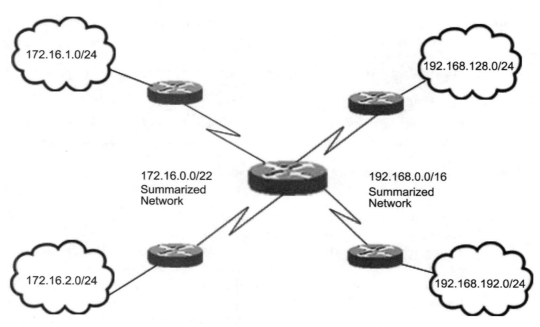

FIGURE 2.11 IP address allocation.

All the information may not be pertinent to every scenario, but most of it is required for you to determine the network design. The pattern of questions may vary, depending on the specifics of a particular technology, for example, Frame-Relay and ATM; which would require additional information like Quality of Service (QoS), numbering of DLCIs, and connectivity of VPIs and VCIs.

One of the most important considerations while designing a WAN is the allocation of the IP addresses. The IP addressing scheme should be done in such a manner that the advantage of prefix routing can be used by all adjacent and contiguous network segments. This saves a lot of bandwidth by not avoiding detailed subnet information over the WAN. Figure 2.11 shows that contiguous networks have been allocated IP addresses that range with the same higher order prefix.

The key design considerations are:

- Collect all information pertaining to the present number of networks, subnets, and hosts with the future requirement. This should include the growth potential in the near future.
- IP addressing should incorporate the requirement for the number of hosts in individual subnets and the total number of hosts in the entire network.
- Interconnectivity of various subnets, according to the customer's requirement.
- To avoid burdening the router, the route summarization should be done at key topological gateways.

■ Determine the routing protocols to be used. Wherever possible, protocols that support VLSM should be used such that you can divide the network into the required number of subnets.

■ Understand the nature and flow of traffic in the network.

The example shown in Figure 2.12 illustrates the key design consideration for ABC Inc. The organization has just merged with another organization, XYZ Inc. Both entities have different classes of addresses, as shown in Figure 2.12.

In Figure 2.12, one of the merging companies has the network addresses of 172.16.1.0/24, 172.16.2.0/24, and so on. XYZ has the network addresses of 192.168.128.0/24, 192.168.192.0/24, and so on. If these network addresses are sent as routing updates to the central router connecting the two entities, the central router will be overwhelmed with the vast amount of information. While designing the merger, the summarization should be planned such that routing updates over the WAN are minimized, and bandwidth utilization is optimized. This in turn brings down the routing cycles required for complex calculation.

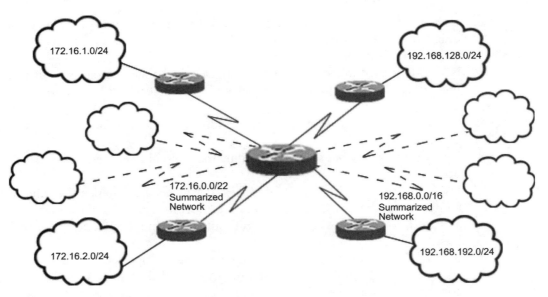

FIGURE 2.12   Designing network for a merger.

Network Address Translation

Network Address Translation (NAT) translates IP addresses of hosts of a network to a different IP address in a different network. NAT reduces the need to use a public IP address for hosts that do not access the Internet. NAT connects hosts having private IP addresses to the public domain or the Internet, by translating the private IP to a registered public IP. A private IP address is used within an organization, while a public IP address is a registered address that is used on the Internet.

For example, using NAT, an organization can use a few registered public IP addresses for all hosts to access the Internet and its resources. The organization is not required to acquire individual registered IP addresses for each host in the organization. The range of private IP addresses as defined in RFC 1918 is listed in Table 2.13.

TABLE 2.13   Range of Private IP Addresses

Class	Address Range	Subnet Mask
A	10.0.0.0–10.255.255.255	255.0.0.0
B	172.16.0.0–172.31.255.255	255.240.0.0
C	192.168.0.0–192.168.255.255	255.255.0.0

The advantages of using NAT are:

- Users with private IP addresses can access the Internet if there is at least one public registered address. This is possible by using NAT, which converts all private addresses to public addresses.
- NAT can be used to reduce the effort required to change IP addresses during a merger. NAT can be used to translate the IP addresses of all hosts in both organizations.
- Users do not need to reconfigure the IP addresses while switching over to a new service provider.
- NAT provides network security by hiding the internal IP address of its hosts in private networks.
- Users can reuse registered IP addresses. If you need to change the internal address of all the hosts, NAT reduces the effort involved. NAT translates the address to a new one at the router without modifying the IP address of each host.
- NAT enables balancing the TCP load by mapping multiple local IP addresses to a single global IP NAT. This feature of NAT is called TCP load sharing. This concept is discussed later in the chapter.

The disadvantages of NAT are:

- NAT increases delay in transmission. This is because each time a packet needs to be translated, the CPU of a router refers the database to check the IP address of the source packet.

■ Applications that require physical addresses of hosts do not function in a NAT environment. This is because NAT hides the actual addresses and replaces them with the translated addresses.

Before configuring NAT, take a quick look at some terminology pertaining to NAT:

Inside Local address: Refers to the same host on the inside network. This address is not always legitimate and refers to the private addresses defined in RFC 1918.

Inside Global address: Refers to a registered, legitimate address to which the inside address gets translated. The ISP provides these addresses.

Outside Local address: Refers to the IP address of an outside host as it appears to the inside network. This address need not be legitimate.

Outside Global address: Refers to the Internet address, that is, the outside address as it appears to the Internet community. This address is assigned to a host on the outside network.

Figure 2.13 depicts a network setup where NAT is used.

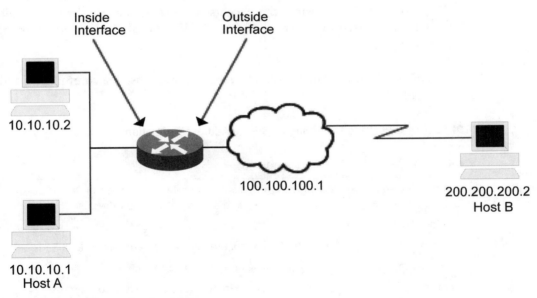

FIGURE 2.13 Network using NAT.

Static Translation

Static translation refers to a one-to-one mapping of the Inside Local address and the Inside Global address. This is helpful when a host in the internal network is accessed from outside.

Referring to the scenario in Figure 2.13, the sequence of NAT is captured as:

1. IP address of the end-user at Host A is 10.10.10.1. Host A attempts to open a connection with Host B with the IP address 200.200.200.2. Host A is on a private IP (RFC 1918) and would not be routed. This host needs a public IP address to connect to host B, which has a public IP.
2. Router checks its NAT table when it receives the first packet on its inside interface. The router translates the Inside Local address to an Inside Global address. This is done assuming that a static translation entry for 10.10.10.1 is defined in the NAT table.
3. NAT router translates the Inside Local address of 10.10.10.1 to the Inside Global address of 100.100.100.1. Host B communicates with this Inside Global address.
4. Host B with the global IP address 200.200.200.2 replies to the connection request and sends the packets to the Global IP address 100.100.100.1.
5. NAT router receives packets from Host B with the Global address 200.200.200.2, routed for Host A with the Inside Global address 100.100.100.1; it checks the NAT table. The router translates the entry to 10.10.10.1 and forwards the packets to Host A, because the NAT table has a translation entry between Inside Local and Inside Global addresses.

To configure static inside source address translation, use the commands listed in Table 2.14.

TABLE 2.14 Commands to Configure Source Address Translation

Command	Explanation
ip nat inside source static local-ip global-ip	Performs static translation between an Inside Local address and an Inside Global address
interface type number	Specifies the inside interface
ip nat inside	Configures the interface as connected to the inside
interface type number	Specifies the outside interface
ip nat outside	Marks the interface as connected to the outside

The global command that defines the source address 10.10.10.1 to be translated to 100.100.100.1 is:

```
NATrouter(config)# ip nat inside source static 10.10.10.1 100.100.100.1
```

When NAT is defined on the global configuration, the inside and outside interfaces should be defined to specify the direction of translation. The interface commands that define the Inside Local address on the interface of Ethernet 0 are:

```
NATrouter(config)# interface e0
NATrouter(config-if)# ip address 10.10.10.1
NATrouter(config-if)# ip nat inside
```

The interface command that defines the Inside Local address on the interface of Serial 0 is:

```
NATrouter(config)# interface s0
NATrouter(config-if)# ip address 100.100.100.1
NATrouter(config-if)# ip nat outside
```

Dynamic Translation

In dynamic translation, the inside address is mapped onto a pool of Inside Global addresses. Consider the example discussed in static translation. Host A makes a connection to Host B, and the sequence of dynamic NAT is captured as:

1. The IP address of the end-user at Host A is 10.10.10.1. Host A attempts to open a connection with Host B, with the IP address 200.200.200.2. Host A does not have a legitimate IP address to connect to host B, which is connected to the Internet.
2. When the router receives the first packet on its inside interface, the router checks its NAT table and translates the Inside Local address to an address defined in the Inside Global address pool.
3. The NAT router translates the Inside Local address 10.10.10.1 to the pool of Inside Global addresses 100.100.100.1. The destination Host B sees and communicates with this address.
4. Host B with the Outside Global IP address 200.200.200.2 replies to the connection request and sends packets to 100.100.100.1.
5. When the NAT router receives packets from Host B with the Outside Global address 200.200.200.2 routed for Host A having the Inside Global address of 100.100.100.1, the router checks the NAT table and translates the address back to 10.10.10.1 and forwards the packets to Host A.

Use the commands listed in Table 2.15 to configure dynamic inside source address translation.

TABLE 2.15 Source Address Translation Commands

Command	Explanation
ip nat pool name start-ip end-ip {netmask netmask | prefix-length prefix-length}	Defines a pool of global addresses to be allocated as needed
access-list access-list-number permit source [source-wildcard]	Defines a standard access list
ip nat outside source list access-list-number pool name	Establishes dynamic outside source translation, specifying the access list defined in the prior step
interface type number	Specifies the inside interface
ip nat inside	Configures the interface as connected to the inside
interface type number	Specifies the outside interface
ip nat outside	Marks the interface as connected to the outside

Using the example of Host A and Host B, all source addresses passing the access list 1 (having a source address from 10.10.10.0/24) are translated to an address from the pool named pool100. The pool contains addresses from 100.100.100.1 to 100.100.100.5. Here, the command shows configuration of dynamic inside source translation.

```
NATrouter(config)# ip nat pool pool100 100.100.100.1 100.100.100.5
  netmask 255.255.255.0
NATrouter(config)# ip nat inside source list 1 pool pool100
```

The interface command invokes the Inside Global address on the interface of Serial 0, as shown here.

```
!
NATrouter(config)# interface s0
NATrouter(config-if)# ip address 100.100.100.10
NATrouter(config-if)# ip nat outside
```

The interface command invokes the Inside Local address on the interface of Ethernet 0 as shown here.

```
NATrouter(config)# interface e0
NATrouter(config-if)# ip address 10.10.10.1
NATrouter(config-if)# ip nat inside
!
access-list 1 permit 10.10.10.0 0.0.0.255
```

Overload

In certain situations, mapping one-to-one addresses may not be an economical solution for connecting to the Internet because you require many registered Inside Global addresses. Instead, you can define a pool of Inside Global addresses to be used by many Inside Local users and conserve registered addresses. This is made possible by using the overload feature of NAT, also known as Port Address Translation (PAT). When NAT overloading is enabled, the router maintains a table of TCP and UDP ports along with the IP addresses of the Inside Local and Inside Global addresses. Theoretically, it is possible to have 65,535 translations per IP address, but Cisco recommends not more that 4,000 translations.

Consider an example of Host A and Host B. Host A makes a connection to Host B. The sequence of NAT is:

1. The IP address of the end-user at Host A is 10.10.10.1. Host A attempts to open a connection with Host B, with the IP address 200.200.200.2. Host A does not have a legitimate IP address to connect to Host B, which is connected on the Internet.
2. The router checks its NAT table when it receives the first packet on its inside interface. In overloading, several Inside Local addresses can be translated to different ports of the same Inside Global address. If overloading is enabled and another translation is active, the router reuses the global address from that translation. The router also updates its NAT database accordingly, such that the address can translate back. This type of entry is called an extended entry.
3. The router replaces the Inside Local source address 10.10.10.1 with the selected Global address, say 100.100.100.1, and forwards the packet. Host B sees and communicates with this address.
4. Host B with Outside Global IP address 200.200.200.2 replies to the connection request and sends packets to 100.100.100.1.
5. The router checks its NAT table when the NAT router receives packets from Host B with the Outside Global address of 200.200.200.2 directed for

Host A with the Inside Global address 100.100.100.1. The router looks for the destination IP and port and translates accordingly.

Table 2.16 lists the commands to configure NAT overloading.

TABLE 2.16 NAT Overloading Commands

Commands	Explanation
ip nat pool name start-ip end-ip {netmask netmask | prefix-length prefix-length}	Defines a pool of global addresses to be allocated as needed
access-list access-list-number permit source [source-wildcard]	Defines a standard access list
ip nat inside source list access-list-number pool name overload	Establishes dynamic source translation, identifying the access list defined in the prior step
interface type number	Specifies the inside interface
ip nat inside	Configures the interface as connected to the inside
interface type number	Specifies the outside interface
ip nat outside	Configures the interface as connected to the outside

To translate different Inside Local addresses to the same Inside Global address having a different port, the set of commands on the router is:

```
NATrouter(config)# ip nat pool pool100 100.100.100.1 100.100.100.5
   netmask 255.255.255.0
NATrouter(config)# ip nat inside source list 1 pool pool100 overload
```

After NAT is defined on the global configuration, the inside and outside interfaces need to be defined in order to specify the direction of translation.

The interface command here defines the Inside Local address on the interface of Ethernet 0:

```
NATrouter(config)# interface e0
NATrouter(config-if)# ip address 10.10.10.1
NATrouter(config-if)# ip nat inside
```

The interface command defines the Inside Global address on the interface of Serial 0 as listed:

```
NATrouter(config)# interface s0
NATrouter(config-if)# ip address 100.100.100.10
NATrouter(config-if)# ip nat outside
access-list 1 permit 10.10.10.0 0.0.0.255
```

Table 2.17 shows the NAT overload translation table of two different hosts.

TABLE 2.17   NAT Overload Translation

Protocol	Inside Local IP Address: Port	Inside Global IP Address: Port	Outside Global IP Address: Port
TCP	10.10.10.1 : 1011	100.100.100.1 : 1011	200.200.200.2 : 21
TCP	10.10.10.2 : 1012	100.100.100.1 : 1012	200.200.200.3 : 21

Translating Overlapping Addresses

NAT translates IP addresses, usually unregistered, to officially assigned IP addresses. At times, the chosen IP address is already officially in use by some other organization.

Table 2.18 lists the steps to be followed for configuring Static Overlapping Address Translation:

TABLE 2.18   Configuring Static Overlapping Address Translation

Command	Explanation
ip nat outside source static global-ip local-ip	Establishes static translation between an Outside Local address and an Outside Global address
interface type number	Specifies the inside interface
ip nat inside	Marks the interface as connected to the inside
interface type number	Specifies the outside interface
ip nat outside	Marks the interface as connected to the outside

Table 2.19 lists the steps to be followed for configuring Dynamic Overlapping Address Translation.

TABLE 2.19 Configuring Dynamic Overlapping Address Translation

Command	Explanation
ip nat pool name start-ip end-ip {netmask netmask | prefix-length prefix-length}	Defines a pool of local addresses to be allocated as needed
access-list access-list-number permit source [source-wildcard]	Defines a standard access list
ip nat outside source list access-list-number pool name	Establishes dynamic outside source translation, specifying the access list defined in the prior step
interface type number	Specifies the inside interface
ip nat inside	Marks the interface as connected to the inside
interface type number	Specifies the outside interface
ip nat outside	Marks the interface as connected to the outside

TCP Load Distribution

There are situations where multiple hosts from the outside network need to communicate with a single host having a valid inside network IP address. This can lead to heavy traffic. Using NAT, it is possible to configure a virtual host in the inside network that coordinates load sharing with other real hosts. The destination address is matched with an access list and is replaced with an address from the rotary pool on a round-robin basis.

Listing 2.1 shows the configuration sequence.

Listing 2.1 Configuration Sequence

```
ip nat pool name start-ip end-ip {netmask netmask | prefix-length
  prefix-length} type rotary
access-list access-list-number permit source [source-wildcard]
ip nat inside destination list access-list-number pool name
interface type number
ip nat outside
```

Table 2.20 shows the explanation for the configuration sequence commands.

TABLE 2.20   Commands of Configuration Sequence

Command	Explanation
ip nat pool name start-ip end-ip {netmask netmask | prefix-length prefix-length} type rotary	Defines the address pool containing the addresses of the real hosts
access-list access-list-number permit source [source-wildcard]	Defines an access list for the address of the virtual host
ip nat inside destination list access-list-number pool name	Establishes dynamic inside destination translation, identifying the access list defined in the prior step
interface type number	Specifies the inside interface
ip nat inside	Configures the interface as connected to the inside
interface type number	Specifies the outside interface
ip nat outside	Configures the interface as connected to the outside

For example, Host B establishes a connection with a virtual host, as shown in Figure 2.14. In Figure 2.14:

■ The Inside Local address is 10.10.10.2
■ The Inside Global address is 100.100.100.1
■ The Outside Global address is 200.200.200.2

In Figure 2.14, the user at Host B with the IP address 200.200.200.2 attempts to open a connection with a host with a virtual IP address 100.100.100.1. The virtual address needs to be translated to the IP address of a real host. The sequence of steps is:

1. The user on Host B with IP address 200.200.200.2 (Outside Global) opens a connection with a virtual host having the IP address of 100.100.100.1 (Inside Global).
2. The NAT router creates a translation and maps it to a real host having the IP address of 10.10.10.2 (Inside Local) when the router receives the request.

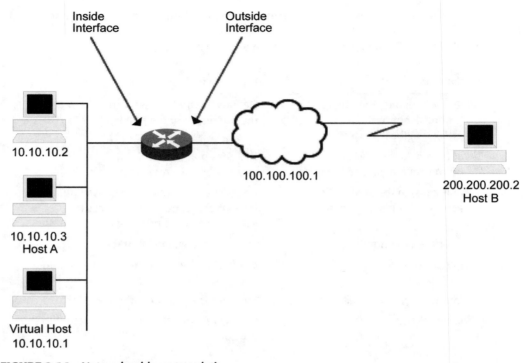

FIGURE 2.14   Network address translation.

3. The Host 10.10.10.2 receives the packet and establishes the connection.
4. The NAT router updates the NAT table with the Inside Local IP address and port numbers, as well as the Inside Global IP address and port numbers. The router translates the address of the real host to that of the virtual host.

After the next request for establishing a connection to 100.100.100.1, the NAT router translates the address to a different Inside Local address. As a result, load distribution is achieved.

SUMMARY

In this chapter, you learned about the two important concepts of the OSI Network layer: IP addressing and routing. We also reminded you about different routing concepts, such as classless and classful routing protocols. In the next chapter, we move on to static and dynamic routing.

POINTS TO REMEMBER

- An IP address is a unique 32-bit Layer 3 address with two components: host and network addresses.
- There are five IP address classes: A, B, C, D, and E.
- A Class A address is used for large organizations with networks supporting large numbers of end users.
- The Class B address format is used for networks of mid-sized organizations.
- The Class C address format is used for small organizations with networks supporting a large number of users.
- Classes D and E are used for experimental purposes.
- Binary representation of IP addresses and subnet masks performs a logical AND operation to find the network portion of the network.
- VLSM divides a network into subnets by assigning different subnet masks.
- VLSM conserves IP addresses and available address space. This is called "subnetting or supernetting a subnet." Subnetting creates a smaller number of networks from one big network to reduce network traffic, thereby increasing performance.
- Subnetworks are easier to manage than one big network, and security policies can be fine-tuned to suit the requirement of each subnet.
- CIDR or prefix routing is a method to replace multiple routes by a single route entry in a top-level global routing table.
- IP v6 is backward compatible with IP v4 and incorporates features like extension headers, flow labeling, authentication, and security checks of IP packets.
- Routing protocols that do not send subnet masks along with their routing updates are called classful routing protocols.
- Classless routing splits a network into multiple subnets and routes the same, using a protocol that carries subnet information.
- Routing protocols that transfer information about the prefix length during routing updates are called classless routing protocols.
- NAT connects hosts having private IP addresses to the public domain or the Internet by translating the private IP used within an organization to a registered public IP that is used on the Internet.
- Static translation refers to one-to-one mapping of Inside Local addresses and the Global address.
- Dynamic translation refers to the mapping of Inside addresses onto a pool of Inside Global addresses.
- In TCP load distribution, the destination address is matched with an access list and is replaced with an address from the rotary pool on a round-robin basis.

3 Static and Dynamic Routing

STATIC ROUTING

Static routes are user-defined routes, configured to route a packet to a specific network destination by directing the packet to the interface of a router. This makes it possible to route the packets through a specific and manually controlled path. Static routes are manually configured, and this is reflected in the routing table and takes precedence over routes chosen by dynamic routing protocols.

Periodic routing updates are not present in static routing. As a result, neighboring routers do not exchange routing information and have no way to communicate any changes in the network topology.

Static routing has many advantages. You have complete control of the network, because the path to be followed in reaching a destination is manually configured with the help of static routes. This feature of static routes increases manageability in small networks and adds some level of security.

The disadvantage of using static routing is that maintaining networks manually can be tedious and prone to human error. Neighboring routers do not exchange routing updates, as in the case with dynamic routing, and as a result, you are responsible for updating routing tables each time there is a change in the network topology. This process can be cumbersome if it is a large network or a network where the requirements change frequently.

Routing updates between participating routers are performed by broadcast or multicast traffic, if they are running a dynamic routing protocol.

Configuring Static Routes

Static routes are configured to reduce the network overhead created with the introduction of dynamic routing. We will discuss dynamic routing later in the chapter. Static routes are also used when two Autonomous Systems (ASs) do not need to exchange their entire routing tables but only information pertaining to a few routes. In Figure 3.1, two ASs are connected over a serial link.

In Figure 3.1, each AS uses a dynamic routing protocol for exchanging network information. There is, however, no routing protocol running between two ASs. Instead, each AS uses a static route pointing to the summarized address of the other AS. As a result, each AS has routing policies confined to its own administrative domain, thereby increasing the security of the network.

The command for configuring static routes in Cisco routers is:

```
ip route destination_network destination_mask address/interface
[distance] [tag tag] [permanent]
```

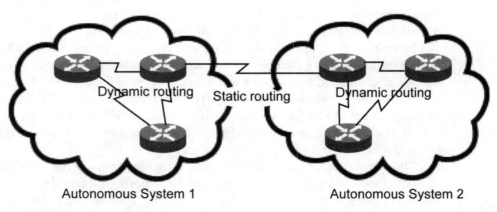

FIGURE 3.1 ASs using static routing.

Table 3.1 explains each of the fields used in the above command.

TABLE 3.1 Commands for Configuring Static Routes

Field	Explanation
ip route	Configures static routes in Cisco routers
destination_network	Specifies the address of the destination network or subnet
destination_mask	Specifies the subnet mask of the destination network or subnet
Address	Specifies the IP address of the next-hop router for reaching the destination address
Interface	Is specified as an alternative to the next-hop IP address
Distance	Specifies the optional administrative distance assigned to the particular route
Tag	Specifies the optional value used along with route maps
Permanent	Specifies that the route will not be removed even if the associated interface with the route goes down

Remember that a static route should be redistributed to advertise to other routers.

The next-hop router must be in the routing table before it is defined in the static route. Usually it is directly connected to the router.

Figure 3.2 shows the configuration of static routes.

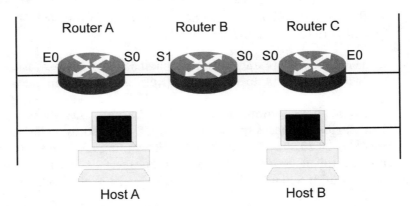

FIGURE 3.2 Configuration of static routes.

In Figure 3.2, three routers, Router A, Router B, and Router C, are connected to each other. There is no routing protocol running in any of these routers. In this example, connectivity is being established between Host A and Host B using static routes. Table 3.2 gives the interface configuration of Router A, Router B, and Router C.

TABLE 3.2 Interface Configuration of Routers

Router A	Router B	Router C
Ethernet 0 = 10.10.10.1	Serial0 = 200.200.200.2	Ethernet 0 = 20.20.20.2
Serial0 = 100.100.100.1	Serial1 = 100.100.100.2	Serial0 = 200.200.200.1

Looking at Figure 3.2, Router A must know the next-hop address in order to reach networks 20.20.20.0 and 200.200.200.0. Router A has no way to gain this knowledge automatically, because there is no routing protocol in use. The only way to ensure that Router A knows the next-hop address to reach networks 20.20.20.0 and 200.200.200.0 is by configuring the next-hop address manually. To configure Router A, the commands are:

```
RouterA(config)# ip route 20.20.20.0 255.255.255.0 Serial 0
RouterA(config)# ip route 200.200.200.0 255.255.255.0 Serial 0
```

Router B should know the next-hop address to 20.20.20.0 and 10.10.10.0. To configure Router B, the commands are:

```
RouterB(config)# ip route 20.20.20.0 255.255.255.0 Serial 0
RouterB(config)# ip route 10.10.10.0 255.255.255.0 Serial 1
```

Router C should know the next-hop address to 100.100.100.0 and 10.10.10.0. To configure Router C, the commands are:

```
RouterC(config)# ip route 100.100.100.0 255.255.255.0 Serial 0
RouterC(config)# ip route 10.10.10.0 255.255.255.0 Serial 0
```

In the above commands, when Host A wants to establish a connection with Host C, Host A checks the destination network address. The network address of Host C is 20.20.20.0. Router A checks its routing table and finds that there is an entry for this network with Serial 0 as the interface. Router A then forwards the packets out of Serial 0. This process is repeated as the packet moves across each router.

Removing Existing Static Routes

You need to remove certain static routes to update the routing table. There could be different reasons for removing or modifying a static route:

- Destination network has changed or is no longer valid.
- Next-hop address has changed or is no longer valid.
- The path to reach a particular destination needs to be modified.
- Static route needs to be removed to incorporate dynamic routing protocols.

Taking the example of Router A, Router B, and Router C, as shown in Figure 3.2, use the following command to remove the static route 20.20.20.0 from the routing table of Router A:

```
RouterA(config)# no ip route 20.20.20.0
```

The Routing Table

A routing table is a table stored in a router or any other internetworking Layer 3 device, and it tracks the routes to a particular network destination. The manual static route entries get reflected in the routing table.

The most important information stored in the routing table is:

- The next-hop address corresponding to a destination network
- The routing protocol used to reach the destination
- The metrics associated with the routes

A routing metric enables a routing algorithm to choose one path over the other in order to reach a destination. Routing tables contain crucial information and play an important role for troubleshooting a networking problem. Listings 3.1, 3.2, and 3.3 show the routing tables for Routers A, B, and C after all the static routes have been added.

LISTING 3.1   Routing Table of Router A after Static Routes Have Been Added

```
RouterA# show ip route
Codes: C - connected, S - static, I - IGRP, R - RIP, M -
   Mobile, B - BGP, D - EIGRP, EX - EIGRP external, O - OSPF,
   IA - OSPF inter area, E1- OSPF external type 1, E1 - OSPF
   external type 2, E - EGP, i - IS_IS, L1 - IS_IS Level -1,
   L2 - IS_IS level-2, * - candidate default
```

```
Gateway of last resort is not set
  S  20.20.20.0        [1/0]       100.100.100.2
  S  200.200.200.0     [1/0]       100.100.100.2
  C  10.10.10.0 is directly connected, Ethernet 0
  C  100.100.100.0 is directly connected, Serial 0
```

LISTING 3.2   Routing Table of Router B after Static Routes Have Been Added

```
RouterB# show ip route
Codes: C - connected, S - static, I - IGRP, R - RIP, M -
  Mobile, B - BGP, D - EIGRP, EX - EIGRP external, O - OSPF,
  IA - OSPF inter area, E1- OSPF external type 1, E1 - OSPF
  external type 2, E - EGP, i - IS_IS, L1 - IS_IS Level -1,
  L2 - IS_IS level-2, * - candidate default

Gateway of last resort is not set
  S  20.20.20.0        [1/0]       200.200.200.3
  S  10.10.10.0        [1/0]       100.100.100.1
  C  200.200.200.0 is directly connected, Serial 0
  C  100.100.100.0 is directly connected, Serial 1
```

LISTING 3.3   Routing Table of Router C after Static Routes Have Been Added

```
RouterC# show ip route
  Codes: C - connected, S - static, I - IGRP, R - RIP, M -
  Mobile, B - BGP, D - EIGRP, EX - EIGRP external, O - OSPF,
  IA - OSPF inter area, E1- OSPF external type 1, E1 - OSPF
  external type 2, E - EGP, i - IS_IS, L1 - IS_IS Level -1,
  L2 - IS_IS level-2, * - candidate default

Gateway of last resort is not set
  S  10.10.10.0        [1/0]       200.200.200.2
  S  100.100.100.0     [1/0]       200.200.200.2
  C  20.20.20.0 is directly connected, Ethernet 0
  C  200.200.200.0 is directly connected, Serial 0
```

Listing 3.4 shows the routing table for Router A after the static route 20.20.20.0 has been removed.

LISTING 3.4   Routing Table of Router A after Static Route Has Been Removed

```
RouterA# show ip route
  Codes: C - connected, S - static, I - IGRP, R - RIP, M -
  Mobile, B - BGP, D - EIGRP, EX - EIGRP external, O - OSPF,
  IA - OSPF inter area, E1- OSPF external type 1, E1 - OSPF
  external type 2, E - EGP, i - IS_IS, L1 - IS_IS Level -1,
  L2 - IS_IS level-2, * - candidate default
```

```
Gateway of last resort is not set
  S  200.200.200.0    [1/0]       100.100.100.2
  C  10.10.10.0 is directly connected, Ethernet 0
  C  100.100.100.0 is directly connected, Serial 0
```

Default Routes

In static routes, each network address is defined separately. For example, networks using stub routers have the same next-hop address. In such a situation, you are not required to define each network separately. Instead, one default network address can be defined.

When a default route is used, the router takes the default route if it finds no specific entry in the routing table for a destination network. If no default network address is defined and the router finds no entry in the routing table for the desired network, the packet is dropped.

In stub networks, routing updates are not sent to stub routers. This causes downstream routers to have a limited knowledge of the network. To overcome this problem, default routes are defined in stub routers.

In Figure 3.3, when Host A needs to send a packet to Host B, the packet is sent to the router R1. The router checks its routing table for an entry pointing to the

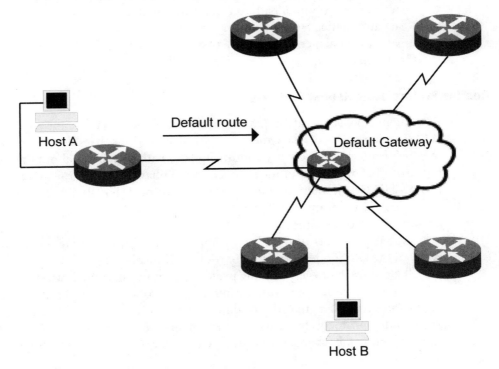

FIGURE 3.3 Transfer of data packet via default route.

destination network. It finds no specific entry for the destination network; instead there is a default route pointing to the WAN cloud. This means that the packet is forwarded to the default address.

Default routes are simple to implement in a small network. In addition, default routes reduce the network overhead, as well as the memory use and CPU cycles of the router.

DYNAMIC ROUTING

Dynamic routing protocols use an algorithm to enable routers to discover routes automatically to different destinations and to share the information with other routers. Dynamic routing protocols determine the best and the next-best paths to any destination networks. The most important feature of dynamic routing protocols is the ability to react to topological changes. Some dynamic routing protocols are even capable of load balancing between equal cost links.

Dynamic routing protocols can:

- Send and receive information about different networks
- Determine the optimal path to reach a destination
- React automatically to topological changes
- Ensure load balances when more than one path exists between the source and the destination

Routing Protocols and Routed Protocols

Routing protocols operate in Layer 3 of the OSI model to route the routed protocols. Routing protocols have their own set of algorithms that describe the method in which routers will exchange routing updates with each other about the available networks. Examples of routing protocols are Enhanced Interior Gateway Routing Protocol (EIGRP), Interior Gateway Routing Protocol (IGRP), Open Shortest Path First (OSPF), and Routing Information Protocol (RIP). Routing protocols are also known as network protocols.

Routed protocols deal with user applications. They operate at the top four layers of the OSI model. Examples of routed protocols are IP, IPX, AppleTalk, and Novell Netware. Not every routing protocol supports all routed protocols.

A routing protocol supports various routed protocols and is used to send routing updates to routers in order to find the best path for transporting data. It supports routed protocols by providing a mechanism for sharing routing information between routers. Exchanges of routes are required to update and maintain routing tables.

Table 3.3 provides a list of routing protocols, along with the routed protocols they support.

TABLE 3.3   Routing Protocols and Corresponding Supported Routed Protocols

Routing Protocols	Routed Protocols
RIP	IP, IPX
EIGRP	IP, IPX, AppleTalk
IGRP	IP
OSPF	IP
IS_IS	IP
NLSP	IPX
AURP	AppleTalk
RTMP	AppleTalk

A routing protocol determines the optimal path among alternative paths to any destination. After all routers running the routing protocols are synchronized, the routers start routing the packets. Each router updates its routing table accordingly, after learning the routes from peer routers that are participating in the process.

Interior Gateway Protocols

Routing protocols that operate and administer within an organization and are within the administrative domain of one or more administrators are called Interior Gateway Protocols (IGPs). These protocols function in Layer 3 of the OSI model. IGPs can be categorized into distance vector protocols and link state protocols. Examples of IGPs are RIP, IGRP, EIGRP, and OSPF.

Distance Vector Protocols

Distance vector protocols were created for small networks. RIP v2 and IGRP are examples of distance vector protocols. Routers using distance vector protocols send all routing updates to directly connected neighbors at periodic intervals. The periodic update is sent after a timer of between 30 and 90 seconds expires.

Distance vector protocols follow the split-horizon rule. According to this rule, routing updates of networks are not sent out through the same interface from where they are received. This avoids routing loops. Distance vector protocols use

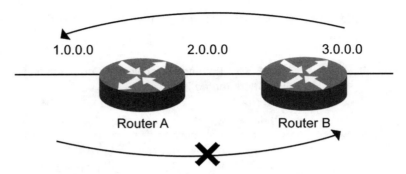

FIGURE 3.4 Split-horizon rule of distance vector protocols.

hop-counts as the metric used to select the optimal path. Figure 3.4 illustrates the split-horizon rule.

Distance vector protocols are more prone to routing loops as compared to link state protocols because they take longer to reach convergence. Distance vector protocols are computationally simpler. Lowest hop-counts do not necessarily signify the optimal path to reach a particular network. It is possible for a distance vector protocol to choose a lower hop-count slow link, such as 64 Kbps, over a higher hop-count faster link, such as E1.

Figure 3.5 shows three routers, each of which knows of the two directly connected networks at initiation. In Figure 3.5, Router A knows of networks 1.0.0.0 and 2.0.0.0, but it is not aware of networks 3.0.0.0 and 4.0.0.0. After exchanging routing updates, the routers know the hop-count to reach these remote networks. Router A's routing table shows a metric of 2 to reach network 4.0.0.0.

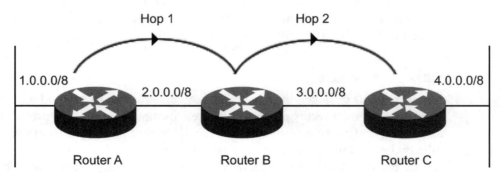

FIGURE 3.5 Distance vector protocols use hop-count for metric.

Apart from the split-horizon rule, distance vector protocols use the following techniques:

Split-horizon with poison reverse: Refers to the routing protocol advertising all routes out of an interface except for those learned from previous updates from the same interface.

Route poisoning: Indicates that a network is unreachable. Routers remove the entry from the routing table after the timer expires.

Hold-down timers: Refers to the period when a new route is detected and the router does not accept or advertise any information about the route in order to avoid routing loops.

Triggered update: Is sent each time there is a topology change. This update does not wait for the update timer to expire. Triggered updates are also called flash updates.

Count to infinity: Refers to the technique used to avoid routing loops, when distance vector protocols use a predefined maximum hop-count, beyond which the network is considered unreachable.

Figure 3.6 displays the route poisoning technique of distance vector protocols. In Figure 3.6, Router B updates the failure of subnet 3.0.0.0 to Router A with a hop-count of infinity to avoid routing loops. When Router A receives this update, it will not route any packets to network 3.0.0.0, because it has an unreachable hop-count.

A disadvantage of distance vector protocol is that it can take up to 90 seconds to realize and announce a link failure. A distance vector protocol also sends out the entire routing table instead of incremental updates. This leads to an increase in the overhead of the router and the network.

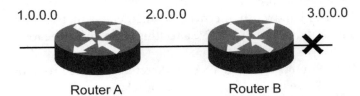

1.0.0.0   2.0.0.0   3.0.0.0

Router A   Router B

FIGURE 3.6 Route poisoning technique of distance vector protocols.

Link State Protocols

The link state protocol is a routing protocol whereby each router broadcasts or multicasts routing information about all its neighbors to all nodes in the internetwork. This makes it possible for each router to have the same image of the network. Link state protocols are also called shortest path or distributed database protocols.

A link state protocol is less prone to routing loops, as it reaches convergence quickly. Different protocols have different update timers, as shown in Table 3.4.

TABLE 3.4 Link State Protocols and Corresponding Update Timers

Link State Protocol	Update Timer
OSPF	Sends incremental updates after every network change. If no update is sent for 30 minutes, it sends a compressed version of the table.
IS_IS	Sends incremental updates after every network change. If no update is sent for 15 minutes, it sends a compressed version of the table.

A link state protocol establishes communication with its adjacent router, which is called adjacency. Two routers forming an adjacency should have the same network and subnet mask and be reachable by Layer 3. Two routers that have formed an adjacency are called neighbors. This neighbor relationship is maintained by exchanging messages called "Hello packets." Link state routing protocols are used in large networks and have better scalability, as compared to distance vector protocols.

Each time a link state router detects that a route is missing in its routing table, it sends an incremental update. This incremental update leads to faster convergence time. In OSPF, these incremental updates are called Link State Advertisements (LSAs).

A metric is used in link state protocols for selecting the path to a network. OSPF uses bandwidth and delay as the default metric. Link state protocols use topological databases to derive a routing table, using the following procedure:

1. Routers exchange LSAs to directly connected routers and form adjacency. These routers form a neighbor relationship.
2. Each router keeps track of the neighbor's name, link status, and cost of the path.
3. All the routers build a topological database.
4. Using metrics, routers calculate the shortest and the most efficient path.
5. Each router builds its own routing table.

Table 3.5 compares distance vector and link state protocols.

TABLE 3.5 Comparison of Distance Vector and Link State Protocols

Distance Vector Protocol	Link State Protocol
Builds the network topology using indirect information from neighbors	Builds the network topology using direct information from neighbors
Selects the path using hop-count as metric	Selects paths using metrics such as bandwidth and delay
Sends routing updates after fixed intervals, usually 30 seconds	Sends routing updates only when there is a topology change, resulting in faster convergence

A third type of protocol combines the best features of both distance vector and link state protocol. This protocol is known as balanced hybrid protocol. EIGRP is an example of a balanced hybrid protocol.

Exterior Gateway Protocols

Exterior Gateway Protocols (EGP) run between ASs and exchange routing information across organizations. These protocols are highly complex because they are managed by different administrators and have individual routing policies. Border Gateway Protocol (BGP) is an example of EGP.

Routing information is shared among different ASs. As a result, individual security and routing policies should be carefully analyzed before implementing EGP. The organizations sharing information using EGP should know each other and agree on the routes to be shared.

Figure 3.7 depicts three ASs, AS1, AS2, and AS3. These ASs use IGPs for exchanging routes within their own boundaries. However, EGP is required to share information with other ASs.

Routing Convergence

Routing convergence is defined as the time taken by all the routers in the network to synchronize their database after a change has taken place in the network topology. During this time, each router recalculates the optimal routes, and when all synchronize, they agree on the optimal routes. The speed and capability of agreeing to the changed topology depends on the algorithm of the routing protocol.

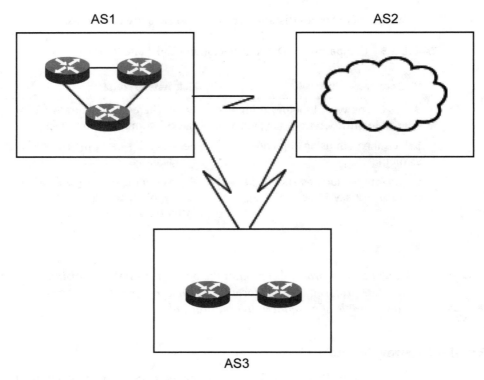

FIGURE 3.7   Border Gateway Protocol.

Delays in convergence can lead to routing loops. This is because the more time that is taken for convergence, the higher the chances of a routing loop.

Each protocol tries to decrease the time taken to synchronize and agree on the changed topology. All routers handle convergence differently. The following section describes the steps for convergence of different routing protocols.

Routing Information Protocol

RIP is a distance vector dynamic routing protocol, which is suitable for small homogeneous networks. The steps for convergence in the RIP algorithm are:

1. A router running RIP sends a flash update when there is a topology change. This is called triggered update with poison reverse.
2. The recipient routers put the affected route in their routing tables in hold-down state and send the flash updates.
3. The originating router again queries its neighbors for alternate paths. If a route exists, the neighbor sends it across. If none exists, then the poisoned route is sent.

4. The router that originated the flash update now installs the best route in its routing table.
5. Other routers, which are still in a state of hold-down, ignore information pertaining to alternate paths. These routes accept the new route once the hold-down timer expires.

Interior Gateway Routing Protocol

IGRP is a Cisco proprietary distance vector dynamic routing protocol. It was designed in the 1980s for ASs that contain large, complex networks. The steps for convergence in IGRP are:

1. A router running the IGRP protocol sends a triggered update with poison reverse when there is a topology change.
2. The recipient routers put the affected route in their routing tables in hold-down and send the flash update.
3. The originating router again queries its neighbors for the alternate path. If one exists, the neighbor sends it across, or else it sends the poisoned route.
4. The router that originated the flash update installs the best route in its routing table.
5. Other routers, which are still in a state of hold-down, ignore information pertaining to alternate paths. These routes accept the new route once the hold-down timer expires.

Enhanced Interior Gateway Routing Protocol

EIGRP is a hybrid protocol that is proprietary to Cisco. The following steps describe convergence in EIGRP:

1. A router running the EIGRP protocol checks for a feasible successor in its topology table when there is a change in topology.
2. If there is a feasible successor, then convergence occurs immediately. If not, then the router changes itself to active state.
3. The originating router again queries its neighbors for alternate paths. If a route exists, the neighbor sends it across as a successor. This is updated in the routing database.
4. A flash update on the new successor is sent out to the network.

Open Shortest Path First

OSPF is a protocol that supports very large networks and is not proprietary to Cisco. It supports IP subnetting and tagging of routes. The steps for convergence for OSPF are:

1. A router running OSPF protocol sends LSAs to neighbors when there is a topology change.
2. Routers update their routing table on receiving the LSAs.

Route Summarization

Route summarization refers to the consolidation of advertised subnetwork addresses such that a single summary route is advertised to other areas by an area border router. Route summarization occurs at major network boundaries to save bandwidth. Outside networks only see the summarized address and cannot see the internal individual subnetworks.

Route summarization in RIP v1 and IGRP is automatic and cannot be disabled, because these do not send the subnet mask information. Route summarization can be enabled or disabled in RIP v2 and EIGRP as they carry the subnet mask information and allow the use of VLSM and summarization. Figure 3.8 shows the benefits of using route summarization.

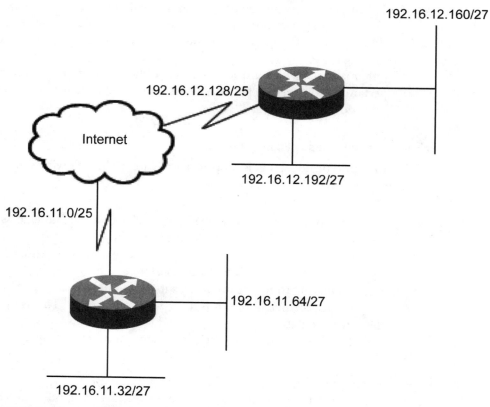

FIGURE 3.8   Benefits of route summarization.

In Figure 3.8, two routers are connected to the Internet. One of the routers has networks 192.16.12.160/27 and 192.16.12.192/27 that are summarized to 192.16.12. 128/25. The other router has networks 192.16.11.64/27 and 192.16.11.32/27 that are summarized to 192.16.11.0/25. It is possible to summarize subnets and consolidate them into one summarized address.

The advantages of route summarization are:

- Summary addresses consolidate several subnets and reduce the size of the routing table. This reduces the overhead of the routers and makes it more efficient.
- Internal individual networks are not known outside the networks. Only the summarized address is seen from outside. This increases security.
- Summarization enables scalability of networks.

Figure 3.9 shows how routing summarization is used by different Internet Service Providers (ISPs) while exchanging routing updates.

In Figure 3.9, the networks of 200.16.1.0/24, 200.16.2.0/24, and 200.16.3.0/24 that belong to different customers are summarized to 200.16.0.0/22 by ISP 1 and sent as routing updates to ISP 2.

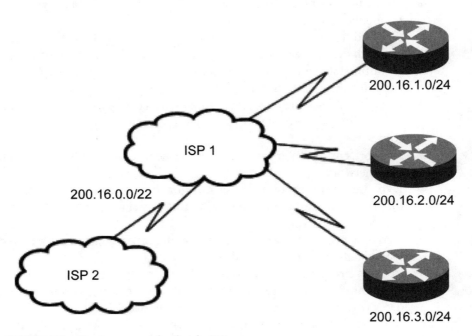

FIGURE 3.9 Route summarization in ISPs.

Scalability of Routing Protocols

Route summarization decreases the routers' overhead. This allows the network a high degree of scalability. Scalability of routing protocols refers to that feature by which the size of the network can expand with increasing requirements without a significant decrease in performance. This would not have been possible without the networks being summarized as the routes move up the top hierarchical layers.

SUMMARY

In this chapter, we have discussed the two types of routing, static and dynamic. We have also discussed how static routing is configured manually, whereas dynamic routing automatically reflects changes in the routing table. In the next chapter we will discuss Routing Information Protocol.

POINTS TO REMEMBER

- Static routing is configured manually for an entry into the routing table.
- Dynamic routing automatically modifies the routing table to reflect changes in the network topology.
- Static routes are user-defined routes, configured to route a packet to a specific network destination by directing the packet to the interface of a router.
- A routing table is a table stored in a router, or any other Layer 3 device, to track the routes to a particular network destination.
- Routing tables store the next-hop address corresponding to destination network, routing protocol used to reach destination, and metrics associated with routes.
- Networks using stub routers have the same next-hop address and do not define each network separately.
- Dynamic routing protocols use an algorithm for routers to discover routes to different destinations automatically, and they share the information with other routers.
- Dynamic routing protocols react to topological changes and are even capable of load balancing between equal cost links.
- Routing protocols such as EIGRP, IGRP, OSPF, and RIP have their own set of algorithms that describe the method in which routers exchange routing updates with each other about the available networks.
- IGPs operate and administer within an organization and are within the administrative domain of one or more administrators.

- IGPs can be categorized into distance vector protocols and link state protocols.
- Routers using distance vector protocols send the entire routing updates to directly connected neighbors at periodic intervals.
- Distance vector protocols follow the split-horizon rule to avoid routing loops.
- In link state protocol, each router broadcasts or multicasts routing information about all its neighbors to all nodes in the internetwork.
- EGPs run between ASs and exchange routing information across organizations.
- Routing convergence is the time taken by all routers in the network to synchronize their database after a change in the network topology.
- Route summarization is the consolidation of advertised subnetwork addresses such that a single summary route is advertised to other areas by an area border router.
- Route summarization decreases the routers' overhead. This allows the networks a high degree of scalability.
- Scalability of routing protocols refers to that feature by which the size of the network can expand with increasing requirement without a significant decrease in performance.

4 | Routing Information Protocol

IN THIS CHAPTER

- Operation and Configuration of RIP
- RIP Versions
- Global Commands to Enable RIP

INTRODUCTION TO RIP

Routing Information Protocol (RIP) is a distance vector protocol that follows the Bellman-Ford algorithm. RIP was developed at Xerox Network Systems (XNS)™ and later was introduced into the Berkeley Software Design (BSD) Unix™ TCP/IP protocol suite. Today, RIP is supported by most operating systems, including Windows™ and Unix.

RIP OPERATION

RIP is suitable for small, homogeneous networks and is relatively easy to implement. It has certain limitations that render it unsuitable for large and complex networks:

- Does not support classless routing (leads to waste of address space)
- Cannot be used in networks that require CIDR and VLSM
- Is not suitable for networks that require more than 15 hops
- Convergence is slow and is prone to routing loops
- Uses broadcasts for routing updates, causing an increase in network traffic

RIP uses hop-count for its metric. A hop-count of 16 signifies an unreachable network. RIP sends "Request" and "Response" messages during routing updates.

RIP uses UDP port 520 for all messages. RIP sends out Request messages using broadcasts when the RIP process begins.

Neighboring routers send Response messages containing their routing tables after receiving updates. On receiving the Response from its neighbor, a router checks to see whether the update is new or not. If the update is found to be new, then it is entered in the routing table along with the address of the advertising router. If the entry already exists for the particular network, then the updates are ignored unless one is received with a lower hop-count.

A hop-count between 0 and 15 is considered to be valid. A count of 16 would indicate a network that is unreachable.

RIP uses the major classful network number for route summarization, because it does not carry subnet mask routing information. Subnet masks enable the efficient utilization of the IP addressing scheme. The lack of subnet mask information carried by RIP means that a router must assume that the subnet mask with which it is configured is effective for all the subnets.

When multiple paths can be used to reach a destination network, the router makes a choice based on the reliability of the routing information source, known as its Administrative Distance. The higher the value of Administrative Distance, the lower the reliability. RIP has an administrative distance of 120.

RIP is defined in Request For Comment (RFC) 1058 and 1723.

RIP Timers

Timers are a property of protocols that determine several variables, such as the frequency of routing updates and the amount of time for a route to become invalid. RIP uses the timers depicted in Table 4.1.

TABLE 4.1 Types of RIP Timers

Timer	Description
Update Timer	Refers to the period after which the router sends the entire routing table to all neighbors. By default, the Update Timer is 30 seconds.

Table 4.1   *(continued)*

Timer	Description
Hold-down Timer	Refers to the period the router waits to receive information about an invalid route, ignoring any updates about the route. By default, the Hold-down Timer is 180 seconds.
Invalid Timer	Refers to the period after which the router waits without getting any response from a particular network, before advertising to its neighbors that the route is invalid. The default value of the Invalid Timer is 90 seconds.
Flush Timer	Refers to the period a router waits after a route has been declared invalid before it flushes the route out of the routing table. By default, the Flush Interval is 240 seconds.

RIP Synchronization

All routers need to understand the network topology in order to perform routing operations. This is possible only if the routers participating in RIP are able to communicate with each other. This communication is known as RIP synchronization.

The steps for the synchronization of RIP are:

1. Each router advertises its directly connected networks to it neighbors using RIP updates. RIP v1 uses broadcasts and RIP v2 uses multicasts for sending routing updates.
2. The router builds its own routing table on receiving these routing updates.
3. A metric is used to select a path from different alternatives for reaching a particular destination network.
4. RIP considers the network to be unreachable if the hop-count is more than 15.

The frequency of sending routing updates to the neighbors is 30 seconds. The RIP routing updates include the IP destination address, metric, next-hop address, timers, flags, hold-downs, split-horizon information, and poison reverse updates. Flags indicate whether the route has been recently changed or not.

RIP VERSIONS

RIP has two versions, RIP v1 and RIP v2. Networks that implement RIP v1 often encounter classful routing loops, because RIP v1 does not carry subnet information. RIP v2 was developed to address this shortcoming of RIP v1. It was designed to carry subnet mask information, by supporting VLSM.

RIP v1

RIP v1 is the first version of RIP and is widely used in a large number of networks. RIP v1 is ideal for small networks that do not need to carry subnet information. Table 4.2 depicts the features of RIP v1.

TABLE 4.2   Features of RIP v1

Hop-count Features	RIP v1 Values
Type of Protocol	Distance vector
Authentication	No
Updates	Broadcast
Subnet Information	Classful
Update Timer	30 seconds
Hold-down Timer	180 seconds
Metric	Hop-count
Count to Infinity	16
Protocol	UDP port 520
Flash or Triggered Updates	Yes

To understand the features of RIP v1, we will look at the IP RIP v1 packet format. Figure 4.1 shows the different fields of the IP RIP v1 packet format, along with the corresponding byte lengths.

1 Byte	1 Byte	2 Bytes	2 Bytes	2 Bytes	4 Bytes	4 Bytes	4 Bytes	4 Bytes
Command	Version Number	Unused	Address Family Identifier	Unused	Address	Unused	Unused	Metric

FIGURE 4.1   IP RIP v1 packet format.

Table 4.3 lists the fields of the IP RIP v1 packet format depicted in Figure 4.1.

TABLE 4.3   Fields of IP RIP v1 Packet Format

Field	Explanation
Command	Indicates whether the packet is meant to be a request or a response
Version Number	Denotes the different RIP versions
Unused	Value equal to zero
Address Family Identifier (AFI)	Indicates address family of the routed protocol; for example, the AFI for IP is 2
Address	Specifies address of the routed protocol; for example, for IP, it is the IP address
Metric	Indicates the best path to reach a network

RIP v2

RIP v2 is the second version of RIP, and it addresses the shortcomings of RIP v1. RIP v2 has added security features, such as plain text and Message Digest 5 (MD5) authentication. In addition, RIP v2 provides support for VLSM, allowing it to perform classless routing.

RIP v2 supports CIDR. RIP v2 uses multicast addresses for routing updates, thereby reducing network traffic and increasing network performance. RIP v2 also provides route summarization. Table 4.4 discusses the features of RIP v2.

TABLE 4.4   Features of RIP v2

Features	RIP v2 Values
Type of Protocol	Distance vector
Subnet Information	Classless
Update Timer	30 seconds
Hold-down Timer	180 seconds
Metric	Hop-count
Count to Infinity	16
Routing Updates Mode	Multicast; address 224.0.0.9
Protocol	UDP port 520
Flash or Triggered Updates	Yes

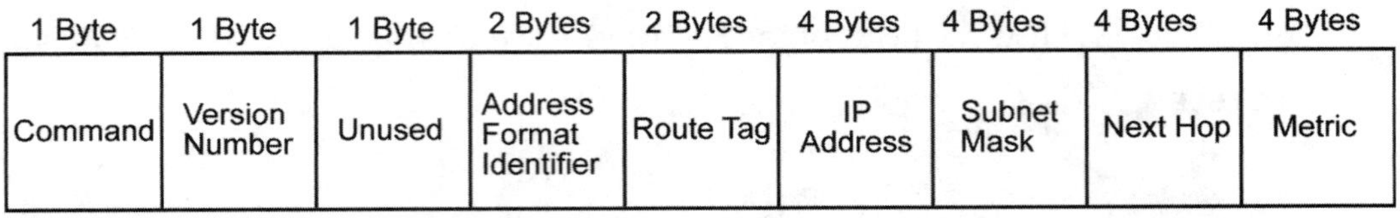

1 Byte	1 Byte	1 Byte	2 Bytes	2 Bytes	4 Bytes	4 Bytes	4 Bytes	4 Bytes
Command	Version Number	Unused	Address Format Identifier	Route Tag	IP Address	Subnet Mask	Next Hop	Metric

FIGURE 4.2   RIP v2 packet format.

To understand the features of RIP v2, we will look at the IP RIP v2 packet format. Figure 4.2 shows the different fields of IP RIP v2 packet format, along with their corresponding byte lengths.

The description of each of the fields in IP RIP v2 is given in Table 4.5.

TABLE 4.5   Fields of RIP v2 Packet Format

Next-hop Field	Description
Command	Indicates whether the packet is meant to be a request or a response
Version Number	Denotes the RIP version
Unused	Denotes a value equal to zero
Address Family Identifier (AFI)	Indicates the address family of the routed protocol; for example, the AFI for IP is 2
Route Tag	Contains tag information like whether a route is learnt from RIP or from any other protocol
IP Address	Specifies the address of the routed protocol; for example, for IP, it is the IP address
Subnet Mask	Contains subnet mask information
Next-hop	Contains the next-hop information that is reflected in the routing table

RIP v1 and RIP v2 share the same database. Cisco routers send version 1 updates but listen to both versions 1 and 2. These default configurations can be changed on a global or interface basis.

Advantages of RIP v2 over RIP v1

RIP v1 does not carry subnet information, and as a result, it does not support VLSM. This leads to classful routing loops in noncontiguous networks. This shortcoming is addressed in RIP v2, which carries subnet mask information and supports VLSM. RIP v2 can route packets to noncontiguous networks, overcoming the shortcomings of RIP v1.

When different subnets of the same classful network lie on the same side of a router, they are known as contiguous networks. When these subnets lie on different sides of the router, they form a noncontiguous network. For example, Router A, Router B, and Router C are connected such that Router B has noncontiguous networks, 180.1.1.0/24 and 180.1.2.0/24, on either side.

RIP v2 overcomes the limitations of RIP v1. If RIP v1 is used as the routing protocol in each of the three routers, it would lead to a routing loop. RIP v1 does not carry subnet mask information, and as Router B does not have any interface in the 180.1.x.0 subnet, it sees a classful summary address of 180.1.0.0/16. Routing loops occur because Router B does not know that to route packets to 180.1.1.0/24, the next-hop router is Router A, while for routing packets to 180.1.2.0/24, the next-hop router is Router C.

Listing 4.1 shows the routing table of Router B from Figure 4.3, when RIP v1 is being used as the routing protocol. The network 180.1.0.0 has two different next-hop addresses.

LISTING 4.1   Routing Table of Router B with RIP v1

```
RouterB# show ip route
Codes: C - connected, S - static, I - IGRP, R - RIP, M - mobile, B - BGP
   D - EIGRP, EX - EIGRP external, O - OSPF, IA - OSPF inter area
   N1 - OSPF NSSA external type 1, N2 - OSPF NSSA external type 2
   E1 - OSPF external type 1, E2 - OSPF external type 2, E - EGP
   i - IS_IS, L1 - IS_IS level-1, L2 - IS_IS level-2, * - candidate
      default
   U - per-user static route, o - ODR

Gateway of last resort is not set
   C  1.1.1.0/16 is directly connected, Serial0
   C  1.2.2.0/16 is directly connected, Serial1
   R  180.1.0.0 [120/1] via 1.1.1.1, 00:01:14, Serial0
      [120/1] via 1.2.2.2, 00:01:14, Serial1
```

Listing 4.2 shows the routing table of Router B when RIP v2 is the routing protocol. In this scenario, Router B knows that in order to reach 180.1.1.0/24, the next-hop router is Router A, and for reaching network 180.1.2.0/24, the next-hop router is Router C.

LISTING 4.2   Routing Table of Router B with RIP v2

```
RouterB# show ip route
Codes: C - connected, S - static, I - IGRP, R - RIP, M - mobile, B - BGP
   D - EIGRP, EX - EIGRP external, O - OSPF, IA - OSPF inter area
```

```
N1 - OSPF NSSA external type 1, N2 - OSPF NSSA external type 2
E1 - OSPF external type 1, E2 - OSPF external type 2, E - EGP
i - IS_IS, L1 - IS_IS level-1, L2 - IS_IS level-2, * - candidate
  default
U - per-user static route, o - ODR

Gateway of last resort is not set
  C  1.1.1.0/16 is directly connected, Serial0
  C  1.2.2.0/16 is directly connected, Serial1
  R  180.1.1.0 [120/1] via 1.1.1.1, 00:01:14, Serial0
  R  180.1.2.0[120/1] via 1.2.2.2, 00:01:14, Serial1
```

In Listing 4.2, networks 180.1.1.0 and 180.1.2.0 have different and unambiguous next-hop addresses.

CONVERGENCE OF RIP

Each time there is a change in the network topology, all routers under RIP must agree on the changed topology by exchanging routing information. Convergence under RIP is complete when all the routers agree on the optimal routes that are available. Convergence is defined as the time taken for every router to synchronize their database after a change in network topology. During this time, each router recalculates the optimal routes, and finally, when all synchronize, they agree on the routes.

Consider an example of three routers, Router 1, Router 2, and Router 3 connected with each other, as shown in Figure 4.3.

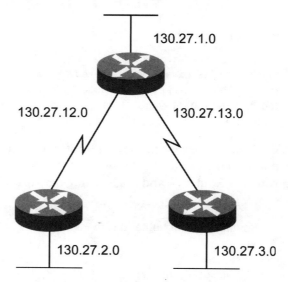

FIGURE 4.3 Router 2 with hop-count 0.

In Figure 4.3, the routers are running RIP. The network addresses of the connected interfaces are also shown in the figure. The routing table shows only the directly connected networks with the metric, 0, during initialization. The routing table of Router 2 shows networks, 130.27.2.0 and 137.27.12.0, with the metric 0. Table 4.6 shows the routing table of Router 2.

TABLE 4.6 Routing Table of Router 2 with Hop-count 0

Network	Metric
130.27.2.0	0
130.27.12.0	0

RIP gathers network information from other routes when it exchanges routing information during the updates. The entries of networks 130.27.1.0 and 130.27.13.0 will appear in the routing table of Router 2 but with a hop-count of 1, as shown in Figure 4.4. In Figure 4.4, 130.27.3.0 (the network beyond Router 3) has a metric of 2, because it has to make two hops to reach that network. Table 4.7 shows the routing table of Router 2.

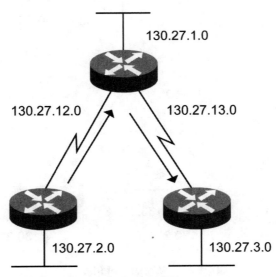

FIGURE 4.4 Router 2 with Hop-count of 1.

TABLE 4.7 Routing Table of Router 2 with Hop-count 1

Network	Metric
130.27.2.0	0
130.27.12.0	0
130.27.1.0	1
130.27.13.0	1
130.27.3.0	2

Router 2 and Router 3 are connected, as shown in Figure 4.5. In Figure 4.5, Router 2 receives routing information of network 130.27.3.0 directly from Router 3, as well as through Router 1. The metrics of these two alternate paths are different, and the path with the lower metric is installed in the routing table. In this case, network 130.27.3.0 is installed in the routing table of Router 2 with metric 1, as shown in Table 4.8.

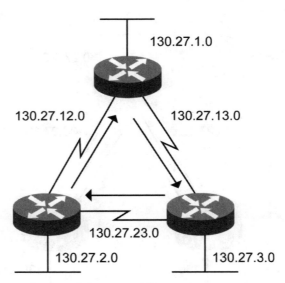

FIGURE 4.5 Working of RIP.

TABLE 4.8   Routing Table of Router 2

Network	Metric
130.27.2.0	0
130.27.12.0	0
130.27.23.0	0
130.27.1.0	1
130.27.13.0	1
130.27.3.0	1

Consider a situation where the network 130.27.3.0 goes down, as shown in Figure 4.6. This causes Router 3 to send routing updates with a metric of 16 (infinity) to Router 2.

In Figure 4.6, Router 2 does not know of the change, and updates are sent to Router 3; it still marks routes to network 130.27.3.0 with a metric of 1. This causes Router 3 to mark network 130.27.3.0 with a metric of 2 with a next-hop address pointing to Router 2. Meanwhile, Router 2 receives the routing updates from Router 3 and sees that network 130.27.3.0 has been given a metric of 16 with a next-hop address

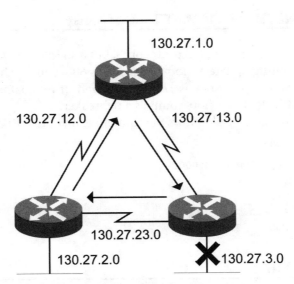

FIGURE 4.6   Router 3 sends router updates.

point to Router 3. This leads to a situation where Router 2 and Router 3 have a next-hop address pointing to each other, causing a routing loop.

The routing loop is avoided using the split-horizon feature, which is used in the RIP algorithm. Using split-horizon, Router 2 does not include network 130.27.3.0 in the routing updates to Router 3, because it has learned about the network from Router 3. In case of split-horizon with poison reverse, Router 2 sends updates on network 130.27.3.0 to Router 3—but with a metric of 16 (infinity). Table 4.9 shows the routing tables of Router 3 and Router 2.

TABLE 4.9 Routing Tables of Routers 3 and 2

Router	Network	Next-hop Router	Metric
3	130.27.3.0	Router 2	2
2	130.27.3.0	Router 3	16

When a router learns about a network that has gone down, it ignores new information for the period set by the Hold-down Timer. This is useful for reaching convergence in a network that contains multiple paths. As a result, Router 2 gets information that network 130.27.3.0 has gone down from Router 3; it ignores information about the network from any other router until the Hold-down Timer expires.

RIP AND INTERNETWORK PACKET EXCHANGE

RIP not only routes the routed protocol IP, but also Internetwork Packet Exchange (IPX). The routing update period of IPX RIP is 60 seconds. The metric used in IPX RIP is ticks and hops. A tick is 1/18 of a second. If two paths have the same tick count, then IPX RIP uses hop-count as a tiebreaker.

The following example shows the configuration of IPX RIP on the serial port of a router:

```
ipx routing 0000.0000.0002
interface Serial0
ipx network 200
ipx router rip
network 200
```

RIP CONFIGURATION

You can configure RIP in a router with the help of certain commands. This section details the processes and commands required to configure the different features of

RIP. All features and configurations are not mandatory. They are used per the requirement.

Global Commands to Enable RIP

To enable RIP, you can use the following global commands:

```
Router(config)# router rip
Router(config-router)# network ip-address
```

Table 4.10 lists the explanation of the global commands.

TABLE 4.10   Global Commands

Commands	Explanation
Router(config)# router rip	Enables RIP
Router(config-router)# network ip-address	Associates the network to the RIP routing process

For example, three routers, R1, R2, and R3, are connected, as shown in Figure 4.7. In Figure 4.7, the global command for enabling RIP on R1 is:

```
R1(config)# router rip
R1(config-router)# network 150.150.0.0
```

The global command for enabling RIP on R2 is:

```
R2(config)# router rip
R2(config-router)# network 150.150.0.0
```

The global command for enabling RIP on R3 is:

```
R3(config)# router rip
R3(config-router)# network 150.150.0.0
```

150.150.1.0/24   150.150.2.0/24   150.150.3.0/24   150.150.4.0/24

FIGURE 4.7   Three routers connected in RIP.

The routing table of R1 is shown in Listing 4.3.

LISTING 4.3 Routing Table of R1

```
R1#sh ip route
Codes: C - connected, S - static, I - IGRP, R - RIP, M - mobile, B - BGP
   D - EIGRP, EX - EIGRP external, O - OSPF, IA - OSPF inter area
   N1 - OSPF NSSA external type 1, N2 - OSPF NSSA external type 2
   E1 - OSPF external type 1, E2 - OSPF external type 2, E - EGP
   i - IS_IS, L1 - IS_IS level-1, L2 - IS_IS level-2, * - candidate
      default
   U - per-user static route, o - ODR

Gateway of last resort is not set
   R   150.150.3.0/24 [120/1] via 150.150.2.2, 03:20:04, Serial0
   R   150.150.3.0/24 [120/2] via 150.150.2.2, 03:20:04, Serial0
   C   150.150.1.0/24 is directly connected, Ethernet0
   C   150.150.2.0/24 is directly connected, Serial0
```

The routing table of R2 is shown in Listing 4.4.

LISTING 4.4 Routing Table of R2

```
R2#sh ip route
Codes: C - connected, S - static, I - IGRP, R - RIP, M - mobile, B - BGP
   D - EIGRP, EX - EIGRP external, O - OSPF, IA - OSPF inter area
   N1 - OSPF NSSA external type 1, N2 - OSPF NSSA external type 2
   E1 - OSPF external type 1, E2 - OSPF external type 2, E - EGP
   i - IS_IS, L1 - IS_IS level-1, L2 - IS_IS level-2, * - candidate
      default
   U - per-user static route, o - ODR

Gateway of last resort is not set
   R   150.150.1.0/24 [120/1] via 150.150.2.1, 03:20:04, Serial1
   R   50.150.4.0/24 [120/1] via 150.150.3.3, 03:20:04, Serial0
   C   150.150.2.0/24 is directly connected, Serial1
   C   150.150.3.0/24 is directly connected, Serial0
```

The routing table of R3 is shown in Listing 4.5.

LISTING 4.5 Routing Table of R3

```
R3#sh ip route
Codes: C - connected, S - static, I - IGRP, R - RIP, M - mobile, B - BGP
   D - EIGRP, EX - EIGRP external, O - OSPF, IA - OSPF inter area
```

```
N1 - OSPF NSSA external type 1, N2 - OSPF NSSA external type 2
E1 - OSPF external type 1, E2 - OSPF external type 2, E - EGP
i - IS_IS, L1 - IS_IS level-1, L2 - IS_IS level-2, * - candidate
  default
U - per-user static route, o - ODR

Gateway of last resort is not set
  R   150.150.1.0/24 [120/2] via 150.150.3.2, 03:20:04, Serial1
  R   150.150.2.0/24 [120/1] via 150.150.3.2, 03:20:04, Serial1
  C   150.150.4.0/24 is directly connected, Ethernet0
  C   150.150.3.0/24 is directly connected, Serial1
```

Passive-interface and Neighbor Commands

By default, RIP works in the broadcast mode. In non-broadcast networks such as frame-relay (FR), it can be configured to send routing updates using the multicast mode. To stop RIP from sending broadcasts out of a particular interface, the passive-interface command is used:

```
Router(config-router)# passive-interface interface
```

The neighbor command is then used to invoke multicast communication with a particular neighbor:

```
Router(config-router)# neighbor ip-address
```

Figure 4.8 shows four routers, R1, R2, R3, and R4.

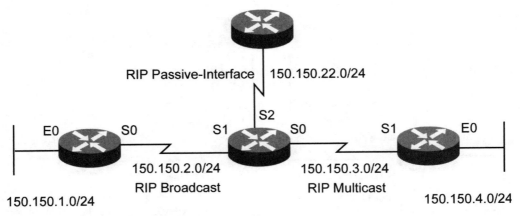

FIGURE 4.8 Passive-interface and neighbor commands in RIP.

In Figure 4.8, R2 sends RIP broadcasts to R1, and RIP multicasts to R3 and sends no updates to R4. Table 4.11 explains the passive-interface and neighbor configuration commands.

TABLE 4.11   RIP neighbor and passive-interface Commands

Commands	Explanation
Router(config-router)# neighbor ip-address	Defines the neighbor to whom multicast routing updates are to be sent
Router(config-router)# passive-interface interface	Identifies the interface from which no broadcasts are to be sent

The configuration for passive-interface and neighbor commands is shown in Listing 4.6.

LISTING 4.6   Configuration for passive-interface and neighbor Commands

```
R2(config)# router rip
R2(config-router)# network 150.150.0.0
R2(config-router)# passive-interface Serial2
R2(config-router)# neighbor 150.150.3.3
```

RIP v1 and RIP v2 can co-exist in a network as long as two neighboring routers have the same version of RIP on both sides of the network link. Figure 4.9 depicts such a scenario, where there are four routers, R1, R2, R3, and R4.

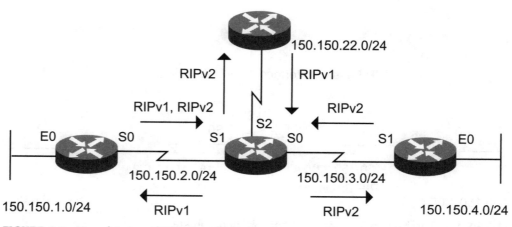

FIGURE 4.9   Co-existence of RIP v1 and RIP v2.

In Figure 4.9, R2 has the default RIP configuration with R1; it sends and receives RIP v2 with R3, while it sends RIP v2 and receives RIP v1 with R4. The corresponding commands to implement the same are shown in Listing 4.7.

LISTING 4.7   Commands to Send and Receive Updates

```
Router(config-router)# version {1 | 2}
Router(config-if)# ip rip send version 1
Router(config-if)# ip rip send version 2
Router(config-if)# ip rip send version 1 2
Router(config-if)# ip rip receive version 1
Router(config-if)# ip rip receive version 2
Router(config-if)# ip rip receive version 1 2
```

Table 4.12 explains the commands used for the enabling of sending and receiving updates of RIP v1 and RIP v2.

TABLE 4.12   Commands to Enable Sending and Receiving Updates

Command	Explanation
Router(config-router)# version {1 | 2}	Defines the default version of RIP that the router sends and receives
Router(config-if)# ip rip send version 1	Configures a particular interface to send RIP v1 packets only
Router(config-if)# ip rip send version 2	Configures a particular interface to send RIP v2 packets only
Router(config-if)# ip rip send version 1 2	Configures a particular interface to send both RIP v1 and RIP v2 packets
Router(config-if)# ip rip receive version 1	Configures a particular interface to receive RIP v1 packets only
Router(config-if)# ip rip receive version 2	Configures a particular interface to receive RIP v2 packets only
Router(config-if)# ip rip receive version 1 2	Configures a particular interface to receive both RIP v1 and RIP v2 packets.

R1 has the default configuration of RIP. The configuration of R1 is:

```
R1(config)# router rip
R1(config-router)# network 150.150.
```

The versions of RIP are defined specifically on each interface, according to Figure 4.10. The configuration of R2 is shown in Listing 4.8.

LISTING 4.8   Configuration of R2

```
R2(config)# router rip
R2(config-router)# network 150.150.0.0
R2(config)# interface Serial1
R2(config-if)# ip rip send version 1
R2(config-if)# ip rip receive version 1 2
R2(config)# interface Serial0
R2(config-if)# ip rip send version 2
R2(config-if)# ip rip receive version 2
R2(config)# interface Serial2
R2(config-if)# ip rip send version 2
R2(config-if)# ip rip receive version 1
```

RIP v2 has been defined under the routing process instead of being defined separately. This configuration of R3 is:

```
R3(config)# router rip
R3(config-router)# version 2
R3(config-router)# network 150.150.0.0
```

R4 works under the default configuration, as by default RIP sends RIP v1 and receives RIP v1 and RIP v2. The configuration of R4 is:

```
R4(config)# router rip
R4(config-router)# network 150.150.0.0
```

Authentication Commands

RIP v1 does not support authentication. RIP v2 supports plain text and Message Digest 5 (MD5) authentication. This security feature prevents any router from participating in the RIP network without your prior knowledge and permission. This is because a password is required to participate in a RIP network.

In order to configure RIP authentication, a key-chain containing keys for the interfaces needs to be configured. Although the default mode of authentication is plain text, it is best to use MD5 because it encrypts the password.

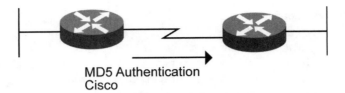

FIGURE 4.10 Authentication keys do not match in RIP.

Consider an example of two routers, R1 and R2, that are connected over a serial link. In Figure 4.10, R1 and R2 do not exchange routing information, as the authentication keys do not match.

In Figure 4.11, R1 and R2 exchange routing updates after their authentication keys match.

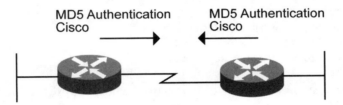

FIGURE 4.11 Authentication keys match in RIP.

Table 4.13 shows the authentication commands in RIP.

TABLE 4.13 Authentication Commands

Commands	Explanation
Router(config-if)# ip rip authentication key-chain name-of-key-chain	Enables RIP authentication
Router(config-if)# ip rip authentication mode {text | md5}	Configures the type of authentication
Router(config)# key chain name-of-the-key-chain	Creates a key-chain
Router(config-keychain)# key number	Creates a key inside the key-chain

LISTING 4.9 Configuration of R1

```
R1(config)# interface serial 0
R1(config-if)# ip rip authentication key-chain newchainR1
R1(config-if)# ip rip authentication mode md5
R1(config-if)# ip rip send version 2
R1(config-if)# ip rip receive version 2
R1(config)# key chain newchain
R1(config-keychain)# key cisco
```

The configuration of R2 is shown in Listing 4.10.

LISTING 4.10 Configuration of R2

```
R2(config)# interface serial 0
R2(config-if)# ip rip authentication key-chain newchainR2
R2(config-if)# ip rip authentication mode md5
R2(config-if)# ip rip send version 2
R2(config-if)# ip rip receive version 2
R2(config)# key chain newchain
R2(config-keychain)# key cisco
```

RIP v2 supports route summarization, which suppresses specific subnets and creates a summarized network instead. By default, RIP summarizes classful networks while crossing classful network boundaries. This feature of RIP can be disabled with the no auto-summary command.

Apart from the automatic summary feature of RIP, it is also possible to summarize to a specified interface using the ip summary-address rip command. However, automatic summarization will prevail if both automatic summarization and the summary address features are enabled at the same time.

Summary address that is configured on the interface will prevail over automatic summarization only if:

- Split-horizon is disabled on the interface that is summarizing the address
- Interface summary address that is configured shares the same classful network address with that of the IP address of the interface

Split-horizon is enabled on any interface by default. It needs to be disabled on certain topologies, such as frame-relay networks.

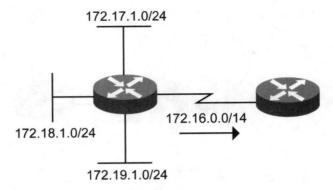

172.17.1.0/24

172.16.0.0/14

172.18.1.0/24

172.19.1.0/24

FIGURE 4.12 Summarizing at supernet.

RIP routing protocol maintains its own routing database. All the entries in the routing database may not find a place in the routing table. Each child route that is summarized into a summary address is kept in the routing database. For some reason, if the last of the child routes is removed from the routing database, the summary address is also removed:

interface Ethernet0/0

ip address 100.1.1.1 255.255.255.0

ip summary-address rip 100.2.0.0 255.255.0.0

no ip split-horizon

The summary address cannot be configured if:

■ Supernet summarization, using prefixes less than classful networks, is not allowed.
■ Summarizing at different subnets of the same classful networks is not allowed.

Figure 4.12 depicts how summarizing is not allowed at the superset, if the prefix is less than the classful network.

R1 summarizes to an invalid supernet address. The configuration of R1 is shown in Listing 4.11.

LISTING 4.11 Configuration of R1

```
interface Serial0
.
.
ip summary-address rip 172.16.0.0 255.252.0.0
interface Serial1
ip address 172.17.1.1 255.255.255.0
```

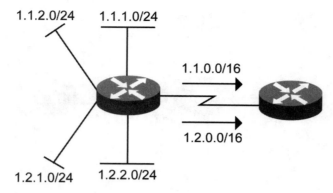

FIGURE 4.13 Summarizing at subnets.

```
interface Serial2
ip address 172.18.1.1 255.255.255.0
interface Serial3
ip address 172.19.1.1 255.255.255.0
```

Figure 4.13 depicts how summarizing is not allowed at different subnets of the same classful networks.

Each summarized address does not have a unique classful address. This summarization is invalid. Configuration of R1 is shown in Listing 4.12.

LISTING 4.12 Configuration of R1

```
interface Serial0
 .
 .
ip summary-address rip 1.1.0.0 255.255.0.0
ip summary-address rip 1.2.0.0 255.255.0.0
interface Serial1
ip address 1.1.1.1 255.255.255.0
interface Serial2
ip address 1.1.2.1 255.255.255.0
interface Serial3
ip address 1.2.1.1 255.255.255.0
interface Serial4
ip address 1.2.2.1 255.255.255.0
```

Distance vector protocols follow the split-horizon rule. Using this rule, protocols do not send routing updates of networks from the same interface where they are received. This avoids routing loops in networks that use broadcast-type networks.

When routers are connected using RIP in a non-broadcast type network, it is best to disable split-horizon, so that routing updates can reach the routers.

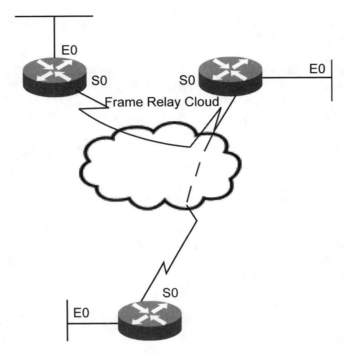

FIGURE 4.14   Split-horizon prevents routing loops.

Be careful while disabling split-horizon, as it may lead to routing loops.

Figure 4.14 depicts a scenario where three routers, R1, R2, and R3, are connected over frame-relay.

In Figure 4.14, R1, R2, and R3 run on RIP. Over the FR cloud, Serial 0 of R1 is connected to R2 and R3. RIP is enabled by default; it prevents R1 from sending routing updates of the routes of the interface from where it received the updates. This would prevent R3 from seeing the networks of R2. In such a situation, split-horizon needs to be disabled.

Table 4.14 shows the commands for disabling and re-enabling split-horizon.

TABLE 4.14   Disabling and Re-enabling Split-horizon

Commands	Explanation
Router(config-if)# no ip split-horizon	Disables the split-horizon
Router(config-if)# ip split-horizon	Re-enables split-horizon

The configurations of all three routers are shown in the following listings. Configuration for R1 is:

```
interface Serial2/0
ip address 130.1.45.1 255.255.255.0
encapsulation frame-relay
(no ip split-horizon)
```

The command no ip split-horizon is by default in FR serial interface, so it will not show in the configuration.

Configuration for R2 is:

```
interface Serial2/0
ip address 130.1.45.2 255.255.255.0
encapsulation frame-relay
ip split-horizon
```

Configuration for R3 is:

```
interface Serial2/0
ip address 130.1.45.3 255.255.255.0
encapsulation frame-relay
ip split-horizon
```

Split-horizon is disabled by default in frame-relay serial interface and should be enabled.

Commands to Change Metrics

You can change the metrics used by RIP. This can be fine-tuned to the extent of changing the metrics for a certain set of routes and applying them on a particular interface. The command for changing the metrics is:

```
Router(config-router)# offset-list {access-list-number | access-list-
name} {in | out} offset [interface-type interface-name]
```

The explanation for this command is given in Table 4.15.

TABLE 4.15   Command to Change Metrics in RIP

Router(config-router)# offset-list {access-list-number | access-list-name} {in | out} offset [interface-type interface-name]	The routing path to a destination network is determined by the metric associated with the route in the routing table. The metrics of the routes learned by RIP can be increased with the help of "offset-list."

Commands to Change Timers

The command for changing the timers is:

```
Router(config-router)# timers basic update invalid holddown flush
[sleeptime]
```

The explanation for this command is given in Table 4.16.

TABLE 4.16   Command to Change Timers in RIP

Commands	Explanation
Router(config-router)# timers basic update invalid holddown flush [sleeptime]	Adjusts the various timers of the RIP routing protocol.

Commands to Adjust Inter-packet Delay

When two routers of different capacity are interconnected, you may need to adjust the delay of RIP packets exchanged between the routers. By default, RIP does not add any delay. You may need to adjust the inter-packet delay when a high-end hub router at the central site communicates with low-end spoke routers at different remote sites.

For example, in Figure 4.15, one high-end hub router, R1, is connected to several low-end spoke routers, R2, R3, and R4. In Figure 4.15, an inter-packet delay is introduced to avoid the situation of flooding low-end routers with routing updates. The command to adjust inter-packet delay is:

```
Router(config-router)# output-delay delay
```

The explanation for this command is shown in Table 4.17.

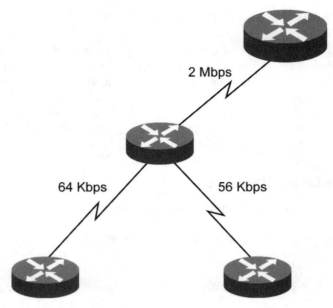

FIGURE 4.15   Inter-packet delay in RIP.

TABLE 4.17   Command to Adjust Inter-packet Delay

Commands	Explanation
Router(config-router)# output-delay delay	Adjusts the inter-delay in RIP updates

Commands for Monitoring and Troubleshooting RIP

It is important to know the different monitoring and troubleshooting tips to maintain a network running RIP routing protocol.

The command show ip interface is used to know the details of all the interfaces of the router. Listing 4.13 shows the output for this command.

LISTING 4.13   Output for show ip interface Command

```
Router# show ip interface
Ethernet0 is up, line protocol is up
Internet address is 10.0.1.10, subnet mask is 255.255.255.0 Broadcast
address is 255.255.255.255
Address determined by setup command
```

```
MTU is 1500 bytes
Helper address is not set
Directed broadcast forwarding is enabled
Multicast groups joined: 224.0.0.1 224.0.0.2
Outgoing access list is not set
Inbound access list is not set
Proxy ARP is enabled
Security level is default
Split-horizon is enabled
ICMP redirects are always sent
ICMP unreachables are always sent
ICMP mask replies are never sent
IP fast switching is enabled
IP fast switching on the same interface is disabled
IP SSE switching is disabled
Router Discovery is disabled
IP accounting is disabled
TCP/IP header compression is disabled
Probe proxy name replies are disabled
Gateway Discovery is disabled
Serial0 is up, line protocol is up
Internet address is 200.2.2.1, subnet mask is 255.255.255.0
Broadcast address is 255.255.255.255
Address determined by setup command
MTU is 1500 bytes
Helper address is not set
Directed broadcast forwarding is enabled
Multicast groups joined: 224.0.0.1 224.0.0.2
Outgoing access list is not set
Inbound  access list is not set
Proxy ARP is enabled
Security level is default
Split-horizon is enabled
ICMP redirects are always sent
ICMP unreachables are always sent
ICMP mask replies are never sent
IP fast switching is enabled
IP fast switching on the same interface is disabled
IP SSE switching is disabled
Router Discovery is disabled
IP accounting is disabled
TCP/IP header compression is disabled
Probe proxy name replies are disabled
Gateway Discovery is disabled
```

Listing 4.14 shows a sample output from the show ip interface brief command.

LISTING 4.14   Output for show ip interface brief Command

```
Router# show ip interface brief
Interface    IP-address     OK?  Method    Status      Protocol
Ethernet0    10.0.1.10      YES  manual    up          up
Serial0      200.2.2.1      YES  manual    up          up
```

The command show ip protocols generate detailed information on the routing protocols present in the router. The output includes timers, RIP version, the networks it is routing, gateway, and summary information. Listing 4.15 shows the output for the show ip protocols command:

LISTING 4.15   Output for show ip protocols Command

```
Router# show ip protocols
Routing Protocol is "rip"
Sending updates every 30 seconds, next due in 8 seconds
Invalid after 180 seconds, hold down 180, flushed after 240
Outgoing update filter list for all interfaces is
Incoming update filter list for all interfaces is
Redistributing: rip
Routing for Networks:
10.0.0.0
```

Routing Information Sources:

```
Gateway    Distance    Last Update
10.1.1.1   120         0:10:00
```

Default version control: send version 2, receive version 2

```
Interface Send Recv Triggered RIP Key-chain
Ethernet2 2 2
Ethernet3 2 2
Ethernet4 2 2
Ethernet5 2 2
Automatic network summarization is not in effect
Address Summarization:
```

The routing table is shown using the show ip route protocol. This can be further filtered to give the output of RIP only, as shown in the sample output in Listing 4.16.

LISTING 4.16 Output for show ip route Command

```
Router# show ip route rip
Router# sh ip route
Codes: C - connected, S - static, I - IGRP, R - RIP, M - mobile, B -
  BGP
D - EIGRP, EX - EIGRP external, O - OSPF, IA - OSPF inter area
N1 - OSPF NSSA external type 1, N2 - OSPF NSSA external type 2
E1 - OSPF external type 1, E2 - OSPF external type 2, E - EGP
i - IS_IS, L1 - IS_IS level-1, L2 - IS_IS level-2, ia - IS_IS inter
  area
* - candidate default, U - per-user static route, o - ODR
P - periodic downloaded static route

Gateway of last resort is not set.
   R   150.150.1.0/24 [120/1] via 150.150.2.1, 03:20:04, Serial1
   R   150.150.4.0/24 [120/1] via 150.150.3.3, 03:20:04, Serial0
```

A sample output of the ip rip database command is shown in Listing 4.17.

LISTING 4.17 Output of ip rip database Command

```
Router# show ip rip database
172.16.0.0/16     auto-summary
172.16.0.0/16
[1] via 172.16.1.2, 00:00:23 (permanent), Serial1
* Triggered Routes:
- [1] via 172.16.1.2, Serial1/0
172.20.0.0/16     auto-summary
172.20.0.0/16
[1] via 172.16.1.2, 00:02:45 (permanent), Serial1/0
* Triggered Routes:
- [1] via 172.16.1.2, Serial1/0
```

At times, monitoring is not enough to trace a problem. In such circumstances you need to debug the different events of RIP as shown in Listing 4.18.

LISTING 4.18 Output for debug Command

```
Router# debug ip rip events
RIP: received v1 triggered request from 10.1.1.2 on Serial1
RIP: start retransmit timer of 10.1.1.2
RIP: received v1 triggered ack from 10.1.1.2 on Serial1
RIP: Stopped retrans timer for 10.1.1.2
RIP: sending v1 ack to 10.1.1.2 via Serial1
```

```
Router# undebug ip rip
```

Debugging uses a number of router resources, and it adversely affects network performance. It should be used only when required and preferably when the network is not in use.

SUMMARY

In this chapter, we discussed the operation and configuration of Routing Information Protocol (RIP). We also discussed the two versions of RIP—RIP v1 and RIP v2. In the next chapter, we will learn about Interior Gateway Routing Protocol.

POINTS TO REMEMBER

- RIP is suitable for small homogeneous networks and has two versions—RIP v1 and RIP v2.
- RIP routing updates include IP destination address, metric, next-hop address, timers, flags, hold-downs, split-horizon information, and poison reverse updates.
- RIP v1 uses the major classful network number for route summarization, because it does not carry subnet mask routing information.
- RIP v2 has added security features, such as plain text and MD5 authentication.
- RIP v2 provides support for VLSM, allowing it to perform classless routing and avoid classful routing loops.
- Convergence is the time taken for every router to synchronize its database after a change in network topology.
- RIP not only routes the routed protocol IP, but also IPX.
- The IPX RIP routing update period is 60 seconds.
- IPX RIP uses ticks—1/18 of a second—and hops.
- Router(config)# router rip enables RIP.
- Router(config-router)# network ip-address associates the network to the RIP routing process.
- Router(config-router)# neighbor ip-address defines the neighbor to whom multicast routing updates are to be sent.
- The Router(config-router)# passive-interface interface command identifies the interface from which no broadcasts are to be sent.

- Router(config-if)# ip rip authentication key-chain name-of-key-chain enables RIP authentication.
- The Router(config-if)# ip rip authentication mode {text | md5} command configures the type of authentication.
- The Router(config)# key-chain name-of-the-key-chain command creates a key-chain.
- The Router(config-key-chain)# key number command creates a key inside the key-chain.
- Router(config-if)# no ip split-horizon disables the split-horizon.
- Router(config-if)# ip split-horizon re-enables split-horizon.
- Router(config-router)# offset-list {access-list-number | access-list-name} {in | out} offset [interface-type interface-name] is the command used to change metrics in RIP.
- Router(config-router)# timers basic update invalid holddown flush [sleeptime] is the command used to change timers in RIP.
- Router(config-router)# output-delay delay adjusts the inter-delay in RIP updates.
- Router# show ip interface shows the details of all the interfaces of the router.

5 | Interior Gateway Routing Protocol

IN THIS CHAPTER

- IGRP Operation
- IGRP Configuration Commands
- IGRP Design Considerations

INTRODUCTION TO IGRP

Interior Gateway Routing Protocol (IGRP) is a distance vector protocol, which is suitable for medium- and large-sized networks. IGRP advertises the entire or a part of the routing table to neighbors via broadcasts. The metrics used in IGRP are a combination of delay, bandwidth, reliability, load, and Maximum Transmission Unit (MTU).

IGRP OPERATION

Being a distance vector protocol, IGRP advertises the entire or a part of the routing table to neighbors every 90 seconds via broadcasts. If a route remains unreachable even after three consecutive updates, it is declared inaccessible. The inaccessible route is not removed from the routing table even at this stage. It is only deleted from the routing table if it fails to respond to seven consecutive routing updates.

The routes that are advertised by IGRP can be classified into:

Interior routes: For subnets connected to the interface of the router.

System routes: For networks from routers within the same AS. These networks do not include subnetworks.

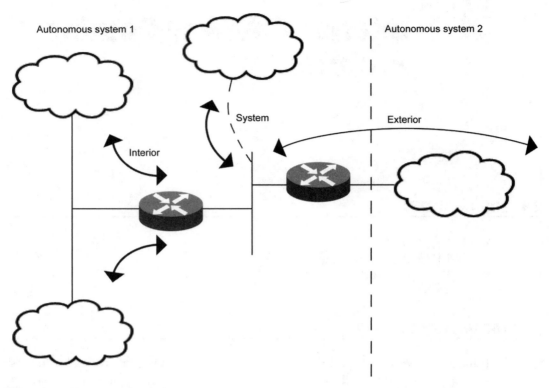

FIGURE 5.1 IGRP has interior, system, and exterior routes.

Exterior routes: From networks outside ASs. These routes are used while defining a default-gateway using "Gateway of Last Resort."

Figure 5.1 shows the interior, system, and exterior routes in a network using IGRP.

We will discuss Gateway of Last Resort later in the chapter.

IGRP FEATURES

Being a distance vector protocol, IGRP has features such as split-horizon, split-horizon with poison reverse, route poisoning, hold-down timers, triggered updates, and count to infinity. Table 5.1 lists the features of IGRP.

TABLE 5.1 Features of IGRP

Features	IGRP Values
Protocol Type	Distance vector
Subnet Information	Classful
Metric	Composite (bandwidth, delay, reliability, loading)
Count to Infinity	100 by default (Maximum is 255)
Routing Updates Mode	Broadcast
Flash or Triggered Updates	Yes
Load Balancing	Up to four paths, by default
Algorithm	Bellman-Ford

In addition to the features mentioned in Table 5.1, another important characteristic of IGRP is that it does not support discontiguous networks. A discontiguous network occurs when one major network address is separated by another major network address.

Autonomous System Number Feature

An AS was traditionally used when different routing protocols were run on the same network. Typically, each organization is assigned its own Autonomous System Number (ASN). If an organization does not have its own registered ASN, it can use any ASN out of the private range assigned by the Internet Assigned Number Authority (IANA) 64512–65534.

Unlike RIP, IGRP uses this feature of ASN. Since RIP does not use the ASN feature, it does not limit its routing update broadcasts within any boundary. RIP broadcasts are sent to all routers, even when it is not required. Figure 5.2 shows a RIP network that does not use the ASN feature and therefore broadcasts updates to AS1 and AS2.

The update broadcasts are not sent beyond the boundary of the IGRP process. This reduces broadcasts, thereby increasing network performance. The ASN feature also builds up network security, because only the routers of the same AS exchange routing updates. Figure 5.3 shows that IGRP broadcasts of one AS do not go to another AS.

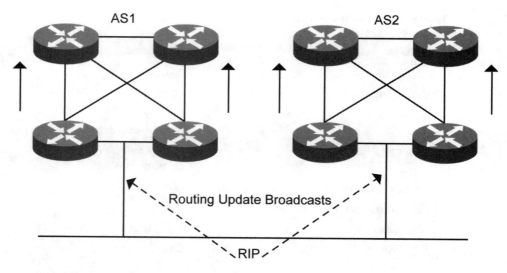

FIGURE 5.2 RIP does not use the ASN feature.

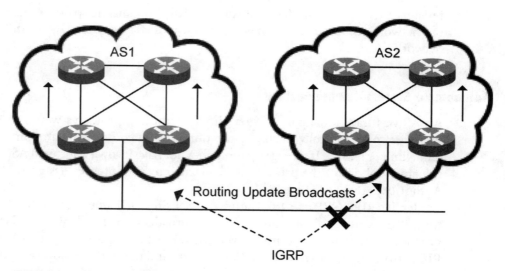

FIGURE 5.3 Using the ASN feature in IGRP.

IGRP Timers

IGRP requires timers for the different routing operations. These timers are set to default values that can be changed, if required. Table 5.2 lists the timers used by IGRP.

TABLE 5.2   IGRP Timers

Timer	Value
Update Timer	Default value is 90 seconds
Invalid Timer	Value is three times the update timer. The default value is 270 seconds.
Hold-down Timer	Value is 10 seconds more than invalid timer. The default value is 280 seconds.
Flush Timer	Value is seven times the update timer. The default value is 630 seconds.

IGRP Metrics

IGRP uses a composite metric to calculate the best path to reach a destination by making a distinction between slow and fast links. To calculate the metric, IGRP uses a combination of:

Bandwidth: Indicates the throughput capacity of a given network. It is the difference between the highest and lowest frequencies available for network signals. Bandwidth can take a value that ranges from 1.2 Kbps to 10 Gbps. For example, the bandwidth for Ethernet could be 10 Mbps, and 64 Kbps for lease lines.

Delay: Denotes the time taken for a packet to move from source to destination. In the equation for calculating the metric, delay is expressed in units of 10 microseconds. It also indicates the time taken between the initiation of a transaction by a sender and the first response received by the sender.

Load: Denotes the amount of router resource that is used. Load is measured on a scale of 1-255.

Reliability: Denotes the ratio of expected keepalives to received keepalives from a link. If the ratio is high, the line is reliable. Reliability is expressed on a scale of 1-225.

MTU: Denotes the maximum packet size that an interface can handle. MTU is expressed in bytes.

IGRP Message Format

The message format of IGRP is depicted in Figure 5.4.

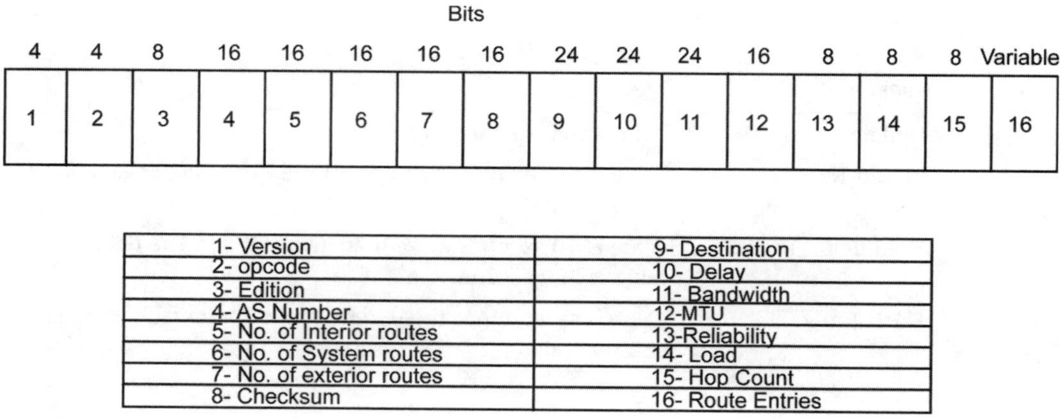

Bits

4	4	8	16	16	16	16	16	24	24	24	16	8	8	8	Variable
1	2	3	4	5	6	7	8	9	10	11	12	13	14	15	16

1- Version	9- Destination
2- opcode	10- Delay
3- Edition	11- Bandwidth
4- AS Number	12-MTU
5- No. of Interior routes	13-Reliability
6- No. of System routes	14- Load
7- No. of exterior routes	15- Hop Count
8- Checksum	16- Route Entries

FIGURE 5.4   IGRP message format.

The descriptions of the various fields of the message packet formats are listed in Table 5.3.

TABLE 5.3   IGRP Message Format Fields

Field	Description
Version	Sets value to 0x01 always.
Opcode	Sets value to 0x01 for a request and 0x02 for an update.
Edition	Increments counter by the sender such that the receiving router knows that it always keeps the latest version of the routing updates.
Autonomous System Number	Specifies IGRP process ID number.
Number of Interior Routes	Indicates the number of routes that are subnets of a directly connected network
Number of System Routes	Indicates the number of routes for networks from routers within the same AS.
Number of Exterior Routes	Indicates the number of routes from networks outside the ASs.
Checksum	Is calculated from an algorithm based on the header and the entries.
Destination	Indicates the destination network.
Delay	Expresses delay as a multiple of 10 microseconds.

TABLE 5.3   *(continued)*

Field	Description
Bandwidth	Defines IGRP bandwidth.
MTU	Expresses MTU in bytes.
Reliability	Specifies error rates of the link.
Load	Specifies load on the link.
Hop-count	Is a number between 0x00 and 0Xff.
Route entries	Are variable in nature.

Delay and bandwidth values are 24-bit with IGRP. For EIGRP, these values are 32-bit values (256*IGRP value).

IGRP CONFIGURATION

IGRP constitutes some mandatory and optional configuration commands. Mandatory commands should be present for the proper functioning of IGRP. Optional commands are used, depending on specific requirements. In this section, we will look at the different scenarios where mandatory and optional configuration commands are used.

IGRP Minimum Configuration

The minimum configuration commands required to enable the IGRP routing process and to define the networks associated with the process are:

```
Router(config)# router igrp as-number
Router(config-router)# network network-number
```

Table 5.4 lists and explains the minimum configuration commands.

TABLE 5.4   Minimum Commands to Enable IGRP Routing

Command	Explanation
Router(config)# router igrp as-number	Enables the IGRP routing process for a specific ASN
Router(config-router)# network network-number	Defines the network under the routing process

Minimum configuration commands are used in the global configuration mode.

IGRP Classful Routing Configuration

IGRP is a classful routing protocol, that is, the network number is defined in a classful format.

Consider a router with subnets 172.16.4.0/24 and 172.16.5.0/24 as the networks on which IGRP is going to send routing updates. The network command should define the classful network address 172.16.0.0 and not 172.16.4.0 or 172.16.5.0. Figure 5.5 illustrates the classful nature of IGRP.

Routers A, B, and C are connected to each other, as shown in Figure 5.5. Router A has connected the subnets of 172.16.1.0 and 172.16.2.0. When all three routers are configured in IGRP and exchange routing information, Router A receives the route to the Ethernet segment of Router C. Route 172.20.1.0/24 belongs to an entirely different classful network than any of its connected subnets. Router A installs the classful network 172.20.0.0/16 and not 172.16.1.0/24 in its routing table. The routing table of Router A shows the different subnets of 172.16.0.0 but shows only the major classful network of 172.20.0.0. The configuration for Router A is shown in Listing 5.1.

LISTING 5.1   Configuration for Router A

```
interface FastEthernet0/0
ip address 172.16.1.1 255.255.255.0
interface Serial0/0
ip address 172.16.2.1 255.255.255.0
router igrp 1
network 172.16.0.0
```

The routing table for Router A is shown in Listing 5.2.

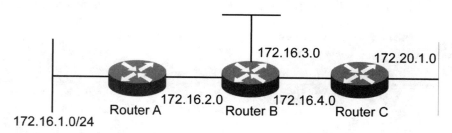

FIGURE 5.5   Classful nature of IGRP.

LISTING 5.2   Routing Table for Router A

```
RouterA# sh ip route
Codes: C - connected, S - static, I - IGRP, R - RIP, M - mobile, B - BGP
  D - EIGRP, EX - EIGRP external, O - OSPF, IA - OSPF inter area
  N1 - OSPF NSSA external type 1, N2 - OSPF NSSA external type 2
  E1 - OSPF external type 1, E2 - OSPF external type 2, E - EGP
  i - IS_IS, L1 - IS_IS level-1, L2 - IS_IS level-2, ia - IS_IS inter
    area
  * - candidate default, U - per-user static route, o - ODR
  P - periodic downloaded static route

Gateway of last resort is not set
  I 172.16.3.0/24 [100/12100] via 172.16.2.2, 00:00:50, Serial0/0
  I 172.16.4.0/24 [100/12100] via 172.16.2.2, 00:00:50, Serial0/0
  I 172.20.0.0/16 [100/12100] via 172.16.2.2, 00:00:50, Serial0/0
  C 172.16.1.0 is directly connected, FastEthernet0/0
  C 172.16.2.0 is directly connected, Serial0/0
```

The configuration for Router B is shown in Listing 5.3

LISTING 5.3   Configuration for Router B

```
interface Serial0/0
ip address 172.16.2.2 255.255.255.0
interface Serial0/1
ip address 172.16.3.2 255.255.255.0
interface Serial0/2
ip address 172.16.4.2 255.255.255.0
router igrp 1
network 172.16.0.0
```

The routing table for Router B is shown in Listing 5.4.

LISTING 5.4   Routing Table for Router B

```
RouterB# sh ip route
Codes: C - connected, S - static, I - IGRP, R - RIP, M - mobile, B - BGP
  D - EIGRP, EX - EIGRP external, O - OSPF, IA - OSPF inter area
  N1 - OSPF NSSA external type 1, N2 - OSPF NSSA external type 2
  E1 - OSPF external type 1, E2 - OSPF external type 2, E - EGP
  i - IS_IS, L1 - IS_IS level-1, L2 - IS_IS level-2, ia - IS_IS inter
    area
  * - candidate default, U - per-user static route, o - ODR
  P - periodic downloaded static route
```

```
Gateway of last resort is not set
  I 172.16.1.0/24 [100/12100] via 172.16.2.1, 00:00:50, Serial0/0
  I 172.20.0.0/16 [100/12100] via 172.16.4.3, 00:00:50, Serial0/2
  C 172.16.2.0 is directly connected, Serial0/0
  C 172.16.3.0 is directly connected, Serial0/1
  C 172.16.4.0 is directly connected, Serial0/2
```

The configuration for Router C is shown in Listing 5.5.

LISTING 5.5   Configuration for Router C

```
interface Serial0/0
ip address 172.16.4.3 255.255.255.0
interface FastEthernet0/0
ip address 172.20.4.3 255.255.255.0
router igrp 1
network 172.16.0.0
network 172.20.0.0
```

The routing table for Router C is shown in Listing 5.6

LISTING 5.6   Routing Table for Router C

```
RouterC #sh ip route
Codes: C - connected, S - static, I - IGRP, R - RIP, M - mobile, B - BGP
  D - EIGRP, EX - EIGRP external, O - OSPF, IA - OSPF inter area
  N1 - OSPF NSSA external type 1, N2 - OSPF NSSA external type 2
  E1 - OSPF external type 1, E2 - OSPF external type 2, E - EGP
  i - IS_IS, L1 - IS_IS level-1, L2 - IS_IS level-2, ia - IS_IS inter
    area
  * - candidate default, U - per-user static route, o - ODR
  P - periodic downloaded static route

Gateway of last resort is not set
  I 172.16.1.0/24 [100/12100] via 172.16.4.2, 00:00:50, Serial0/0
  I 172.16.2.0/24 [100/12100] via 172.16.4.2, 00:00:50, Serial0/0
  I 172.16.3.0/24 [100/12100] via 172.16.4.2, 00:00:50, Serial0/0
  C 172.16.4.0 is directly connected, Serial0/0
  C 172.20.4.0 is directly connected, FastEthernet0/0
```

IGRP Load Balancing

IGRP can load balance between alternate paths when there are multiple paths to a given destination. IGRP can perform load balancing over four paths. If there are more than four routes to the same destination network address, the best four paths are selected as the feasible paths. This load balancing feature of IGRP increases performance by increasing the throughput.

The variance command enables load balancing between alternate paths with different metrics.

```
Router(config-router)# variance value
```

A variance value is a multiplier that, when multiplied with the metric of the primary path, gives a value that should be greater than or equal to the metric of the destination through the alternate path. The default value of variance is 1. This means that IGRP will load balance only among paths having equal costs.

The default value of variance is 1 for equal-cost load balancing. Changing the default value will set up the router for unequal-cost load balancing.

Consider a scenario where two routers, R1 and R2, are connected via two serial links of different capacity—64 Kbps and 2 Mbps. This is depicted in Figure 5.6.

The configuration of R1 is shown in Listing 5.7.

LISTING 5.7 Configuration for R1

```
interface FastEthernet0
ip address 1.1.3.1 255.255.255.0
interface Serial0
ip address 1.1.2.1 255.255.255.0
interface Serial1
ip address 1.1.1.1 255.255.255.0
router igrp 1
network 1.0.0.0
```

The routing table for R1 is shown in Listing 5.8.

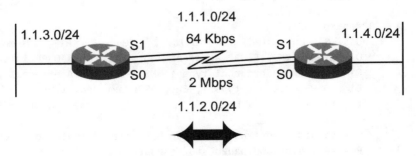

FIGURE 5.6 Load balancing in IGRP.

LISTING 5.8   Routing Table for R1

```
R1#sh ip route
Codes: C - connected, S - static, I - IGRP, R - RIP, M - mobile, B - BGP
   D - EIGRP, EX - EIGRP external, O - OSPF, IA - OSPF inter area
   N1 - OSPF NSSA external type 1, N2 - OSPF NSSA external type 2
   E1 - OSPF external type 1, E2 - OSPF external type 2, E - EGP
   i - IS_IS, L1 - IS_IS level-1, L2 - IS_IS level-2, ia - IS_IS inter
     area
   * - candidate default, U - per-user static route, o - ODR
   P - periodic downloaded static route

Gateway of last resort is not set
   I 1.1.4.0/24 [100/10576] via 1.1.2.2, 00:00:50, Serial0
   C 1.1.3.0 is directly connected, FastEthernet0
   C 1.1.2.0 is directly connected, Serial0
   C 1.1.1.0 is directly connected, Serial1
```

The configuration for R2 is shown in Listing 5.9.

LISTING 5.9   Configuration for R2

```
interface FastEthernet0
ip address 1.1.4.2 255.255.255.0
interface Serial0
ip address 1.1.2.2 255.255.255.0
interface Serial1
ip address 1.1.1.2 255.255.255.0
router igrp 1
network 1.0.0.0
```

The routing table for R2 is shown in Listing 5.10.

LISTING 5.10   Routing Table for R2

```
R2#sh ip route
Codes: C - connected, S - static, I - IGRP, R - RIP, M - mobile, B - BGP
   D - EIGRP, EX - EIGRP external, O - OSPF, IA - OSPF inter area
   N1 - OSPF NSSA external type 1, N2 - OSPF NSSA external type 2
   E1 - OSPF external type 1, E2 - OSPF external type 2, E - EGP
   i - IS_IS, L1 - IS_IS level-1, L2 - IS_IS level-2, ia - IS_IS inter
     area
   * - candidate default, U - per-user static route, o - ODR
   P - periodic downloaded static route
```

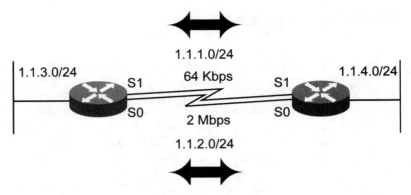

FIGURE 5.7 Unequal-cost load balancing in IGRP.

```
Gateway of last resort is not set
 I 1.1.1.0/24 [100/10576] via 1.1.2.1, 00:00:50, Serial0
 C 1.1.4.0 is directly connected, FastEthernet0
 C 1.1.2.0 is directly connected, Serial0
 C 1.1.1.0 is directly connected, Serial1
```

IGRP can perform unequal-cost load balancing using variance, thereby using both the links instead of one. Figure 5.7 shows two routers connected via serial links of varying capacity.

Figure 5.7 defines a variance parameter to load balance the routers. This takes care of the unequal-cost load balancing. The configuration for R1 is shown in Listing 5.11.

LISTING 5.11 Configuration for R1

```
interface FastEthernet0
ip address 1.1.3.1 255.255.255.0
interface Serial0
ip address 1.1.2.1 255.255.255.0
interface Serial1
ip address 1.1.1.1 255.255.255.0
router igrp 1
network 1.0.0.0
variance 10
```

The routing table for R1 is shown in Listing 5.12.

LISTING 5.12 Routing Table for R1

```
R1#sh ip route
Codes: C - connected, S - static, I - IGRP, R - RIP, M - mobile, B - BGP
   D - EIGRP, EX - EIGRP external, O - OSPF, IA - OSPF inter area
   N1 - OSPF NSSA external type 1, N2 - OSPF NSSA external type 2
   E1 - OSPF external type 1, E2 - OSPF external type 2, E - EGP
   i - IS_IS, L1 - IS_IS level-1, L2 - IS_IS level-2, ia - IS_IS inter area
   * - candidate default, U - per-user static route, o - ODR
   P - periodic downloaded static route

Gateway of last resort is not set
   I 1.1.4.0/24 [100/10576] via 1.1.2.2, 00:00:50, Serial0
   [100/12100] via 1.1.1.2, 00:00:50, Serial1
   C 1.1.3.0 is directly connected, FastEthernet0
   C 1.1.2.0 is directly connected, Serial0
   C 1.1.1.0 is directly connected, Serial1
```

The configuration for R2 is shown Listing 5.13.

LISTING 5.13 Configuration for R2

```
interface FastEthernet0
ip address 1.1.4.2 255.255.255.0
interface Serial0
ip address 1.1.2.2 255.255.255.0
interface Serial1
ip address 1.1.1.2 255.255.255.0
router igrp 1
network 1.0.0.0
variance 10
```

The routing table for R2 is shown in Listing 5.14.

LISTING 5.14 Routing Table for R2

```
R2#sh ip route
Codes: C - connected, S - static, I - IGRP, R - RIP, M - mobile, B - BGP
   D - EIGRP, EX - EIGRP external, O - OSPF, IA - OSPF inter area
   N1 - OSPF NSSA external type 1, N2 - OSPF NSSA external type 2
   E1 - OSPF external type 1, E2 - OSPF external type 2, E - EGP
   i - IS_IS, L1 - IS_IS level-1, L2 - IS_IS level-2, ia - IS_IS inter area
   * - candidate default, U - per-user ,static route, o - ODR
   P - periodic downloaded static route
```

```
Gateway of last resort is not set
  I 1.1.1.0/24 [100/10576] via 1.1.2.1, 00:00:50, Serial0
  [100/12100] via 1.1.1.1, 00:00:50, Serial1
  C 1.1.4.0 is directly connected, FastEthernet0
  C 1.1.2.0 is directly connected, Serial0
  C 1.1.1.0 is directly connected, Serial1
```

The variance command is used to determine the feasibility of a potential alternate route into the routing table. If the next router is closer to the destination than the present router and has a path whose metric is less than or equal to the variance, it is called feasible path. To perform load balancing between multiple alternate paths:

- Metric of the alternate path should be less than or equal to the variance. The metric of the alternate path should also be greater than or equal to the variance multiplied with the metric of the local path.
- Next-hop router should be closer to the destination than the present router; that is, the metric to the destination from the next-hop router should be less than that of the present router.

Figure 5.8 shows three connected routers, R1, R2, and R3. R1 has two paths to reach network 1.1.1.0/24. One of them is a direct path, while the other is through R2.

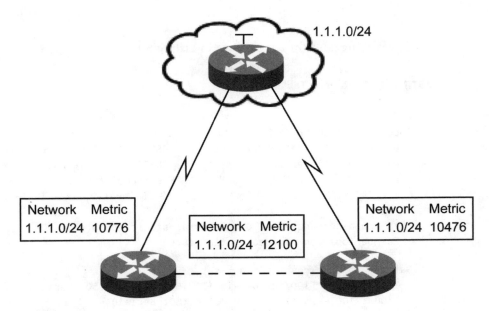

FIGURE 5.8 IGRP feasible successor.

In Figure 5.8, the metric for reaching 1.1.1.0/24 from R2 is less than that of the metric used to reach the same network from R1. This satisfies one of the conditions required for an alternate path to form a feasible successor.

If R1 defines a variance of 2, that is, a multiplying factor, then 10776 * 2 = 21552 should be greater than the metric to reach 1.1.1.0/24 from R1 through the alternate path, which is also true. Both the conditions to become a feasible successor are met, and R1 will load balance the traffic to 1.1.1.0/24 through both the paths.

The configuration for R1 is shown in Listing 5.15.

LISTING 5.15   Configuration for R1

```
interface FastEthernet0
ip address 172.16.1.1 255.255.255.0
interface Serial0
ip address 172.16.2.1 255.255.255.0
interface Serial1
ip address 172.16.3.1 255.255.255.0
router igrp 10
network 172.16.0.0
variance 2
.
.
.
```

The routing table for R1 is shown in Listing 5.16.

LISTING 5.16   Routing Table for R1

```
R1#sh ip route
Codes: C - connected, S - static, I - IGRP, R - RIP, M - mobile, B - BGP
   D - EIGRP, EX - EIGRP external, O - OSPF, IA - OSPF inter area
   N1 - OSPF NSSA external type 1, N2 - OSPF NSSA external type 2
   E1 - OSPF external type 1, E2 - OSPF external type 2, E - EGP
   i - IS_IS, L1 - IS_IS level-1, L2 - IS_IS level-2, ia - IS_IS inter
     area
   * - candidate default, U - per-user static route, o - ODR
   P - periodic downloaded static route

Gateway of last resort is not set
   I 1.1.1.0/24 [100/10776] via 172.16.3.3, 00:01:50, Serial1
   [100/12100] via 172.16.2.2, 00:01:50, Serial0
   C 172.16.1.0 is directly connected, FastEthernet0
```

```
C 172.16.2.0 is directly connected, Serial0
C 172.16.3.0 is directly connected, Serial1
```

The traffic-share balanced command is used to control and distribute traffic among different alternate paths of unequal cost.

```
Router(config-router)# traffic-share balanced
```

This command is required to balance the traffic among different paths of unequal costs.

IGRP Maximum Paths

IGRP can load balance between two points up to four links using the variance parameter. If the difference between metric of multiple links falls within the variance specified, IGRP will use up to a maximum of four parallel links.

Engineering IGRP Metrics

The various metrics in IGRP can be changed, if you need to change the default route selection process to provide more flexibility. When traffic is routed over multiple paths in IGRP, the lower the metric value, the higher the traffic carried.

The command used to change the value of different metric weights is:

```
Router(config-router)# metric weights tos k1 k2 k3 k4 k5
```

The formula used to calculate the metric in IGRP is:

$$(K_1 * Bw) + (K_2 * Bw)/(256\text{-}Load) + (K_3 * Delay) * (K_5/(Reliability + K_4))$$

The default values for weights are displayed in Table 5.5.

TABLE 5.5 Default Values of IGRP Weights

Weight	Default Value
K_1	1
K_2	0
K_3	1
K_4	0
K_5	0

Substituting the default value into the equation, the metric used in IGRP becomes "bandwidth plus delay." Table 5.6 lists the values of bandwidth and delay of some common physical media.

TABLE 5.6 Bandwidth and Delay Values of Common Physical Media

Physical media	Bandwidth (Kbps)	Delay (uS)
ATM 622 Mbps	622000	100
ATM 155 Mbps	155000	100
Fast Ethernet	100000	100
FDDI	100000	100
HSSI	45045	20000
Token Ring	16000	630
Ethernet	10000	1000
T1	1544	20000
E1	2048	20000
DS0	64	20000
Serial line (56 Kbps)	56	20000

You should be careful while modifying the bandwidth and delay values, because it can degrade the network performance and may also result in a routing loop.

Suppressing IGRP Advertisements

There are instances when you do not want to send routing updates to a specific neighbor or interface. The passive-interface command is used to control the number of interfaces over which the IGRP routing updates are going to be sent. Passive-interface can receive routing updates. Consider the following example of suppressing IGRP advertisements.

In Figure 5.9, R1 has two connected interfaces in the network, 172.16.1.0/24 and 172.16.2.0/24.

In Figure 5.9, when the classful network 172.16.0.0 is defined under the IGRP routing process, it sends routing updates through both the Ethernet and the serial interfaces. After R1 and R2 exchange routing information, the routing table of R1 will contain a route to the Ethernet interface of R2. Similarly, the routing table of R2 will have a route to 172.16.1.0/24.

The configuration for R1 is shown in Listing 5.17.

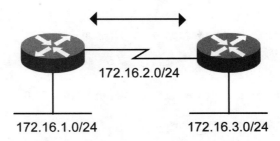

FIGURE 5.9 Classful IGRP routing process.

LISTING 5.17 Configuration for R1

```
interface FastEthernet0/0
ip address 172.16.1.1 255.255.255.0
interface Serial0/0
ip address 172.16.2.1 255.255.255.0
router igrp 1
network 172.16.0.0
```

The routing table for R1 is shown in Listing 5.18.

LISTING 5.18 Routing Table for R1

```
R1#sh ip route
Codes: C - connected, S - static, I - IGRP, R - RIP, M - mobile, B - BGP
   D - EIGRP, EX - EIGRP external, O - OSPF, IA - OSPF inter area
   N1 - OSPF NSSA external type 1, N2 - OSPF NSSA external type 2
   E1 - OSPF external type 1, E2 - OSPF external type 2, E - EGP
   i - IS_IS, L1 - IS_IS level-1, L2 - IS_IS level-2, ia - IS_IS inter
     area
   * - candidate default, U - per-user static route, o - ODR
   P - periodic downloaded static route

Gateway of last resort is not set
   I 172.16.3.0/24 [100/12100] via 172.16.2.2, 00:00:50, Serial0/0
   C 172.16.1.0 is directly connected, FastEthernet0/0
   C 172.16.2.0 is directly connected, Serial0/0
```

The configuration for R2 is shown in Listing 5.19.

LISTING 5.19 Configuration for R2

```
interface FastEthernet0/0
ip address 172.16.3.2 255.255.255.0
```

```
interface Serial0/0
ip address 172.16.2.2 255.255.255.0
router igrp 1
network 172.16.0.0
```

The routing table for R2 is shown in Listing 5.20.

LISTING 5.20 Routing Table for R2

```
R2#sh ip route
Codes: C - connected, S - static, I - IGRP, R - RIP, M - mobile, B - BGP
   D - EIGRP, EX - EIGRP external, O - OSPF, IA - OSPF inter area
   N1 - OSPF NSSA external type 1, N2 - OSPF NSSA external type 2
   E1 - OSPF external type 1, E2 - OSPF external type 2, E - EGP
   i - IS_IS, L1 - IS_IS level-1, L2 - IS_IS level-2, ia - IS_IS inter area
   * - candidate default, U - per-user static route, o - ODR
   P - periodic downloaded static route

Gateway of last resort is not set
   I 172.16.1.0/24 [100/12100] via 172.16.2.1, 00:00:50, Serial0/0
   C 172.16.3.0 is directly connected, Fastethernet0/0
   C 172.16.2.0 is directly connected, Serial0/0
```

If the serial interfaces of the two routers are defined as passive-interfaces, there will be no exchange of routing information between them. The route to the 172.16.3.0/24 is no longer present in the routing table for R1. Figure 5.10 shows the working of the passive-interface command.

The configuration of R1 is shown in Listing 5.21.

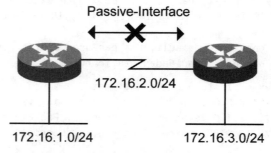

FIGURE 5.10 Working of passive-interface command.

LISTING 5.21 Configuration of R1

```
interface FastEthernet0/0
ip address 172.16.1.1 255.255.255.0
```

```
interface Serial0/0
ip address 172.16.2.1 255.255.255.0
router igrp 1
passive-interface serial 0/0
network 172.16.0.0
```

The routing table of R1 is shown in Listing 5.22.

LISTING 5.22   Routing Table of R1

```
R1#sh ip route
Codes: C - connected, S - static, I - IGRP, R - RIP, M - mobile, B - BGP
   D - EIGRP, EX - EIGRP external, O - OSPF, IA - OSPF inter area
   N1 - OSPF NSSA external type 1, N2 - OSPF NSSA external type 2
   E1 - OSPF external type 1, E2 - OSPF external type 2, E - EGP
   i - IS_IS, L1 - IS_IS level-1, L2 - IS_IS level-2, ia - IS_IS inter area
   * - candidate default, U - per-user static route, o - ODR
   P - periodic downloaded static route

Gateway of last resort is not set
   C 172.16.1.0 is directly connected, FastEthernet0/0
   C 172.16.2.0 is directly connected, Serial0/0
```

The configuration of R2 is shown in Listing 5.23.

LISTING 5.23   Configuration of R2

```
interface FastEthernet0/0
ip address 172.16.3.2 255.255.255.0
interface Serial0/0
ip address 172.16.2.2 255.255.255.0
router igrp 1
passive-interface serial 0/0
network 172.16.0.0
```

The routing table of R2 is shown in Listing 5.24.

LISTING 5.24   Routing Table of R2

```
R2#sh ip route
Codes: C - connected, S - static, I - IGRP, R - RIP, M - mobile, B - BGP
   D - EIGRP, EX - EIGRP external, O - OSPF, IA - OSPF inter area
   N1 - OSPF NSSA external type 1, N2 - OSPF NSSA external type 2
   E1 - OSPF external type 1, E2 - OSPF external type 2, E - EGP
   i - IS_IS, L1 - IS_IS level-1, L2 - IS_IS level-2, ia - IS_IS inter area
   * - candidate default, U - per-user static route, o - ODR
   P - periodic downloaded static route
```

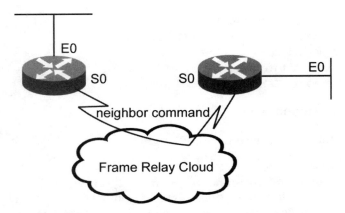

FIGURE 5.11 Use of neighbor command in IGRP.

```
Gateway of last resort is not set
  C 172.16.3.0 is directly connected, Fastethernet0/0
  C 172.16.2.0 is directly connected, Serial0/0
```

Unicast Updates in IGRP

IGRP sends routing updates through broadcasts. However, if the need arises, unicast updates can also be sent using the neighbor command. IGRP defines a neighbor in non-broadcast networks such as Frame-Relay (FR). The command is:

```
Router(config-router)# neighbor ip-address
```

With the neighbor command, IGRP sends unicast routing updates instead of broadcasts. Figure 5.11 shows two routers, R1 and R2, connected over FR with their serial interfaces S0 and S0, respectively.

In Figure 5.11, the neighbor command enables the neighbor relationship. The configuration of R1 is:

```
router igrp 1
network 172.16.0.0
neighbor 172.16.12.2
```

The configuration of R2 is:

```
router igrp 1
network 172.16.0.0
neighbor 172.16.12.1
```

Modifying IGRP Timers

You may need to change the timers from the default values so that the time taken for network convergence is less. The command used to change the timers is:

```
Router(config-router)# timers basic update invalid holddown flush
(sleeptime)
```

This command adjusts the different timers of the routing process as required and is entered in IGRP configuration mode.

When a new route is found, it is placed in such a state that the router does not accept any information about the route. The length of time that the route is kept in this state is called the hold-down timer. During this period, the router ignores any incoming advertisements about the route from any other router. This is done to avoid any routing loops when convergence takes place.

Gateway of Last Resort

When a packet needs to be sent to a particular destination, the router checks the routing table for an entry of the network. The packet is forwarded to the next-hop router if an entry is found and is dropped if the entry is not found. However, an alternative method would be to have a default network where packets with no entry in the routing table are forwarded. This default network is known as a Gateway of Last Resort.

Candidate default routes can be defined using the ip default-network network command. The network field should be that of an unconnected, classful network that is present in the routing table and may be learned from any dynamic or static routing protocol. The best default route is selected among multiple candidates' default routes.

Consider the example depicted in Figure 5.12. The figure shows three routers, R1, R2, and R3, connected to each other, such that R2 runs IGRP and another classless protocol like OSPF.

In Figure 5.12, R2 generates a default network that is propagated to R1. R1 sees this network as the Gateway of Last Resort and uses it to reach networks that do not have an entry in its routing table. IGRP will not understand the routing table, and it will have no entry of any classless networks in the table. This is because it is a classful routing protocol. It will not be possible to reach any of the classless networks from R1.

This problem can be resolved if R2 generates a default route pointing to a classful network address and advertises the same. R1 receives the advertisement and installs the default route in its routing table. This default route points to the classful network, which is already present in the routing table of R1. This feature enables R1 to send packets to networks that do not have an entry in their routing tables by forwarding the same to the Gateway of Last Resort.

The configuration for R1 is shown in Listing 5.25.

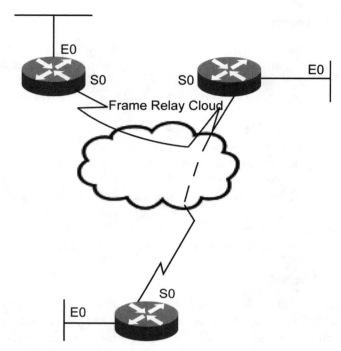

FIGURE 5.12 Gateway of Last Resort.

LISTING 5.25 Configuration for R1

```
interface FastEthernet0
ip address 200.20.10.1 255.255.255.0
interface Serial0
ip address 200.20.20.1 255.255.255.0
.
.
router igrp 1
network 200.20.10.0
network 200.20.20.0
```

The routing table for R1 is shown in Listing 5.26.

LISTING 5.26 Routing Table for R1

```
R1#sh ip route
Codes: C - connected, S - static, I - IGRP, R - RIP, M - mobile, B - BGP
   D - EIGRP, EX - EIGRP external, O - OSPF, IA - OSPF inter area
```

```
N1 - OSPF NSSA external type 1, N2 - OSPF NSSA external type 2
E1 - OSPF external type 1, E2 - OSPF external type 2, E - EGP
i - IS_IS, L1 - IS_IS level-1, L2 - IS_IS level-2, ia - IS_IS inter area
* - candidate default, U - per-user static route, o - ODR
P - periodic downloaded static route

Gateway of last resort is 200.20.20.2 to network 172.20.0.0
   I *172.20.0.0/16 [100/12100] via 200.20.20.2, 00:00:50, Serial0
   I 200.20.30.0 [100/12100] via 200.20.20.2, 00:00:50, Serial0
   C 200.20.10.0 is directly connected, FastEthernet0
   C 200.20.20.0 is directly connected, Serial0
   .
   .
   .
```

The configuration for R2 is shown in Listing 5.27.

LISTING 5.27   Configuration for R2

```
interface Serial0
ip address 200.20.20.2 255.255.255.0
interface Serial1
ip address 200.20.30.2 255.255.255.0
.

.
router ospf 1
.

.
router igrp 1
redistribute ospf 1
network 200.20.20.0
network 200.20.30.0
.

.
ip default-network 172.20.0.0
```

The routing table for R2 is shown in Listing 5.28.

LISTING 5.28   Routing Table for R2

```
R2#sh ip route
Codes: C - connected, S - static, I - IGRP, R - RIP, M - mobile, B - BGP
   D - EIGRP, EX - EIGRP external, O - OSPF, IA - OSPF inter area
   N1 - OSPF NSSA external type 1, N2 - OSPF NSSA external type 2
```

```
          E1 - OSPF external type 1, E2 - OSPF external type 2, E - EGP
          i - IS_IS, L1 - IS_IS level-1, L2 - IS_IS level-2, ia - IS_IS inter area
          * - candidate default, U - per-user static route, o - ODR
          P - periodic downloaded static route

   Gateway of last resort is not set
       O E2172.20.32.0/24 [110/100] via 200.20.30.3, 01:11:24, Serial1
       I 200.20.10.0 [100/12100] via 200.20.20.1, 00:00:50, Serial0
       C 200.20.20.0 is directly connected, Serial0
       C 200.20.30.0 is directly connected, Serial1
       .
       .
       .
```

Commands for Monitoring and Troubleshooting IGRP

Several commands are used for maintaining, monitoring, and troubleshooting IGRP. The show ip interface command is used to know details of all the interfaces of the router:

```
Router# show ip interface
```

The output of Router# show ip interface command is shown in Listing 5.29.

LISTING 5.29   Output for Router# show ip interface Command

```
Ethernet0 is up, line protocol is up
   Internet address is 10.0.1.10, subnet mask is 255.255.255.0
   Broadcast address is 255.255.255.255
   Address determined by setup command
   MTU is 1500 bytes
   Helper address is not set
   Directed broadcast forwarding is enabled
   Multicast groups joined: 224.0.0.1 224.0.0.2
   Outgoing access list is not set
   Inbound access list is not set
   Proxy ARP is enabled
   Security level is default
   Split horizon is enabled
   ICMP redirects are always sent
   ICMP unreachables are always sent
   ICMP mask replies are never sent
   IP fast switching is enabled
   IP fast switching on the same interface is disabled
   IP SSE switching is disabled
```

```
  Router Discovery is disabled
  IP accounting is disabled
  TCP/IP header compression is disabled
  Probe proxy name replies are disabled
  Gateway Discovery is disabled

Serial0 is up, line protocol is up
  Internet address is 200.2.2.1, subnet mask is 255.255.255.0
  Broadcast address is 255.255.255.255
  Address determined by setup command
  MTU is 1500 bytes
  Helper address is not set
  Directed broadcast forwarding is enabled
  Multicast groups joined: 224.0.0.1 224.0.0.2
  Outgoing access list is not set
  Inbound access list is not set
  Proxy ARP is enabled
  Security level is default
  Split horizon is enabled
  ICMP redirects are always sent
  ICMP unreachables are always sent
  ICMP mask replies are never sent
  IP fast switching is enabled
  IP fast switching on the same interface is disabled
  IP SSE switching is disabled
  Router Discovery is disabled
  IP accounting is disabled
  TCP/IP header compression is disabled
  Probe proxy name replies are disabled
  Gateway Discovery is disabled
```

The command Router# show ip interface brief lists all the interfaces, as well as IP address and its status:

```
Router# show ip interface brief
```

Listing 5.30 is a sample output for the show ip interface brief command.

LISTING 5.30   Output for show ip interface brief Command

```
Router# show ip interface brief
InterfaceIP-address  OK?  Method Status Protocol
Ethernet0 10.0.1.10  YES  manual up      up
Serial0   200.2.2.1  YES  manual up      up
```

Listing 5.31 is a sample output for the show ip protocol command.

LISTING 5.31 Output for show ip protocol Command

```
Router# show ip protocol
Routing protocol is "igrp 1"
  Sending updates every 90 seconds, next due in 65 seconds
  Invalid after 270 seconds, hold down 280, flushed after 630
  Outgoing update filter list for all interfaces is not set
  Incoming update filter list for all interfaces is not set
  Default network flagged in outgoing updates
  Default networks accepted from incoming updates
  IGRP metric weight K1=1, K2=0, K3=1, K4=0, K5=0
  IGRP maximum hopcount 100
  IGRP maximum metric variance 1
  Redistribution: igrp 1

Routing for Networks:
  1.0.0.0

Routing Information Sources:
  Gateway     Distance    Last Update
  1.1.1.2     100         0:01:50
  Distance: (default is 100)
```

Listing 5.32 is a sample output for the debug ip igrp events command.

LISTING 5.32 Output for debug ip igrp events Command

```
Router# debug ip igrp events
IGRP event debugging is on
HH:MM:SSIGRP: sending update to 255.255.255.255 via FastEthernet0/0
(1.1.1.1)
HH:MM:SSIGRP: update contains 5 interior, 0 system, and 1 exterior
routes
HH:MM:SSIGRP: total routes in update: 6
HH:MM:SSIGRP: received update from 1.1.2.2 on Serial0/0
HH:MM:SSIGRP: update contains 3 interior, 0 system, and 1 exterior
routes
HH:MM:SSIGRP: total routes in update: 4
Router# undebug ip igrp events
```

Listing 5.33 is a sample output for the debug ip igrp transactions command.

LISTING 5.33   Output for debug ip igrp transactions Command

```
Router# debug ip igrp transactions
HH:MM:SSIGRP: received update from 1.1.2.2 on serial0/0
HH:MM:SS  Subnet 1.2.1.0, metric 182671 (neighbor 12100)
HH:MM:SS  Subnet 1.2.2.0, metric 182671 (neighbor 12100)
HH:MM:SS  Subnet 1.2.3.0, metric 182671 (neighbor 12100)
HH:MM:SS  IGRP: sending update to 255.255.255.255 via FastEthernet 0/0
(1.1.1.1)
HH:MM:SS  Subnet 1.2.1.0, metric 182671
HH:MM:SS  Subnet 1.2.2.0, metric 182671
HH:MM:SS  Subnet 1.2.3.0, metric 182671
HH:MM:SS  Subnet 1.2.4.0, metric 182671
HH:MM:SS  IGRP: sending update to 255.255.255.255 via serial0/0
(1.1.2.1)
HH:MM:SS  Subnet 1.2.4.0, metric 182671
HH:MM:SS  IGRP: received update from 1.1.2.2 on serial0/0
HH:MM:SS  Subnet 1.2.1.0, metric 182671 (neighbor 12100)
HH:MM:SS  Subnet 1.2.2.0, metric 182671 (neighbor 12100)
HH:MM:SS  Subnet 1.2.3.0, metric 182671 (neighbor 12100)
Router# undebug ip igrp transactions
```

Other Optional IGRP Tasks

Hold-down timers increase convergence time. The hold-down is enabled by default. If required, it can be disabled using the following command:

```
Router(config-router)# no metric holddown
```

The offset-list command is used to increase the metric of the routes sent or received by IGRP. This command can be applied to a particular set of routes by the use of an access-list. An offset-list can also be applied to all routes sent or received by any particular interface. The syntax for the command offset-list is:

```
Router(config-router)# offset-list [access-list-number | access-list-
name] {in | out} offset [interface-type | interface-number]
```

A distance vector protocol uses hop-counts to define the maximum diameter to which the routing process can extend. This value is 100 in IGRP. This means that an IGRP protocol will not advertise any route that has a hop-count beyond 100. This default value can be increased up to 255 using the command:

```
Router(config-router)# metric maximum-hops number
```

IGRP validates the source address of incoming IGRP routing updates. This security feature can be disabled using the command:

```
Router(config-router)# no validate-update-source
```

This command is used to disable the validation of the source IP address of the router participating in the IGRP network.

The commands used for disabling and enabling split-horizon are:

```
Router(config-if)# no ip split-horizon
Router(config-if)# ip split-horizon
```

When distance vector protocols like IGRP are configured in routers connected in a broadcast network, split-horizon is enabled by default to prevent routing loops. However, in case of non-broadcast packet-switching networks, such as FR, you may need to disable split-horizon because:

- Split-horizon prevents routes from being advertised out of the same interface from where it is learned.
- Interfaces that have subinterfaces do not advertise routes to the secondary network address.

You can connect one interface to two different routers in packet-switching networks. Routing updates from one router will not reach the other unless split-horizon is disabled on the hub router.

NOTE

Enabling and disabling split-horizon for all routing protocols on an interface must be done with caution because this could lead to routing loops.

CAUTION

Consider three routers, R1, R2, and R3, connected over a FR network, as shown in Figure 5.13. In Figure 5.13, three routers, R1, R2, and R3, are running in IGRP. R1 is connected to R2 and R3 via Serial 0, over the FR cloud. IGRP would prevent R1 from sending routing updates from the same interface from where it has received the updates. As a result, R3 is unable to communicate with the networks of R2. Disabling split-horizon on the serial interface of R1 will overcome this problem.

The configuration of R1 is:

```
interface Serial2/0
ip address 130.1.45.1 255.255.255.0
```

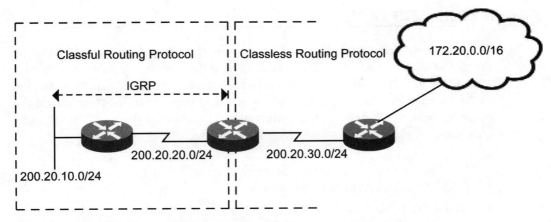

FIGURE 5.13   Split-horizon Disabled in FR Network.

```
encapsulation frame-relay
no ip split-horizon
```

The no ip split-horizon command is in FR serial interface by default and will not be visible in the configuration.

The configuration of R2 is:

```
interface Serial2/0
 ip address 130.1.45.2 255.255.255.0
 encapsulation frame-relay
 ip split-horizon
```

The command ip split-horizon disables split-horizon by default in FR serial interface.

The configuration of R3 is:

```
interface Serial2/0
 ip address 130.1.45.3 255.255.255.0
 encapsulation frame-relay
 ip split-horizon
```

The command ip split-horizon disables split-horizon by default in FR serial interface and has to be enabled.

DESIGN CONSIDERATIONS

There are some characteristics and parameters that should be considered while designing an IGRP network. IGRP is used on a flat network topology. Also note that by default, IGRP has a maximum hop-count of 100, which can be extended to 255. You must also make note of the method used by IGRP to calculate the optimal path while designing the network. The other features of IGRP to be considered are:

Load Balancing: Use this feature where IGRP balances load over equal-cost paths, to enhance the overall performance of the network.

Scalability: Consider the size, growth, and scalability of the network while deploying IGRP in the network.

IGRP recalculates the entire routing table during convergence. This leads to increased CPU cycles and memory usage and as a result, puts a limit to the scalability of the network. A lot of bandwidth is also consumed during the update process.

ADVANTAGES OF IGRP OVER RIP

RIP and IGRP are both distance vector protocols. They are similar in the way they function. However, IGRP has overcome a few drawbacks of RIP:

- RIP uses hop-count as a metric. IGRP uses composite metric for calculating the best path to reach a destination. This enables IGRP to make a distinction between slow and fast links while selecting the best path to reach a destination.
- RIP has a maximum hop-count of 15, rendering it incapable for use in large networks where the number of hop-counts is more than 15. IGRP can be used in such networks because it can be configured to scale up to 255 hop-counts.
- RIP does not support the ASN feature. IGRP has an ASN that makes it possible to run different routing systems on the same network. This is particularly needed when several organizations converge over the same network.
- RIP and IGRP employ different methods to handle default networks. IGRP has several candidate default networks and chooses the one with the lowest metric as the default route.

Table 5.7 draws a comparison between the RIP and IGRP routing protocols.

TABLE 5.7   Comparing RIP and IGRP

Feature	RIP	IGRP
Hop-Count	15	100
Update Timer	30 seconds	90 seconds
Metric	Hop-count	Bandwidth, delay, loading, reliability, and MTU
Load Sharing	No	Possible over four equal paths
AS	No	Yes

IGRP has some limitations that are common to distance vector protocols:

- Does not support classless routing and as a result, does not support CIDR
- Does not support subnet information and as a result, VLSM does not work
- Takes more time to achieve convergence because IGRP recalculates the entire table
- Sends the entire routing table during updates, consuming significant amounts of bandwidth in the process
- Cannot be used in routers manufactured by other vendors because it is a Cisco-proprietary protocol

SUMMARY

In this chapter we discussed IGRP's operation and configuration. We also discussed the advantages of IGRP over RIP. In the next chapter, we will learn about EIGRP.

POINTS TO REMEMBER

- Metric in IGRP is a combination of delay, bandwidth, reliability, load, and MTU.
- IGRP has features such as split-horizon, split-horizon with poison reverse, route poisoning, hold-down timers, triggered updates, and count to infinity.

- Routes advertised by IGRP can be classified into interior, system, and exterior routes.
- The ASN feature of IGRP enables only the routers of the same AS to exchange routing updates, thereby enhancing network security.
- Router(config)# router igrp as-number enables IGRP routing process for the particular ASN.
- Router(config-router)# network network-number defines the network under the routing process.
- Router(config-router)# variance value enables load balancing between different alternate paths having different metrics.
- Router(config-router)# traffic-share balanced balances traffic among different paths of unequal costs.
- Router(config-router)# metric weights tos k1 k2 k3 k4 k5 changes the values of different metric weights.
- $(K_1*Bw) + (K_2*Bw)/(256-Load) + (K_3*Delay)* (K_5/(Reliability + K_4))$ is the formula used to calculate the IGRP metric.
- Router(config-router)# neighbor ip-address: defines a neighbor in non-broadcast networks.
- Router(config-router)# timers basic update invalid holddown flush (sleeptime) is used to change the timers.
- Router(config-router)# no metric holddown disables hold-down.
- Router(config-router)# offset-list [access-list-number | access-list-name] {in | out} offset [interface-type | interface-number] is the syntax for the offset-list command.
- Router(config-router)# metric maximum-hops number increases hop-count up to 255.
- Router(config-router)# no validate-update-source disables validation of the source address of incoming routing updates.
- Router(config-if)# no ip split-horizon disables split-horizon.
- Router(config-if)# ip split-horizon enables split-horizon.
- Router# show ip interface shows details of the interfaces of the router.
- Router# show ip interface brief lists all the interfaces, IP address, and IP address status of the router.
- You should consider load balancing and scalability features while designing the IGRP network.
- IGRP has a default network, known as a Gateway of Last Resort, to which packets with no entry in the routing table are forwarded.

6 | Enhanced Interior Gateway Protocol

INTRODUCTION TO EIGRP

Enhanced Interior Gateway Routing Protocol (EIGRP) is an enhanced version of IGRP. The working of EIGRP is similar to that of IGRP in many ways, but it uses a better algorithm, Diffusing Update Algorithm (DUAL), for metric calculation. EIGRP addresses many shortcomings of IGRP, the most important being the support of classless addressing. The only disadvantage of EIGRP is that it is a Cisco-proprietary protocol and is not chosen for environments with routers other than Cisco.

EIGRP OPERATION

EIGRP is a Cisco-proprietary routing protocol. This means that EIGRP does not support routers manufactured by any vendor other than Cisco. EIGRP has the following characteristics:

- Supports VLSM
- Achieves faster convergence
- Supports multiple routed protocols such as Internet Protocol (IP), Internetwork Packet Exchange (IPX), and AppleTalk
- Requires lesser bandwidth
- Uses DUAL

- Performs load balancing over paths having unequal paths
- Finds alternate routes

DUAL is explained later in the chapter

The features of EIGRP are shown in Table 6.1.

TABLE 6.1   Features of EIGRP

Features	EIGRP Values
Protocol type	Hybrid
Subnet information	Classless
Routing updates	Sends routing updates when there is a change in topology
Metric	Composite (bandwidth, delay, reliability, load)
Count to infinity	224
Routing updates mode	Multicast; uses address 224.0.0.10
Algorithm	DUAL

EIGRP uses different data packets during the routing operation. They are:

Hello: Sent by a router to its neighbor during the discovery process. These packets are sent using the multicast unreliable delivery method.

Acknowledgment: Sent by an EIGRP router on receipt of Hello packets. The unicast delivery protocol is used to send acknowledgment packets back to the sender of Hello packets.

Routing Updates: Sent in the form of Reliable Transport Protocol (RTP) when there is a change in the network topology. They are sent using multicast or unicast, depending upon the number of routers to which data are to be sent.

Queries: Sent by EIGRP using multicast or unicast reliable delivery protocol.

Replies: Sent by EIGRP using unicast.

Unreliable delivery is discussed in a later section.

Reliable Transport Protocol

Reliability of protocols guarantee that data will not only reach the destination but that they will also be delivered in the correct sequence. RTP ensures that EIGRP is a reliable protocol. This reliability is achieved by the use of acknowledgments sent by the recipient of the data packet. EIGRP sends routing updates, using the multicast address of 224.0.0.10. Each neighbor acknowledges receipt by sending back an acknowledgment using unicast. This ensures that the routing updates have reached each of the EIGRP neighbors.

In reliable transport, the EIGRP neighbor may fail to respond by sending an acknowledgment. In such cases, the packet is retransmitted only to the neighbor that has failed to send the acknowledgment. If the neighbor fails to acknowledge receipt even after 16 tries, it is declared 'dead'.

Each packet has a sequence number assigned by the router, and then incremented as the packet traverses each router. Each packet also contains the last sequence number such that the trail of numbers can be maintained. RTP ensures that the packets are sent to the destination such that the sequence number is maintained.

EIGRP also sends packets that do not use RTP. In such cases, sending the acknowledgment or maintaining the sequence number is not required. This is known as unreliable delivery.

In Figure 6.1, three routers, R1, R2, and R3, are connected, where R1 sends out multicast routing updates to routers R2 and R3.

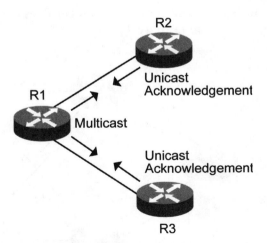

FIGURE 6.1   EIGRP uses RTP to send packets to its neighbors and receive acknowledgments.

In Figure 6.1, R2 and R3 send unicast acknowledgments back to R1 after receiving data packets.

Neighbor Discovery

When a new router joins the EIGRP network, it needs to discover the other EIGRP neighbors for optimal functioning of the routing process. The new router initiates this discovery process by sending multicast Hello packets. These Hello packets are sent across the network over a periodic interval of 5 seconds over fast links, and 60 seconds over slow links.

If a router does not receive a Hello packet for a certain period, the neighbor is considered unreachable. This time period is known as the Hold-time and is normally three times that of the Hello interval. For example, the Hold-time is 15 seconds on fast links and 180 seconds on slow links. The Hello-interval and Hold-time can be modified, if required.

Fast Ethernet is an example of a fast link, while Frame-Relay is a slow link.

A router running EIGRP builds its Neighbor table by sending Hello packets to its neighbors. The Neighbor table contains the IP address of each neighbor from which a router receives Hello packets. The Neighbor table also contains the interface on which it receives these Hello packets.

Figure 6.2 shows several routers exchanging Hello packets in order to build their Neighbor table.

The show ip eigrp neighbors command enables you to see the Neighbor table. The output is shown in Listing 6.1.

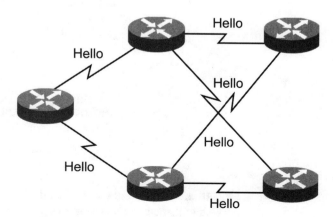

FIGURE 6.2   EIGRP routers exchange Hello packets to establish a Neighbor table.

LISTING 6.1   show ip eigrp neighbors Command

```
p12r1#show ip eigrp neighbors
IP-EIGRP neighbors for process 1
H Address Interface Hold Uptime SRTT RTO Q Seq Type
(sec) (ms) Cnt Num
3 192.12.4.1 Et0 10 01:54:47 80 480 0 2
2 192.12.2.1 Se2 14 01:55:40 26 200 0 5
1 192.12.1.1 Se1 11 01:55:40 62 372 0 6
0 192.12.5.1 Se0 14 01:03:20 19 200 0 97
```

Hybrid, Link State, and Distance Vector Protocols

EIGRP is a hybrid protocol because it displays the characteristics of both distance vector and link state protocols. Distance vector protocols send periodic routing updates of the entire routing table to all directly connected neighbors. These protocols consume a lot of bandwidth and are prone to routing loops. Distance vector protocols also take a long time to attain routing convergence.

Link state protocols send partial routing updates only when there is a change of topology of directly connected neighbors. These protocols consume less bandwidth and reach convergence faster than distance vector protocols. This feature enables link state protocols to provide a loop-free routing environment.

EIGRP uses the same metrics as used by distance vector protocols. However, it does not send the entire routing table to neighbor routers at regular intervals. EIGRP sends partial changed link state information only when there is a topological change. EIGRP uses less bandwidth as compared to other distance vector protocols. By default, EIGRP uses up to 50 percent of the bandwidth on a link. This value can be modified, if required.

EIGRP is a classless routing protocol unlike other distance vector protocols, such as RIP and IGRP.

Diffusing Update Algorithm

EIGRP uses Diffusing Update Algorithm (DUAL) to calculate new routes whenever there is a topological change. This algorithm works on the principle of using multiple routers in a coordinated manner while performing routing calculations. This reduces the time taken to achieve routing convergence and prevents routing loops in the network.

The following terms and concepts will help you understand DUAL of EIGRP:

Adjacency: Virtual link between two routers over which routing information is sent and received. Adjacency is formed by exchanging Hello packets between neighboring routers. Routers exchange routing updates after adjacency is established. A router adds the distance metric to the network as advertised by its neighbor to the cost of the link. This enables the router to access its neighbor. Routers perform this calculation for every route to a destination network.

Feasible Distance: The lowest calculated metric among the different alternate paths to reach a destination network.

Advertised Distance: Distance advertised by a neighboring router.

Feasible Successor: Neighboring router that has an advertised distance that is less than the feasible distance. The feasible successor is closer to the destination than the router. A router can have multiple feasible successors for reaching a destination network.

Successor: The feasible successor that has the lowest metric to reach a destination.

Consider a network with three routers, R1, R2, and R3. Figure 6.3 illustrates the different concepts of DUAL in this network.

In Figure 6.3, R1 and R2 form an adjacency. The metrics to reach Network 1 from R1 through R2 is 45, and through R3, it is 50. The feasible distance is 45. The feasible distance of R1 through R3 to reach Network 1 is 50. R3 has an advertised distance of 30 to reach the same destination. So the advertised distance of R3 to reach Network 1 is less than the feasible distance of R1. This means that R3 is a feasible successor of R1. Similarly, R2 is also a feasible successor.

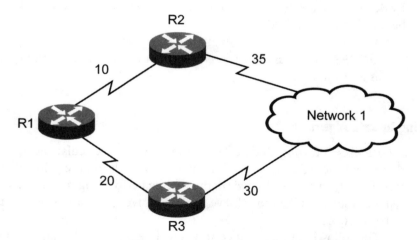

FIGURE 6.3   R1 selects the path to reach Network 1 from the list of two alternate paths based on its feasible distance to reach Network 1.

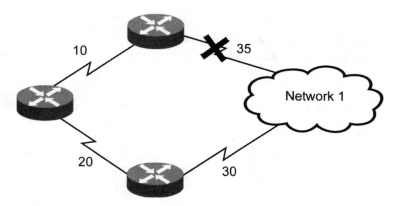

FIGURE 6.4 EIGRP uses DUAL to find an alternate path for reaching a network when the topology changes.

A list of feasible successors is maintained so that a router can select the next successor from the list in case there is a topological change due to a link failure.

Figure 6.4 illustrates how R1 reaches Network 1 even when the link between R2 and Network 1 fails.

EIGRP saves and maintains data in a topological table. This topological table contains:

■ Feasible distance to reach destination network
■ List of all feasible successors
■ Advertised distance to reach the destination for all feasible successors
■ Feasible distance to reach a destination network
■ Interface of the router connecting to all feasible successors

A router will recalculate its feasible successors for reaching the destination when an input event occurs. This recalculation is performed using DUAL. An input event can occur if:

■ Cost of a directly connected link changes
■ Link state of a directly connected interfaces changes
■ Update packet is received
■ Query packet is received
■ Reply packet is received

The process of recalculating the distance to the destination network through each of its feasible successors can lead to any of the following results:

■ If the feasible successor has a lower distance metric to reach a destination than the existing successor, the feasible successor becomes the new successor.

- If the new distance is less than the feasible distance, the feasible distance is updated to the new calculated distance.
- If the new distance is different from the existing distance, then routing updates are sent to all neighbors.

When the DUAL algorithm performs this local computation, the router cannot:

- Change the successor of the router
- Change the advertising distance to the destination network
- Change the feasible distance
- Start another DUAL process

EIGRP Packet Formats

EIGRP packets are encapsulated in IP with the protocol field set to 88. The source IP address of the packet is the interface generating this packet. The destination packet can be multicast or unicast depending on the packet type.

The maximum length of the packet is determined by Maximum Transmission Unit (MTU). The value of MTU varies, according to the technologies used; it has a default value of 1500 for Ethernet. The different fields of the EIGRP packet are shown in Figure 6.5.

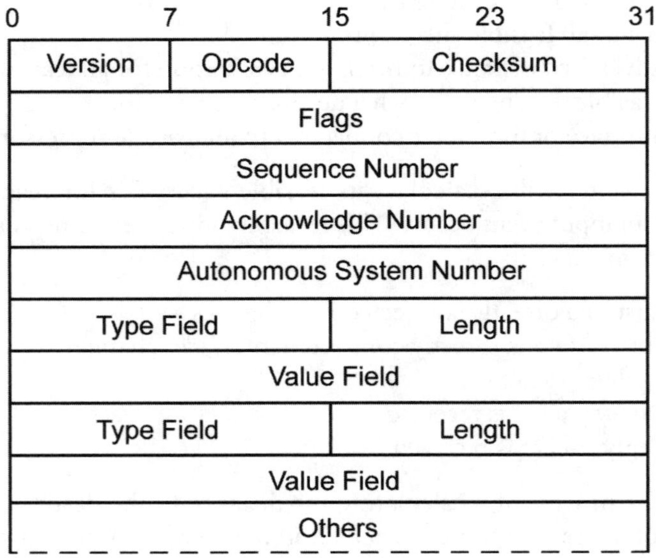

FIGURE 6.5 The EIGRP packet format.

The functions of the key fields of the EIGRP packet are:

Version: Indicates the version of the EIGRP process.

Opcode: Specifies the type of the EIGRP packet. Table 6.2 shows EIGRP packet's Opcode types.

TABLE 6.2   EIGRP Opcode Types

Opcode	Type
1	Update
3	Query
4	Reply
5	Hello
6	IPX, SAP

Checksum: Specifies the checksum of the entire EIGRP packet, excluding the IP header.

Flags: Shows the flag type. An EIGRP packet has two bits. The first bit is called as the init bit and is used in a new neighbor relationship. The second bit is the conditional receive bit and is used in a proprietary reliable multicast algorithm.

Sequence: Is a 32-bit sequence number. It is used to send messages reliably using RTP.

Acknowledgment Number: Sends messages reliably using RTP. It uses a 32-bit sequence number received from a neighbor to which a packet has been sent.

Autonomous System Number: Identifies the EIGRP process that is sending the packet. The destination to which the EIGRP packet is being sent processes the packet only if it has an EIGRP routing process with the same number, or else the packet is rejected.

Type and Length Value Field: Follows the EIGRP header. The values of Type and Length Value (TLV) specific to different protocols are summarized in Table 6.3.

TABLE 6.3   TLV Types Specific to Protocols

Number	TLV Types
General TLV Types	
0x0001	EIGRP parameters (Hello/Hold-time)
0x0003	Sequence
0x0004	Software version
0x0005	Next multicast sequence
TLV Types Specific to IP	
0x0102	IP internal routes
0x0103	IP external routes
TLV Types Specific to AppleTalk	
0x0202	AppleTalk internal route
0x0203	AppleTalk external routes
0x0204	AppleTalk cable configuration
TLV Types Specific to IPX	
0x0302	IPX internal routes
0x0303	IPX external routes

The Opcode field specifies the different types of EIGRP packets. The update information determines whether the route is internal or external.

Internal routes contain information of networks that are learned from within the EIGRP process. External routes contain information of networks that are not learned from within the EIGRP routing process. They are learned from another routing process after being redistributed into the EIGRP process.

The different fields used by EIGRP internal routes are shown in Figure 6.6. The EIGRP internal routes type field is 0x0102. The metric information of EIGRP is similar to that of IGRP. The additional fields used in EIGRP are next-hop and prefix length. The different fields of EIGRP internal routes are:

Next-hop: Specifies the IP address of the interface of the router to which destination-bound packets are forwarded. In most cases, it is the IP address of the router's interface generating the advertisement. This may not be true in case of multi-access networks.

Delay: Is the sum of configured delays.

Bandwidth: Is 2,560,000,000 divided by the lowest configured bandwidth of any interface from the source to the destination along which the packet traverses.

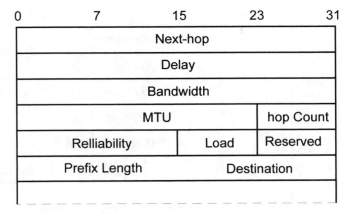

FIGURE 6.6   EIGRP packet format for internal route.

MTU: Is the smallest MTU of any link in the path traversed by the packet from the source to the destination.

Hop-count: Keeps track of the number of hops of the packet in its journey from the source to the destination. It can take any value between 0x01 to 0xFF. The maximum hop-count of EIGRP is 256.

Reliability: Denotes the error rates of the packet as it traverses from the source to the destination. It can accept a value between 0x01 and 0xFF.

Load: Reflects the total outgoing load of the interface along the route and can accept a value between 0x01 and 0xFF.

Reserved: Remains unused.

Prefix length: Specifies the network number of the destination network. It is calculated by performing a logical AND between the IP address and the subnet mask.

Destination: Specifies the destination of the route.

An external EIGRP route is learned from a source that is outside the EIGRP routing process. When routes from another routing process are redistributed into EIGRP, it results in an external route. The destination points to networks that do not belong to the EIGRP AS. Figure 6.7 shows the different fields of an EIGRP external packet.

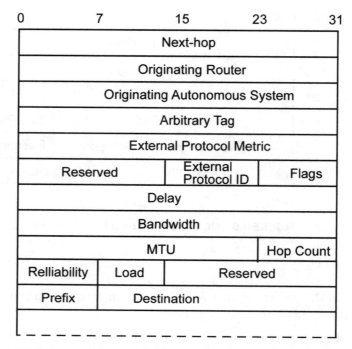

FIGURE 6.7 EIGRP packet format for external route.

The different fields of an EIGRP external packet are:

Next-hop: Denotes the IP address of the interface of the router to which the destination-bound packets are forwarded.

Originating Router: Denotes the ID of the router from which the EIGRP AS has learned about the external routes. It is the router on which redistribution has taken place.

Originating AS: Denotes the ASN of the router that is originating the route.

Arbitrary Tag: Is the tag defined in the route maps.

External Protocol Metric: Refers to the metric that has injected the routes into the EIGRP AS.

Reserved: Remains unused.

External Protocol ID: Denotes the ID of the protocol from which the external routes are learned. A list of IDs for different protocols is given in Table 6.4.

TABLE 6.4   External Protocols and Protocol ID

External Protocol	Protocol ID
IGRP	0x01
EIGRP	0x02
Static route	0x03
RIP	0x04
OSPF	0x06
IS_IS	0x07
EGP	0x08
BGP	0x09
Connected Link	0x0B

Flags: Shows the flag type. EIGRP external routes use two bits. For an external route, the right-most bit is 0x01. For a candidate default route, the second bit is 0x02.

Delay: Denotes the sum of configured delays.

Bandwidth: Is equal to 2,560,000,000 divided by the lowest configured bandwidth of any interface from the source to the destination along which the packet traverses.

MTU: Is the smallest MTU of any link in the path traversed by the packet from the source to the destination.

Hop-count: Keeps track of the number of hops of the packet in its journey from the source to the destination.

Reliability: Denotes the error rates of the packet as it traverses from the source to the destination. It can accept a value between 0x01 and 0xFF.

Load: Reflects the total outgoing load of the interface along the route and can accept a value between 0x01 and 0xFF.

Reserved: Remains unused.

Destination: Specifies the destination of the route.

Prefix length: Specifies the network number of the destination network. It is calculated by performing a logical AND operation between the IP address and the subnet mask.

The IP address interface usually belongs to the router that is advertising the network. However, this is not always true, as in the case of multi-access networks.

Address Aggregation

Address aggregation is a process by which several contiguous network addresses are consolidated into a major network address. A major network address is made of several subnet addresses. Address aggregation is not limited by the boundaries of different classes of IP addresses. An aggregated network can have any subnet mask. Figure 6.8 depicts address aggregation.

Figure 6.8 shows a router that has several LAN interfaces with network addresses of 172.16.192.0/24, 172.16.193.0/24, 172.16.194.0/24, and 172.16.195.0/24. Table 6.5 represents each of these network addresses in both the decimal and binary formats.

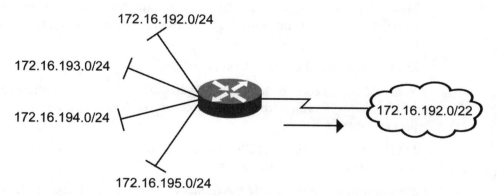

FIGURE 6.8   Several contiguous networks can be represented by a single aggregate address.

TABLE 6.5   Decimal and Binary Formats of Network Addresses

Decimal Format	Binary Format
172.16.192.0/24	10101100.00010000.11000000
172.16.193.0/24	10101100.00010000.11000001
172.16.194.0/24	10101100.00010000.11000010
172.16.195.0/24	10101100.00010000.11000011

To find the aggregate address that can represent all four networks, start from the left-most bit of the first octet and find out to what extent the bit-order is the same for all the networks as you move to the right.

In this example, you can see that the common bits for all the four networks are 10101100.00010000.110000. The decimal equivalent is 172.16.192.0/22. The aggregate address for 172.16.192.0/24, 172.16.193.0/24, 172.16.194.0/24, and 172.16.195.0/24 is 172.16.192.0/22.

This example shows the process by which it is possible to calculate an aggregate address in order to represent several contiguous networks. This concept can also be extended to derive an aggregate address of a group of contiguous aggregates. This is shown in another example, depicted in Figure 6.9.

In Figure 6.9, many groups of networks are connected to a router. Each of these groups sends its aggregate address to the central router. Because these groups of aggregate addresses are contiguous, the central router derives an aggregate and passes it ahead.

The first network has three networks, 172.16.24.0/24, 172.16.25.0/24, and 172.16.26.0/24. The aggregate address of these networks is derived in the same manner, as shown in Figure 6.9. As a result, we get the value 172.16.24.0/21. This aggregate address is sent to the central router. The second network has two networks, 172.16.20.0/24 and 172.16.21.0/24, which are summarized to 172.16.20.0/22. The third network has two networks, 172.16.18.0/24 and 172.16.19.0/24, which are summarized to 172.16.18.0/23. The central router receives three aggregated addresses, 172.16.24.0/21, 172.16.20.0/22, and 172.16.18.0/23.

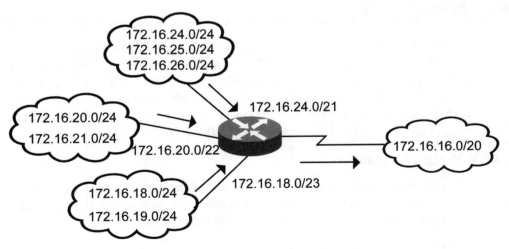

FIGURE 6.9 Aggregate addresses can optimize bandwidth use.

Table 6.6 shows each of these network addresses in both the decimal and binary equivalent.

TABLE 6.6   Formats of the Network Address

Decimal Format	Binary Format
172.16.24.0/21	10101100.00010000.00011
172.16.20.0/22	10101100.00010000.000101
172.16.18.0/23	10101100.00010000.0001001

As you move from the left-most bit to the right-most bit, the common bits are 10101100.00010000.0001, whose decimal equivalent is 172.16.16.0/20.

Decimal Format: 172.16.16.0/20

Binary Format: 10101100.00010000.0001

Address aggregation enables the conservation of bandwidth, by sending only the major network address instead of individual subnets. As a result, routers consume fewer resources. This makes routing efficient and cost-effective because the size of the routing tables is significantly reduced.

Address aggregation is also known as address summarization, and the aggregated address is called summary address.

EIGRP CONFIGURATION

EIGRP constitutes some minimum configuration commands, which are required for the proper functioning of EIGRP. In addition, there are different scenarios where specific configuration commands are required. There are also certain optional configuration commands, which do not significantly impact network performance. In this section, we will look at the configuration commands that are used to configure EIGRP for optimal network performance.

EIGRP Minimum Configuration

The minimum configuration commands define the networks that are included in the EIGRP routing process. These commands are mandatory to enable EIGRP routing and are entered in the global configuration mode:

```
Router(config)# router eigrp autonomous-system
Router(config-router)# network network-number
```

These commands are required to create an EIGRP routing process. Table 6.7 explains the global configuration commands.

TABLE 6.7   Global Configuration Commands

Command	Explanation
Router(config)# router eigrp autonomous-system	Enables the EIGRP routing process. The ASN can be between 1 and 65535.
Router(config-router)# network network-number	Defines the network to be included in the EIGRP routing process.

The ASN need not be assigned by InterNIC. It can be any value between 1 and 65535, but it should be the same for all the EIGRP processes that need to share routing information.

In Figure 6.10, three routers are to be configured to route IP packets using EIGRP. In Figure 6.10, the routers are using an EIGRP ASN equal to 1. The configuration of Router 1 is:

```
router eigrp 1
network 192.16.1.0
network 200.200.1.0
```

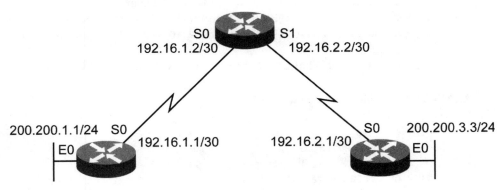

FIGURE 6.10   R1, R2, and R3 route packets using EIGRP.

The configuration of Router 2 is:

```
router eigrp 1
network 192.16.1.0
network 192.16.2.0
```

The configuration of Router 3 is:

```
router eigrp 1
network 192.16.2.0
network 200.200.3.0
```

The routing table of R1 after synchronization is shown in Listing 6.2.

LISTING 6.2 Routing Table for R1

```
R1#sh ip route
Codes: C - connected, S - static, I - IGRP, R - RIP, M - mobile, B - BGP
  D - EIGRP, EX - EIGRP external, O - OSPF, IA - OSPF inter area
  N1 - OSPF NSSA external type 1, N2 - OSPF NSSA external type 2
  E1 - OSPF external type 1, E2 - OSPF external type 2, E - EGP
  i - IS_IS, L1 - IS_IS level-1, L2 - IS_IS level-2, ia - IS_IS inter
    area
  * - candidate default, U - per-user static route, o - ODR
  P - periodic downloaded static route

Gateway of last resort is not set
     192.16.1.0/24 is variably subnetted, 1 subnets
  C  192.16.1.0/30 is directly connected, Serial0
  C  200.200.1.0 is directly connected, Ethernet0
     192.16.2.0/24 is variably subnetted, 1 subnets
  D  192.16.2.0/30 [90/2195456] via 192.16.1.2, 00:02:12, Serial0
  D  200.200.3.0/24 [90/2195456] via 192.16.1.2, 00:02:10, Serial0
```

The routing table of R2 after synchronization is shown in Listing 6.3.

LISTING 6.3 Routing Table for R2

```
R2#sh ip route
Codes: C - connected, S - static, I - IGRP, R - RIP, M - mobile, B - BGP
  D - EIGRP, EX - EIGRP external, O - OSPF, IA - OSPF inter area
  N1 - OSPF NSSA external type 1, N2 - OSPF NSSA external type 2
  E1 - OSPF external type 1, E2 - OSPF external type 2, E - EGP
```

```
   i - IS_IS, L1 - IS_IS level-1, L2 - IS_IS level-2, ia - IS_IS inter
     area
   * - candidate default, U - per-user static route, o - ODR
   P - periodic downloaded static route

Gateway of last resort is not set
     192.16.1.0/24 is variably subnetted, 1 subnets
   C 192.16.1.0/30 is directly connected, Serial0
     192.16.2.0/24 is variably subnetted, 1 subnets
   C 192.16.2.0/30 is directly connected, Serial1
   D 200.200.1.0/24 [90/2195456] via 192.16.1.1, 00:02:12, Serial0
   D 200.200.3.0/24 [90/2195456] via 192.16.2.1, 00:02:10, Serial1
```

The routing table of R3 after synchronization is shown in Listing 6.4.

LISTING 6.4   Routing Table for R3

```
R3#sh ip route
Codes: C - connected, S - static, I - IGRP, R - RIP, M - mobile, B - BGP
   D - EIGRP, EX - EIGRP external, O - OSPF, IA - OSPF inter area
   N1 - OSPF NSSA external type 1, N2 - OSPF NSSA external type 2
   E1 - OSPF external type 1, E2 - OSPF external type 2, E - EGP
   i - IS_IS, L1 - IS_IS level-1, L2 - IS_IS level-2, ia - IS_IS inter
     area
   * - candidate default, U - per-user static route, o - ODR
   P - periodic downloaded static route

Gateway of last resort is not set
   C 200.200.3.0/24 is directly connected, Ethernet0
     192.16.2.0/24 is variably subnetted, 1 subnets
   C 192.16.2.0/30 is directly connected, Serial0
     192.16.1.0/24 is variably subnetted, 1 subnets
   D 192.16.1.0/30 [90/2195456] via 192.16.2.2, 00:02:12, Serial0
   D 200.200.1.0/24 [90/2195456] via 192.16.2.2, 00:02:10, Serial0
```

Changing Adjacency

When an EIGRP routing system is stable, the adjacency formed between the neighbors will not change. However, if the conditions of the links are not stable, the adjacencies continue to vary. A log of all the adjacency changes indicates the condition of the links. The log of adjacency changes is a feature of EIGRP, which is disabled by default. The logging of EIGRP neighbor adjacency changes are enabled by executing the log-neighbor-changes command in the config mode:

```
Router(config)# eigrp log-neighbor-changes
```

Modifying Bandwidth Percent

Like any routing process, EIGRP consumes a part of the link bandwidth that is used for transferring data and for sending routing updates. As a result, the correct balance between the two is vital for the optimal functioning of the network.

By default, EIGRP reserves a maximum of 50 percent for its own packets. If this value needs to be modified, use the command:

```
Router(config-if)# ip bandwidth-percent eigrp percent
```

Modifying Metrics

When there are multiple paths to reach any destination, a routing protocol uses the routing metric to make a choice of one route over the other. The metric is stored in the routing tables. In EIGRP, a metric is calculated using a combination of bandwidth, delay, load, reliability, and MTU.

By default, EIGRP uses predefined constants or weights as the multiplying factor against each of these variables. If it is required to influence the way the metric is calculated, the value of these weights can be modified using this command starting from the router configuration mode:

```
Router(config-router)# metric weights tos k1 k2 k3 k4 k5
```

You can increase the metrics for a particular set of routes both for incoming and outgoing routes from an EIGRP router. This influences the choice of one path over the other. This is crucial while performing route redistribution (to be explained later) between routing protocols.

Avoid changing the metric values unless it is for a specific requirement. Improper use of this command can lead to the erratic behavior of routing protocols.

Defining the Offset Value

The command to define the offset value for a particular set of routes defined by the access-list is:

```
Router(config-router)# offset-list [access-list-number | access-list-
name] {in | out} offset [interface-type interface-name]
```

In the command, the in or out option specifies that the offset should be applied to incoming or outgoing routes.

EIGRP can increase the routes learned by EIGRP using offset-list.

Authenticating EIGRP Routes

Exchanging routing updates between routers can normally be performed without introducing security checks. However, if the need arises, EIGRP is capable of enhancing security during the exchange of routing information between routers.

The secured encrypted authentication prevents exchange of any unauthorized routing messages from unapproved sources. The commands required to enable EIGRP MD5 authentication are executed at the interface that needs to engage in secured updates. This is followed by the creation of a "key-chain," which contains the key, or password, required for authentication before routing updates are sent. The sequence of commands for implementing an EIGRP route authentication is shown in Listing 6.5.

LISTING 6.5 EIGRP Route Authentication Commands

```
Router(config)# interface type number
Router(config-if)# ip authentication mode eigrp autonomous-system md5
Router(config-if)# ip authentication key-chain eigrp autonomous-system
  key-chain
Router(config)# key-chain name-of-chain
Router(config-key-chain)# key number
Router(config-key-chain-key)# key-string text
```

The explanation for the commands in Listing 6.5 is given in Table 6.8.

TABLE 6.8 Commands to Implement EIGRP Route Authentication

Command	Explanation
Router(config)# interface type number	Enables the interface configuration mode
Router(config-if)# ip authentication mode eigrp autonomous-system md5	Enables the MD5 authentication mode for EIGRP updates
Router(config-if)# ip authentication key-chain eigrp autonomous-system key-chain	Links the authentication with its corresponding key-chain
Router(config)# key-chain name-of-chain	Defines the key-chain
Router(config-key-chain)# key number	Defines the key within the key-chain
Router(config-key-chain-key)# key-string text	Defines a key string within the key

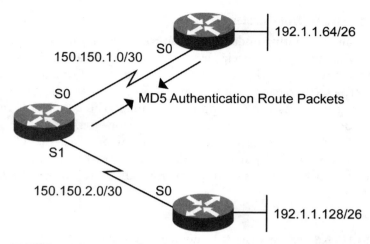

FIGURE 6.11   MD5 authentication is used to exchange routing updates.

The use of MD5 authentication between EIGRP routers is illustrated in Figure 6.11. In Figure 6.11, R1 uses MD5 authentication before exchanging routes with R2, but it does not use such authentication with R3. Authentication between R1 and R2 will be successful only if the keys and not key-chains defined at both the routers match, as shown in the configuration. Configuration for R1 is shown in Listing 6.6.

LISTING 6.6   Configuration for R1

```
Interface serial 0
Ip authentication mode eigrp 1 md5
Ip authentication key-chain R1chain
Key-chain R1chain
Key 1
Key-string commonkey
```

Configuration for R2 is shown in Listing 6.7.

LISTING 6.7   Configuration for R2

```
Interface serial 0
Ip authentication mode eigrp 1 md5
Ip authentication key-chain R2chain
Key-chain R2chain
Key 1
Key-string commonkey
```

Changing the Hello-interval and Hold-time

EIGRP can change the default settings of the hello-interval and hold-time, using the commands:

```
Router(config-if)# ip hello-interval eigrp autonomous-system-number
seconds
Router(config-if)# ip hold-time eigrp autonomous-system-number seconds
```

The functions of these commands are given in Table 6.9.

TABLE 6.9   Commands to Change Hello-interval and Hold-time

Command	Explanation
Router(config-if)# ip hello-interval eigrp autonomous-system-number seconds	Modifies the default hello-interval
Router(config-if)# ip hold-time eigrp autonomous-system-number seconds	Modifies the default hold-time

Disabling and Re-enabling Split-horizon

EIGRP, being a hybrid protocol, uses split-horizon by default. While this is useful for most physical topologies, it may be the cause of failure to advertise routes in non-broadcast networks, such as Frame-Relay. This occurs because in non-broadcast networks the same physical link is used to establish multiple logical connections.

If split-horizon is enabled, the routing update received from one logical connection is not advertised out of the other logical connections because the physical link is the same. In such a situation, it is a good idea to disable split-horizon. The command to disable split-horizon is:

```
Router(config-if)# no ip split-horizon eigrp autonomous-system
Split-horizon can be re-enabled using the command:
Router(config-if)# ip split-horizon eigrp autonomous-system
```

Care should be taken when split-horizon is disabled, as this disabling may lead to routing loops.

In Figure 6.12, routers R1, R2, and R3 are connected over Frame-Relay, where R1 has a common physical link for two different logical connections of R2 and R3.

In Figure 6.12, if split-horizon were enabled, it would prevent R1 from advertising routes learned from R2 to R3, and vice versa. This problem is solved by disabling split-horizon.

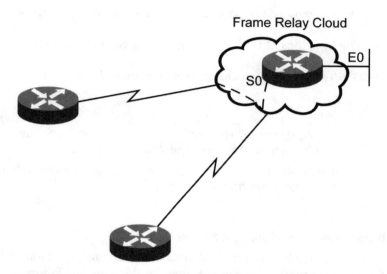

FIGURE 6.12   Split-horizon is disabled in the Frame-Relay hub router R1.

Configuring Stub Routers

You can configure stub routers in EIGRP networks. The central router is known as a hub, and the multiple remote routers are known as spokes. The hub router contains information and performs the routing for all the networks, while spoke routers have only enough information to reach the hub router. Such a type of configuration on the spoke router is called EIGRP stub routing. Figure 6.13 shows a hub-and-spoke connection in which one central router is connected to multiple remote routers.

In Figure 6.13, there is no direct physical connection between any two spokes. Any connectivity between the spokes is established only through the hub router. The command for stub routing is:

```
Router(config-router)# eigrp stub [receive-only | connected | static |
summary]
```

Hub routers, also known as distribution routers, are high-end routers with large processing powers; spoke routers are low-end routers with low processing powers.

NOTE

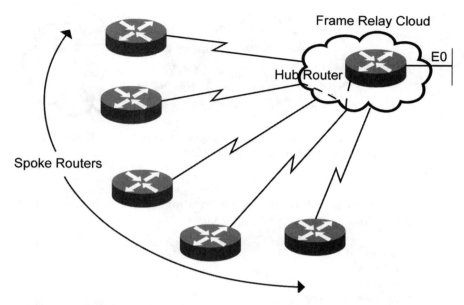

FIGURE 6.13   Multiple spoke routers are connected for EIGRP stub routing.

EIGRP Address Aggregation

EIGRP performs route summarization at network boundaries by default. In this process, the router receives the classful network address for all other networks other than its connected networks. This EIGRP feature of suppressing individual subnets and advertising only the summarized address allows for optimal bandwidth consumption. It can, however, lead to routing loops in certain networking environments. The command used to disable automatic route summarization is:

```
Router(config-router)# no auto-summary
```

Figure 6.14 illustrates a scenario when the default route summarization of EIGRP has to be disabled to avoid routing loops in R1. Here, R2 and R3 have LANs with network addresses, 192.1.1.64/26 and 192.1.1.128/26, respectively.

In Figure 6.14, if the default route summarization were used, the summarized classful address, 192.1.1.0/24, will be generated while crossing the network boundary, and be advertised to R1. This summarized address allows R1 to have ambiguous paths to reach the LANs of R2 and R3. The common entry of 192.1.1.0/24 will confuse R1 into selecting the right path for forwarding packets. This ambiguity can

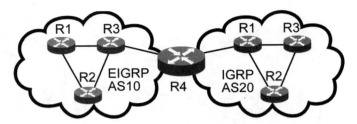

FIGURE 6.14   R1 will face routing loops unless R2 and R3 disable the route summarization.

be avoided by disabling auto-summarization, as shown in the configuration of R2 and R3. Configuration of R2 and R3 is:

```
router eigrp 1
network 150.150.0.0
network 192.1.1.0
no auto-summary
```

Disabling the auto-summarization feature will advertise the individual subnets of 192.1.1.64/26 and 192.1.1.128/26 instead of 192.1.1.0/24. As a result, R1 will know the correct next-hop address for forwarding a packet.

The auto-summarization feature can work in conjunction with another feature of EIGRP that allows it to create summary aggregates by using the command:

```
Router(config-if)# ip summary-address eigrp autonomous-system-number
ip-address mask
```

Under default conditions, the default summarization remains enabled, and generating a summary address using this command is not required. However, if you need to generate a summary aggregate address at a particular interface, the summary address command is used.

In Figure 6.15, router R1 sends out the summary aggregate addresses of 192.1.1.0/24 and 150.0.0.0/8 out of the interface connecting to R4. In Figure 6.15, the configuration command to achieve interface connectivity for R1 is:

```
interface Serial0
ip address 10.1.1.1 255.255.255.0
ip summary-address eigrp 10 192.1.1.0 255.255.255.0
ip summary-address eigrp 10 150.0.0.0 255.0.0.0
```

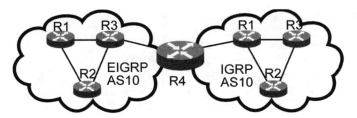

FIGURE 6.15 R1 advertises summary aggregate addresses out of the interface connecting to R4.

EIGRP Verification

The existing configuration of the EIGRP routing protocol should be monitored, maintained, and modified, according to the latest specifications and needs. The most commonly used commands to monitor, manage, debug, and troubleshoot EIGRP are given in Table 6.10.

TABLE 6.10 Commands to Monitor, Manage, Debug, and Troubleshoot EIGRP

Command	Explanation
Router# show ip eigrp neighbor detail	Displays detailed information about the EIGRP neighbor
Router# show ip eigrp interfaces [interface-type | interface-number] [as-number]	Displays information about the interfaces configured in EIGRP
Router# show ip eigrp neighbors [interface-type | interface-number | static]	Displays information about all EIGRP discovered neighbors
Router# show ip eigrp topology [as-number | [ip-address] mask]]	Displays the EIGRP topology table for a particular EIGRP process
Router# show ip eigrp traffic [as-number]	Displays the number of packets sent out and received for an autonomous EIGRP process
Router# clear ip eigrp neighbors [ip-address | interface-type]	Deletes the neighbors from the neighbor table

The output for the show ip interface brief command is shown in Listing 6.8.

LISTING 6.8 Output for the show ip interface brief Command

```
Router# show ip interface brief
Interface IP-Address OK? Method Status Protocol
Serial0 202.11.5.2 YES manual up up
Serial1 202.11.1.2 YES manual up up
Serial2 202.11.2.2 YES manual up up
Serial3 202.11.4.2 YES manual up up
```

The output for the show ip route command is shown in Listing 6.9.

LISTING 6.9 Output for the show ip route Command

```
Router# sh ip route
Codes: C - connected, S - static, I - IGRP, R - RIP, M - mobile, B - BGP
   D - EIGRP, EX - EIGRP external, O - OSPF, IA - OSPF inter area
   N1 - OSPF NSSA external type 1, N2 - OSPF NSSA external type 2
   E1 - OSPF external type 1, E2 - OSPF external type 2, E - EGP
   i - IS_IS, L1 - IS_IS level-1, L2 - IS_IS level-2, ia - IS_IS inter area
   * - candidate default, U - per-user static route, o - ODR
   P - periodic downloaded static route

Gateway of last resort is not set
     202.1.4.0/30 is subnetted, 1 subnets
  D  202.1.4.0 [90/3193456] via 202.11.5.1, 00:02:40, Serial0
     202.2.4.0/30 is subnetted, 1 subnets
  D  202.2.4.0 [90/3193456] via 202.11.5.1, 00:02:46, Serial0
     202.11.4.0/30 is subnetted, 1 subnets
  C  202.11.4.0 is directly connected, Serial3
     202.1.5.0/30 is subnetted, 1 subnets
  D  202.1.5.0 [90/2681456] via 202.11.5.1, 00:02:46, Serial0
     202.2.5.0/30 is subnetted, 1 subnets
  D  202.2.5.0 [90/2681456] via 202.11.5.1, 00:02:47, Serial0
     202.6.5.0/30 is subnetted, 1 subnets
  D  202.6.5.0 [90/2681456] via 202.11.5.1, 00:02:47, Serial0
     202.9.5.0/30 is subnetted, 1 subnets
  D  202.9.5.0 [90/2681456] via 202.11.5.1, 00:02:49, Serial0
     202.11.5.0/30 is subnetted, 1 subnets
  C  202.11.5.0 is directly connected, Serial0
     202.1.1.0/30 is subnetted, 1 subnets
  D  202.1.1.0 [90/3193456] via 202.11.5.1, 00:02:49, Serial0
     202.2.1.0/30 is subnetted, 1 subnets
  D  202.2.1.0 [90/3193456] via 202.11.5.1, 00:02:49, Serial0
     202.11.1.0/30 is subnetted, 1 subnets
  C  202.11.1.0 is directly connected, Serial1
```

```
     202.1.2.0/30 is subnetted, 1 subnets
D    202.1.2.0 [90/3193456] via 202.11.5.1, 00:02:49, Serial0
     202.2.2.0/30 is subnetted, 1 subnets
D    202.2.2.0 [90/3193456] via 202.11.5.1, 00:02:49, Serial0
     202.11.2.0/30 is subnetted, 1 subnets
C    202.11.2.0 is directly connected, Serial2
D    202.1.3.0/24 [90/3219456] via 202.11.5.1, 00:02:49, Serial0
D    202.2.3.0/24 [90/3219456] via 202.11.5.1, 00:02:49, Serial0
D    202.11.3.0/24 [90/2195456] via 202.11.1.1, 00:45:15, Serial1
```

The output for the show ip route eigrp command is shown in Listing 6.10.

LISTING 6.10   Output for the show ip route eigrp Command

```
Router# show ip route eigrp
     202.1.4.0/30 is subnetted, 1 subnets
D    202.1.4.0 [90/3193856] via 202.11.5.1, 00:59:21, Serial0
     202.2.4.0/30 is subnetted, 1 subnets
D    202.2.4.0 [90/3193856] via 202.11.5.1, 00:59:21, Serial0
     202.1.5.0/30 is subnetted, 1 subnets
D    202.1.5.0 [90/2681856] via 202.11.5.1, 00:59:21, Serial0
     202.2.5.0/30 is subnetted, 1 subnets
D    202.2.5.0 [90/2681856] via 202.11.5.1, 00:59:21, Serial0
     202.6.5.0/30 is subnetted, 1 subnets
```

The output for the show ip route connected command is shown in Listing 6.11.

LISTING 6.11   Output for the show ip route Command

```
Router# show ip route connected
     202.11.4.0/30 is subnetted, 1 subnets
C    202.11.4.0 is directly connected, Serial3
     202.11.5.0/30 is subnetted, 1 subnets
C    202.11.5.0 is directly connected, Serial0
     202.11.1.0/30 is subnetted, 1 subnets
C    202.11.1.0 is directly connected, Serial1
     202.11.2.0/30 is subnetted, 1 subnets
C    202.11.2.0 is directly connected, Serial2
```

The output for the show ip protocols command is shown in Listing 6.12.

LISTING 6.12   Output for the show ip protocols Command

```
Router# show ip protocols
  Routing Protocol is "eigrp 1"
```

```
      Outgoing update filter list for all interfaces is not set
      Incoming update filter list for all interfaces is not set
      Default networks flagged in outgoing updates
      Default networks accepted from incoming updates
      EIGRP metric weight K1=1, K2=0, K3=1, K4=0, K5=0
      EIGRP maximum hopcount 100
      EIGRP maximum metric variance 1
      Redistributing: eigrp 1
      Automatic network summarization is not in effect
      Routing for Networks:
      202.11.1.0
      202.11.2.0
      202.11.4.0
      202.11.5.0
Routing Information Sources:

Gateway Distance Last Update
    202.11.5.1 90 00:48:19
    202.11.4.1 90 00:47:22
    202.11.1.1 90 00:47:24
    202.11.2.1 90 00:47:24
Distance: internal 90 external 170
```

The output for the show ip eigrp neighbors command is shown in Listing 6.13.

LISTING 6.13   Output for the show ip eigrp neighbors Command

```
Router# show ip eigrp neighbors
  IP-EIGRP neighbors for process 1
  H Address Interface Hold Uptime SRTT RTO Q Seq Type
  (sec) (ms) Cnt Num
  3 202.11.4.1 Se3 10 00:54:47 80 480 0 2
  2 202.11.2.1 Se2 14 00:55:40 26 202 0 5
  1 202.11.1.1 Se1 11 00:55:40 62 372 0 6
  0 202.11.5.1 Se0 14 01:03:20 19 202 0 97
```

The output for the show ip eigrp topology all-links command is shown in Listing 6.14.

LISTING 6.14   Output for the show ip eigrp topology all-links Command

```
Router# show ip eigrp topology all-links
  IP-EIGRP Topology Table for AS(1)/ID(202.11.5.2)
  Codes: P - Passive, A - Active, U - Update, Q - Query, R - Reply,
  r - reply Status, s - sia Status
```

```
P 202.11.4.0/30, 1 successors, FD is 2169856, serno 24
via Connected, Serial3
P 202.11.5.0/30, 1 successors, FD is 2169856, serno 1
via Connected, Serial0
P 202.9.5.0/30, 1 successors, FD is 2681856, serno 4
via 202.11.5.1 (2681856/2169856), Serial0
P 202.11.1.0/30, 1 successors, FD is 2169856, serno 22
via Connected, Serial1
via 202.11.2.1 (2681856/2169856), Serial2
P 202.11.2.0/30, 1 successors, FD is 2169856, serno 23
via Connected, Serial2
via 202.11.1.1 (2681856/2169856), Serial1
P 202.11.3.0/24, 3 successors, FD is 2195456, serno 27
via 202.11.4.1 (2195456/281600), Serial3
via 202.11.1.1 (2195456/281600), Serial1
via 202.11.2.1 (2195456/281600), Serial2
P 202.2.2.0/30, 1 successors, FD is 3193856, serno 11
via 202.11.5.1 (3193856/2681856), Serial0
—More—
```

The output for the show ip eigrp traffic command is shown in Listing 6.15.

LISTING 6.15   Output for the show ip eigrp traffic Command

```
Router# show ip eigrp traffic
  IP-EIGRP Traffic Statistics for process 1
  Hellos sent/received: 3857/3596
  Updates sent/received: 25/18
  Queries sent/received: 4/0
  Replies sent/received: 0/4
  Acks sent/received: 14/19
  Input queue high water mark 1, 0 drops
  SIA-Queries sent/received: 0/0
  SIA-Replies sent/received: 0/0
```

The output for the show ip eigrp topology active command is shown in Listing 6.16.

LISTING 6.16   Output for the show ip eigrp topology active Command

```
Router# show ip eigrp topology active
  IP-EIGRP Topology Table for AS(1)/ID(10.1.4.2)
  A 20.2.1.0/24, 1 successors, FD is Inaccessible, Q
  1 replies, active 00:01:43, query-origin: Successor Origin
```

```
    via 10.1.3.1 (Infinity/Infinity), Serial1/0
    via 10.1.4.1 (Infinity/Infinity), Serial1/1, serno 146
Remaining replies:
    Via 10.1.5.2, r, Serial1/2
```

DESIGN CONSIDERATIONS

Some of the design considerations while choosing the EIGRP routing protocol are:

- Current number of networks and nodes that will be part of the EIGRP routing process
- Future number of networks and nodes to indicate the scalability
- IP addressing scheme to be deployed
- Summarization to be used
- Load balancing techniques to be used, if any
- Redistribution points
- Security policy
- Bandwidth requirement
- Manageability required
- Routing authentication, if required
- Modification of EIGRP metrics weights, if required
- If the network may add routers from other vendors in future

SUMMARY

In this chapter, you learned about the various concepts of EIGRP. You also learned about the various configuration commands used to enable the EIGRP routing process. In the next chapter, we will introduce you to the different concepts related to OSPF.

POINTS TO REMEMBER

- EIGRP uses different data packets, such as Hello, Acknowledgment, Routing Updates, Queries, and Replies.
- In RTP, if an EIGRP neighbor does not acknowledge a packet, it is retransmitted only to the neighbor.
- The new router initiates itself by sending multicast Hello packets in the network.

- EIGRP is a hybrid protocol because it uses the same metrics used by distance vector protocols but does not send the entire routing table to neighbor routers.
- Adjacency is formed by exchanging Hello packets between neighboring routers.
- Address aggregation is a process by which several contiguous network addresses are consolidated into a major network address, and it enables conservation of bandwidth.
- Router(config)# router eigrp autonomous-system: enables the EIGRP routing process. The ASN can be between 1 and 65535.
- Router(config-router)# network network-number: defines the network to be included in the EIGRP routing process.
- Router(config)# eigrp log-neighbor-changes: enables a change in the adjacency.
- Router(config-if)# ip bandwidth-percent eigrp percent: enables modification of bandwidth percent.
- Router(config-router)# metric weights tos k1 k2 k3 k4 k5: enables modification of metrics.
- Router(config-if)# no ip split-horizon eigrp autonomous-system: disables split-horizon.
- Router(config-router)# eigrp stub [receive-only | connected | static | summary]: configures stub routing

7 Open Shortest Path First Routing Protocol

INTRODUCTION TO OSPF

Open Shortest Path First (OSPF) is a link state routing protocol that runs internally in an Autonomous System (AS). OSPF is based on Djikstra's Shortest Path algorithm. OSPF routers gather link state information and use the shortest path algorithm for calculating the optimal path for routing data packets.

The term "Open" in OSPF denotes that it is an open specification that is available to the public domain. The specifications for OSPF protocol are published as Request for Comments (RFC) 1247. The term "Shortest Path" in OSPF specifies that this protocol is based on Djikstra's algorithm.

OSPF is used in large heterogeneous internetworks. The process of routing updates becomes more tedious and causes degradation of performance as the network grows. The routing protocol divides the large network into smaller segments to overcome this problem. OSPF supports multi-paths, where the shortest routes can be recalculated faster for an area, resulting in the use of minimum network routing traffic.

The OSPF routing protocol keeps track of three tables—the routing table, a table to track directly attached neighbors, and a table to track the topology and design of the whole internetwork. OSPF enables each router in the AS to calculate and update routing information.

Djikstra's shortest path algorithm is used to compute the shortest path between any two nodes in a graphical data structure. The link state protocols like OSPF use the shortest path algorithm to compute the shortest route for forwarding the packets

OSPF TERMINOLOGIES

This section discusses the terminologies used while discussing an OSPF routing protocol. Some of the terms are:

Interface: Physical interface on an OSPF router.

Link: Network or router interface assigned to any given network.

Link State Advertisement (LSA):   OSPF packet containing source, destination, and routing information. This information is advertised to all OSPF routers in a hierarchical area.

Flooding: Periodic updating of topology and routing table information with the sending of LSAs.

Adjacencies: Logical connection between the OSPF router and its Designated Router (DR).

Designated Router: Router used to reduce the number of adjacencies formed in a broadcast network.

Backup Designated Router (BDR):   Router that acts as a standby for DR on broadcast networks. BDR gathers routing information updated from the adjacent OSPF routers and takes the role of DR when the DR goes down.

Autonomous Systems (AS):   Set of routers in the same administrative control using the same protocol for routing process.

Internal Gateway Protocol (IGP):   Routing protocol used in the intranet, within an AS.

Router ID: 32-bit number that uniquely identifies a router in an AS. The router ID is usually an IP address.

Multi-Access/Broadcast Networks: Physical networks that support interconnection of more than two routers that can communicate directly.

Non-Broadcast Multi-Access Networks (NBMA):   Interconnects routers in the OSPF network without having the broadcast capability. NBMA is discussed in detail in a later section.

CONCEPTS OF OSPF

This section discusses the different concepts of OSPF and looks at the importance of Hello packets in OSPF routing protocol. It will also discuss the different types of networks formed.

Hello Packets

Hello packets are sent out periodically by an OSPF router from all the interfaces configured to run the OSPF routing protocol. These Hello packets are crucial in the formation and working of OSPF neighbors. The different parameters contained in Hello packets are:

- Router ID
- Area ID of router interface
- Network mask of router sending Hello packet
- Hello interval
- Dead interval
- Authentication information
- Router priority
- Designated Router (DR) information
- Backup Designated Router (BDR) information
- Router IDs of neighbors
- Other optional characteristics

Hello packets contain information about the router IDs of other neighboring routers. On receiving a Hello packet, adjacency between routers is established by matching parameters at the receiving interface. On the successful formation of adjacency, an entry is made in the neighbor table. This table also records the interface on which the neighbors are located. If the parameters do not match, they are dropped, and the adjacency is not formed.

OSPF routers are identified with the help of a unique IP address known as a router ID. The two rules that govern the router ID are:

- If a router has configured loopback addresses, then the highest IP address among them becomes the router ID for the router.
- If a router has not configured loopback addresses, then the highest IP address of any of its physical interfaces becomes the router ID.

It is best to configure a loopback address so that the router ID is not affected with the changing link status of any of the physical interfaces.

Types of Networks

All neighboring routers do not form adjacency. This formation of a relationship between two neighboring routers depends on the type of network. The different network types are:

- Point-to-point networks
- Broadcast networks

- Non-Broadcast Multi-Access (NBMA) networks
- Point-to-multipoint networks
- Virtual links

Point-to-Point Networks

When two routers have a direct link between themselves, a point-to-point network type is formed. Figure 7.1 shows a point-to-point network where two routers are directly connected over a serial link to one another.

In point-to-point networks, the two neighbors form an adjacency. E1 and T1 links are examples of point-to-point networks.

FIGURE 7.1 Two networks forming a point-to-point network.

Broadcast Networks

All nodes receive a single packet sent from a designated node in broadcast networks. OSPF broadcast type networks elect a DR and a BDR. Figure 7.2 shows a broadcast network where several routers are connected over an Ethernet. Ethernet and token ring are examples of broadcast networks.

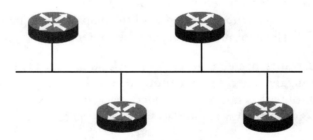

FIGURE 7.2 Multiple routers connected in a broadcast network.

Non-Broadcast Multi-Access Networks

Networks that connect more than two routers without broadcast capability are called Non-Broadcast Multi-Access (NBMA) networks. A packet from one router does not reach all other attached routers because broadcast is not supported in NBMA network. Figure 7.3 shows an NBMA network where several routers are connected over the Frame-Relay cloud.

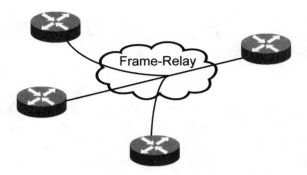

FIGURE 7.3   Multiple routers connected in an NBMA network.

In Figure 7.3, the hub router has one physical connection, but it has three logical connections to each of the other spoke routers. In this configuration, the NBMA network appears to have several point-to-point links. As a result, the NBMA network appears similar to the point-to-multipoint network. However, unlike point-to-multipoint networks, the NBMA elects a DR and BDR. Additional commands are required to configure OSPF in NBMA networks. Examples of NBMA networks are Frame-Relay, ATM, and X.25.

Virtual Links

Virtual links connect OSPF areas that do not have any connection with Area 0. We will discuss the different areas of OSPF routing protocol later in the chapter.

OSPF OPERATION

This section discusses the different steps involved in the operation of the OSPF routing process. Consider a network with routers configured in an OSPF configuration. Hello packets are sent out by each of the OSPF routers from all interfaces. When two routers agree on the parameters specified in the Hello packets, they form adjacencies after which multiple LSAs are sent. LSAs contain information related to router links and connection of different network types. The router notes this information in the link state database. OSPF routers build up identical link state databases.

An OSPF router then calculates loop-free paths to different destinations using Djikstra's algorithm and represents these paths in the form of a graph called the Shortest Path First (SPF) tree. The routing table is built after the SPF tree is completed.

The OSPF routing process only takes place when its database is built up at the startup phase. Once synchronization is reached, OSPF sends out LSAs after every 30 minutes.

OSPF Areas

OSPF is scalable enough for interconnecting large internetworks, which, in turn, is divided further into areas based on the subnet information. These areas are treated as single independent entities capable of communicating with other areas exchanging routing information among them. A network configured with OSPF routing protocol can be divided into single, multiple, and non-standard areas. The following sections discuss each of these areas in further detail.

Areas make OSPF easily scalable and aid in interconnecting large networks.

Single Area

Single area is a logical subdivision of the greater OSPF domain, grouping routers that run OSPF with identical topological databases. Figure 7.4 shows Area 0 representing a single area. Area 0 contains seven routers that maintain a single link state database and updates this database whenever network topology changes. Every other router in any other OSPF area should maintain at least one connection with Area 0.

The limitations of an OSPF single area are:

- Huge size of routing table leading to heavy data exchange even if the entire routing table is not sent to the neighbor routers.
- Huge, unmanageable topological database, leading to heavy data exchange between neighbor routers.
- Frequent use of shortest path algorithm in routers. Even a small change in the network may need recalculation of routing paths.

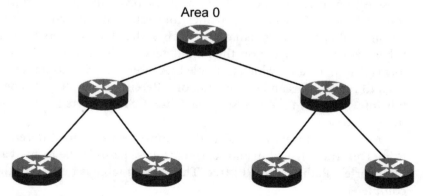

FIGURE 7.4   Routers connected in an OSPF single area network.

- Poor router performance due to heavy data exchange, complex calculations, and more CPU cycles.
- Waste of bandwidth due to LSA.

Area 0 is also known as backbone area or Area 0.0.0.0

NOTE

Multiple Area

OSPF in a single area is not scalable in large networks. To overcome such limitations, very large networks are divided into many manageable areas, as shown in Figure 7.5.

In Figure 7.5, there are three areas, Area 0, Area 20, and Area 80. Router 1 belongs to Area 0, Router 4 and Router 5 belong to Area 20, Router 6 and Router 7 belong Area 80. Router 2 and Router 3 are Area Border Routers (ABRs). Router 2 connects routers in Area 0 and Area 20. Router 3 connects Area 0 with routers in Area 80. Each router in Area 20 and Area 80 is connected to the backbone area through the ABRs, Router 2, and Router 3, respectively.

The advantages of OSPF in multiple areas are:

- Reduced memory overheads because only link state information for routers belonging to the same area are stored.
- Reduced complex recalculations. Even if the network topology changes, only a particular area is affected. You need to recalculate only that particular link state database.
- Smaller and manageable routing tables in OSPF multiple areas.
- Simpler process to exchange routing data as compared to single area.

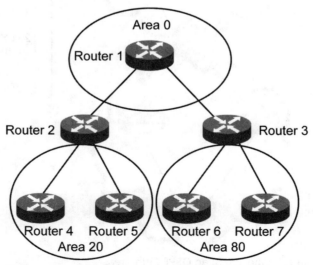

FIGURE 7.5 Routers connected in an OSPF multiple area network.

OSPF multiple areas are complex to implement and configure as compared to OSPF single areas.

Non-standard OSPF Areas

Apart from single area and multiple areas, OSPF defines three more types of areas, which are known as non-standard areas. The different non-standard OSPF areas are:

- Stub Area (SA)
- Totally Stubby Area (TSA)
- Not So Stubby Area (NSSA)

Stub Area

SA does not accept external summary routes, LSA Type 4 packets, and LSA Type 5 packets. SA is configured by a default external route 0.0.0.0. Figure 7.6 shows an OSPF SA.

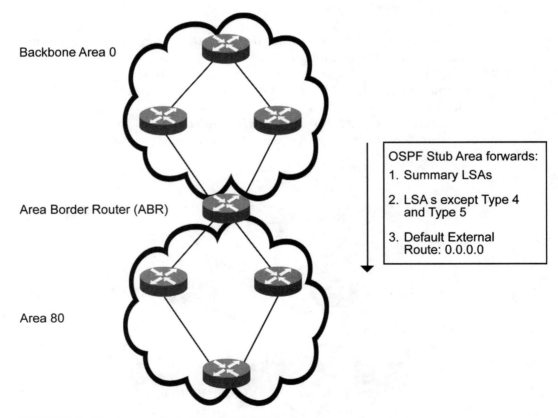

FIGURE 7.6 Routers connected in an OSPF Stub Area.

Instead of passing external summary routes, LSA Type 4 packets and LSA Type 5 packets from other areas are propagated to the SA.

Totally Stubby Area

TSA is Cisco proprietary and is used when there are few networks with limited connectivity with the remaining network. TSA makes link state databases and routing tables as small as possible. The default external route specified in TSA is 0.0.0.0. Figure 7.7 depicts an OSPF TSA.

TSA does not accept external summary LSAs from inside and outside the AS. LSA Types 3, 4, and 5 are also not accepted by the TSA.

TSAs are also used for stable and scalable internetwork configurations.

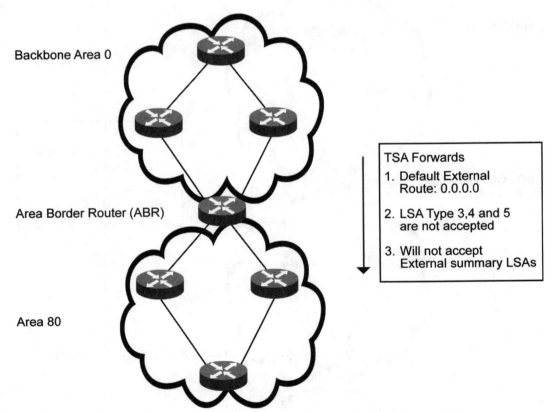

FIGURE 7.7   Routers connected in an OSPF Totally Stub Area.

Not So Stubby Area

NSSA is mainly used in redistribution of routing information. Type 4 and Type 5 LSAs and external route information are not passed in or out of NSSA. NSSA is configured by a default external route 0.0.0.0. Figure 7.8 shows an OSPF NSSA.

NSSA is similar to SA, but it allows Autonomous System Boundary (ASBR) also. In OSPF SA, the ABRs connect multiple OSPF areas. In NSSA, in addition to an ABR, NSSA may have ASBRs, which connect routers in two different ASs. The concept of ASBR is explained later in the chapter.

TIP

Single and multiple areas in OSPF are also known as regular, standard, or ordinary areas.

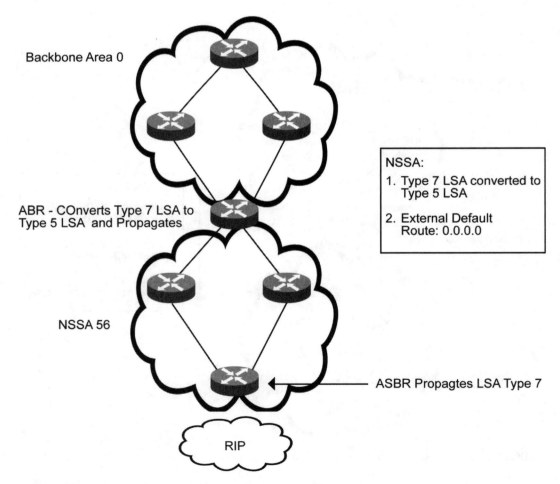

FIGURE 7.8 Routers connected in an OSPF Not So Stubby Area.

Election Process

The formation of adjacencies and LSA transmission between OSPF routers can be tedious and generate heavy traffic when the number of routers increases in the network. This is because each router sends LSAs to its adjacent router, which, in turn, passes on the same to its adjacent neighbors, increasing the total traffic in the network. This situation is avoided with the help of an election process in which a DR and a BDR are elected in multi-access networks.

DR acts as the hub and manages the flooding process in a multi-access network. Full-mesh adjacencies are not required if there is a DR in the network. The prerequisites of the OSPF election process are:

- Hello packets should contain the priority information of the originating router and the DR and BDR interfaces.
- Priority of an OSPF router should not be set to 0. The priority value can be set to influence the selection of the DR.
- DR and BDR are set as 0.0.0.0 when a router initializes.
- Router interfaces in the network should keep note of the IP addresses of DR and BDR in its interface data structure.

The priority of a router can take any value in the range 0–255 with the default value being 1. If the priority is set to 0, the router does not participate in the election process. The command used to set the priority on an interface is ip ospf priority.

When an OSPF router is initialized, it acknowledges the presence of a DR and BDR. A fresh election to dislodge the existing DR is not initiated even if the newly joining router has a higher priority. This prevents an election every time a new router with a higher priority joins the network. As a result, a router with lower priority can be elected as DR.

If the new router detects the absence of even a BDR, an election is held and the router with the highest priority is elected as the new BDR. If there is a tie due to multiple routers having the same priority, then the router with the highest router ID is elected as BDR. If, for some reason, there is no DR, then the BDR is promoted to the role of a DR.

All the other routers participating in the election apart from the elected DR and BDR are known as DRothers. These routers establish adjacencies with the DR and BDR only. DRothers send Hello packets to DR and BDR using the multicast address 224.0.0.6. Updates are flooded to other DRothers using the multicast address of 224.0.0.5.

If there is only one eligible router present in a multi-access network, then it will become the DR. In such a scenario, there will be no BDR present in the network, and all DRothers form adjacencies with the DR. If, however, there is no eligible router in a multi-access network, then no adjacencies will be formed.

Designated Routers

A DR is the leader and representative of all the DRothers in the network. This router reduces the network traffic, because there is no need to form a fully meshed adjacency anymore.

The example shown in Figure 7.9 illustrates the requirement of the DR in an OSPF network. In Figure 7.9, there are six routers in the multi-access network configured to run OSPF routing protocol. The 15 full-meshed adjacencies are formed because each router forms adjacency with the remaining five routers to send LSAs.

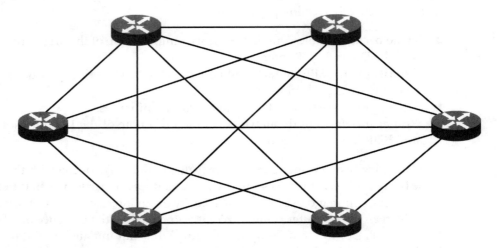

FIGURE 7.9   A network depicting router adjacencies.

For x number of routers, there would be x(x-1)/2 number of adjacencies. Substituting x=6 in the example, we get 6*5/2=15 adjacencies. Each router sends LSAs to other adjacent routers plus one for the network; that is, for x number of routers, each router would send (x-1)+1=x LSAs. As a result, the total number of LSAs in the entire network is x*x=x^2.

If the network were to form full-mesh adjacencies, then the volume of traffic would be enormous as the number of routers increases. Figure 7.10 shows the discussed example with the presence of a DR.

In Figure 7.10, all the routers form adjacency with the DR and not with one another. A DR is said to form a pseudo node to which all the remaining routers connect. The DR sends LSAs to all the other routers, instead of them sending it directly to one another.

All connecting routers within the same IP subnetwork elect their own DR in a multi-access network. A router has multiple interfaces participating in different subnetworks. As a result, it is possible that the router is elected to be the DR

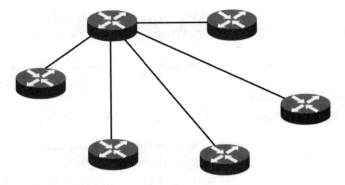

FIGURE 7.10 A network depicting DR adjacencies.

in one subnet and not in the other networks. This scenario is depicted in Figure 7.11. In Figure 7.11, R1 is connected to two different networks in a broadcast network. R1 is the DR for the network to which E0 is connected. However, R1 is not the DR for the network to which the interface E1 is connected.

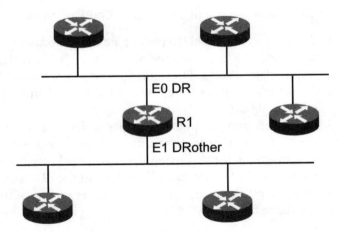

FIGURE 7.11 Election of DRs in a multi-access OSPF network.

Backup Designated Routers

A BDR is elected in a multi-access OSPF network so that the network does not collapse in the event of a DR failure. DRothers form adjacencies with the DR as well as the BDR. DR and BDR are also adjacent to one another. If the DR fails, the BDR is promoted to the role of a DR. The network outage is vastly reduced because fresh elections are not required, and the DRothers have already formed adjacencies with the BDR. In networks that do not have a BDR, another election takes place to elect a new DR. The OSPF network remains unavailable during this period.

Election of BDR is not mandatory in multi-access networks.

DR and BDR Election Process

The election procedure of a DR in an OSPF network is:

1. The routers are eligible to participate in the election process if they indicate their own IP addresses in the DR priority fields in Hello packets.
2. The eligible routers have a priority greater than 0 and form adjacencies with neighbors.
3. The election process identifies all routers that can participate in the DR election from the list of eligible routers.
4. The router having the highest priority is elected as the DR. In case of a tie, the router with the higher router ID is elected as DR.
5. The BDR is elected as the DR if there is no router that has declared itself as the DR.

The election procedure of a BDR in an OSPF network is:

1. The routers are eligible to participate in the election process if they indicate their own IP addresses in the BDR priority fields in Hello packets. A router that declares itself as DR cannot participate in the BDR election process.
2. The eligible routers have a priority greater than 0 and form adjacencies with neighbors.
3. The router with the highest priority is elected as the BDR. In case of a tie, the router with the higher router ID is elected as BDR.
4. The neighbor router having the highest priority becomes the BDR if no router has declared its own IP address in the BDR field.
5. The highest router ID acts as a tiebreaker in case of multiple routers having the same priority.

A router cannot be part of the election eligible list if the priority value is set to 0.

TYPES OF OSPF ROUTERS

This section discusses the different types of OSPF routers. The four types of OSPF routers are Internal Router (IR), Area Border Router (ABR), Backbone Router (BR), and Autonomous System Boundary Router (ASBR).

Internal Routers

IRs belong to the same OSPF area with all directly connected networks and execute only one copy of the OSPF routing algorithm. The IR keeps an updated and accurate link state database for each subnet in that area. Routers with only backbone interfaces are also part of IRs.

Area Border Routers

ABRs connect two or more OSPF areas. These routers execute multiple copies of the OSPF routing algorithm—one copy for each attached OSPF area and another copy for the backbone network. ABRs consolidate the link state database and topological information of their respective areas and distribute them back to the backbone. The backbone further redistributes this link state database information to other areas.

It is mandatory that ABR touch Area 0.

Backbone Routers

Every router within the backbone area that interfaces with an IR or with another OSPF area using ABR is called BR. If all interfaces of a router are connected to the backbone, then it is considered to be an IR.

BR will be inside Area 0.

Autonomous System Boundary Routers

ASBR is any router that is connected to at least one external routing process. This router transmits and receives routing information with routers belonging to other ASs. Each router within the AS knows the ASBR path. The ASBR redistributes link state database information to a separate AS router. ASBRs can function as IRs, ABRs, and BRs.

The different OSPF routers are illustrated in the example shown in Figure 7.12. Figure 7.12 shows a simple OSPF configuration. R1 and R3 that belong to Area 0 and Area 50 are IRs for their respective areas. R2 is an ABR connecting Area 0 to Area 50. R2 also acts as the BR because it connects routers in Area 50 to the backbone area. Link state database information is redistributed to AS1 via R2. This R2

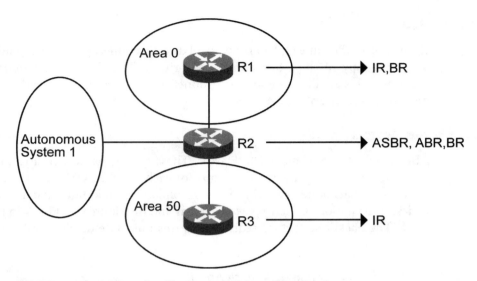

FIGURE 7.12 A network with simple OSPF configuration.

connects one AS with Area 0 and Area 50 to AS1. As a result, R2 acts as the ASBR. R2 also acts as an ASBR, linking two different ASs.

The different OSPF routers are illustrated in a complex OSPF configuration. This scenario is depicted in Figure 7.13.

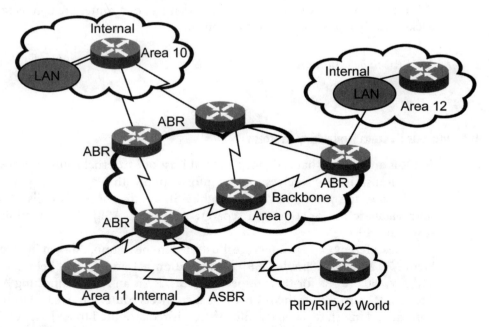

FIGURE 7.13 A network with complex OSPF configuration.

In Figure 7.13, IRs are represented by "Internal" in the respective areas. ABRs connect routers from Area 0 to Area 10, Area 11, and Area 12. The central router that resides inside Area 0 is the BR. The ASBR in Area 11 redistributes and connects Area 11 with another RIP AS.

OSPF PACKET FORMATS

This section discusses the different packet formats in OSPF routing protocol.

General OSPF Packet Format

All OSPF packets consist of eight fields for header and one field for data. Figure 7.14 shows the general OSPF packet format.

The various fields in the general OSPF packet format are:

Version Number: Identifies the version of OSPF used. This field is 1 byte long.

Packet Type: Specifies the packet type used. This field is 1 byte long.

Packet Length: Contains the total OSPF packet length, including the OSPF header in bytes. This field is 2 bytes long.

Router ID: Specifies the source ID of the packet. This field is 4 bytes long.

Area ID: Specifies the area to which the OSPF packet belongs. All OSPF packets belong to a single area. This field is 4 bytes long.

Checksum: Contains the checksum number that specifies if the packet is damaged during transmission. This field is 2 bytes long.

Authentication Type: Contains the authentication type followed by the OSPF packet. This field is 2 bytes long.

Authentication: Contains authentication information. This field is 8 bytes long.

Data: Contains the encapsulated data to be carried forward to the destination.

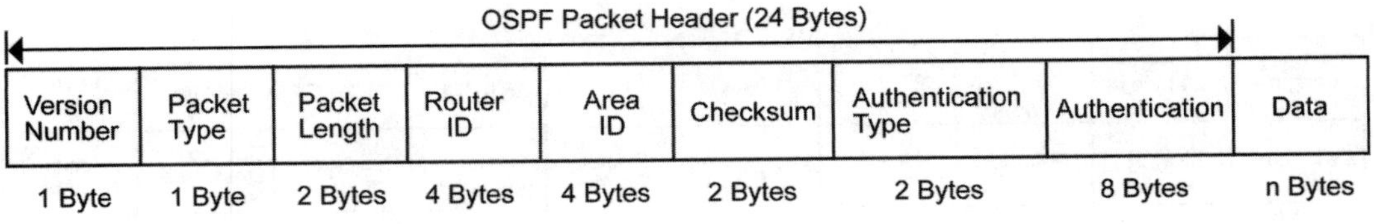

OSPF Packet Header (24 Bytes)

Version Number	Packet Type	Packet Length	Router ID	Area ID	Checksum	Authentication Type	Authentication	Data
1 Byte	1 Byte	2 Bytes	4 Bytes	4 Bytes	2 Bytes	2 Bytes	8 Bytes	n Bytes

FIGURE 7.14 The general OSPF packet format.

There are five different data packet types supported by the OSPF routing protocol. They are listed in Table 7.1.

TABLE 7.1 OSPF Packet Types and Functions

Packet Type	Function
Hello	Establishes and maintains adjacency relationships with neighbors.
Database Description (DBD)	Briefs contents of the database. The DBD packet includes information in the link state database of an OSPF router.
Link State Request (LSR)	Requests for specific part of a neighbor router's link state database. This packet downloads link database contents.
Link State Update (LSU)	Updates link database contents and carries LSAs to neighbor routers.
Link State Acknowledgement (LSA)	Acknowledges receipt of LSAs in neighbor routers.

Hello Packet Format

The packet format of a Hello packet is shown in Figure 7.15. In Figure 7.15, the Hello packet format includes:

Netmask: A 32-bit number indicating the range of IP addresses on the IP network. For example, the network mask for a Class C network can be 0xffffff00.

Hello Interval: The time interval between two Hello packets.

Option: Enables routers to support optional capabilities and helps in communicating their capability to other routers.

Router Priority: The priority of the router that generated this packet.

Length	24 bytes	4 bytes	2 bytes	1 byte	1 byte	4 bytes	4 bytes	4 bytes	4 bytes each
Description	OSPF Packet Header	Netmask	Hello Interval	Option	Router Priority	Dead Interval	Designated Router	Backup Designated Router	Additional Neighbor

FIGURE 7.15 A Hello packet format.

Dead Interval: The time to wait without receiving any Hello packet.

Designated Router: The router ID of the DR.

Backup Designated Router: The router ID of the BDR.

Additional Neighbor: The router ID of the nearest adjacent router that generated this packet.

Database Description Packet Format

The Database Description (DBD) packet format is shown in Figure 7.16. In Figure 7.16, apart from other fields, the DBD packet format includes:

Interface Maximum Transmission Unit: Determines the size of the largest datagram that can be transmitted by an IP interface.

Options: Enables routers to support optional capabilities and helps in communicating their capability to other routers.

DD Sequence Number: Represents the unique 32-bit number that identifies the DBD packet

LSA Header: Discussed in a later section.

```
 0                   1                   2                   3
 0 1 2 3 4 5 6 7 8 9 0 1 2 3 4 5 6 7 8 9 0 1 2 3 4 5 6 7 8 9 0 1
+-+-+-+-+-+-+-+-+-+-+-+-+-+-+-+-+-+-+-+-+-+-+-+-+-+-+-+-+-+-+-+-+
|   Version #   |       2       |         Packet Length         |
+-+-+-+-+-+-+-+-+-+-+-+-+-+-+-+-+-+-+-+-+-+-+-+-+-+-+-+-+-+-+-+-+
|                          Router ID                            |
+-+-+-+-+-+-+-+-+-+-+-+-+-+-+-+-+-+-+-+-+-+-+-+-+-+-+-+-+-+-+-+-+
|                           Area ID                             |
+-+-+-+-+-+-+-+-+-+-+-+-+-+-+-+-+-+-+-+-+-+-+-+-+-+-+-+-+-+-+-+-+
|           Checksum            |             AuthType          |
+-+-+-+-+-+-+-+-+-+-+-+-+-+-+-+-+-+-+-+-+-+-+-+-+-+-+-+-+-+-+-+-+
|                       Authentication                          |
+-+-+-+-+-+-+-+-+-+-+-+-+-+-+-+-+-+-+-+-+-+-+-+-+-+-+-+-+-+-+-+-+
|                       Authentication                          |
+-+-+-+-+-+-+-+-+-+-+-+-+-+-+-+-+-+-+-+-+-+-+-+-+-+-+-+-+-+-+-+-+
|         Interface MTU         |Options |0|0|0|0|R|I|M|MS      |
+-+-+-+-+-+-+-+-+-+-+-+-+-+-+-+-+-+-+-+-+-+-+-+-+-+-+-+-+-+-+-+-+
|                     DD Sequence Number                        |
+-+-+-+-+-+-+-+-+-+-+-+-+-+-+-+-+-+-+-+-+-+-+-+-+-+-+-+-+-+-+-+-+
|                                                               |

                        An LSA Header

                      (LSA Descriptions)

+-+-+-+-+-+-+-+-+-+-+-+-+-+-+-+-+-+-+-+-+-+-+-+-+-+-+-+-+-+-+-+-+
|                                                               |
```

FIGURE 7.16 A DBD packet format.

Link State Request Packet Format

The Link State Request (LSR) packet format is shown in Figure 7.17. In Figure 7.17, apart from other fields, the LSR packet format includes:

LS Type: Specifies the LSA Type. Each type has a different advertisement format.

Link State ID: Specifies the part of the routing environment that is described by the advertisement. For example, for network LSA Type 2, the link state ID is set to the IP address of the network DR.

Advertising Router: Specifies the router ID of the router that created the LSA.

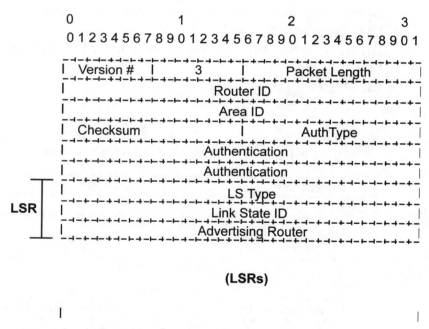

(LSRs)

FIGURE 7.17 A LSR packet format.

Link State Update Packet Format

The Link State Update (LSU) packet format is shown in Figure 7.18. In Figure 7.18, apart from other fields, the LSU packet format includes LSAs. The header format of the LSAs is discussed in the next section.

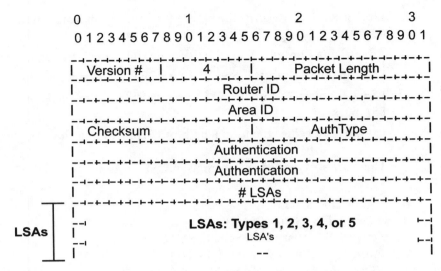

FIGURE 7.18   A LSU packet format.

Link State Acknowledgment Packet Format

The Link State Acknowledgment (LSA) packet format is shown in Figure 7.19. The LSA packet begins with a common 20-byte header that uniquely identifies the LSA. The LSA packet format also includes:

LS Age: Records the time in seconds since the LSA originated.

Options: Specifies the optional capabilities of the routing domain.

LS Type: Specifies the LSA Type. Each LSA Type has a different advertisement format.

```
 0                   1                   2                   3
 0 1 2 3 4 5 6 7 8 9 0 1 2 3 4 5 6 7 8 9 0 1 2 3 4 5 6 7 8 9 0 1
+-+-+-+-+-+-+-+-+-+-+-+-+-+-+-+-+-+-+-+-+-+-+-+-+-+-+-+-+-+-+-+-+
|    LS Age       |    Options      |         LS Type            |
+-+-+-+-+-+-+-+-+-+-+-+-+-+-+-+-+-+-+-+-+-+-+-+-+-+-+-+-+-+-+-+-+
|                      Link State ID                            |
+-+-+-+-+-+-+-+-+-+-+-+-+-+-+-+-+-+-+-+-+-+-+-+-+-+-+-+-+-+-+-+-+
|                    Advertising Router                         |
+-+-+-+-+-+-+-+-+-+-+-+-+-+-+-+-+-+-+-+-+-+-+-+-+-+-+-+-+-+-+-+-+
|                    LS Sequence Number                         |
+-+-+-+-+-+-+-+-+-+-+-+-+-+-+-+-+-+-+-+-+-+-+-+-+-+-+-+-+-+-+-+-+
|    LS Checksum          |          Length                     |
+-+-+-+-+-+-+-+-+-+-+-+-+-+-+-+-+-+-+-+-+-+-+-+-+-+-+-+-+-+-+-+-+
```

FIGURE 7.19   A LSA packet header.

Link State ID: Records the part of the routing environment described by the advertisement.

Advertising Router: Specifies the router ID of the router that created the LSA.

LS Sequence Number: Specifies the sequence number of LSAs.

LS Checksum: Contains an integer specifying the correctness of the OSPF LSA packet.

Length: Specifies the total length of LSA including the 20-byte LSA header.

There are six types of LSA formats in OSPF routing protocol. They are:

- Router Link State (LSA 1)
- Network Link State (LSA 2)
- Network Summary Link State (LSA 3)
- Summary ASBR Link State (LSA 4)
- AS External Link State (LSA 5)
- NSSA External LSA (LSA 6)

These LSA types are discussed in the following sections.

Router Link States

This LSA packet format is also known as LSA Type 1. Router link state is generated and sent by all OSPF routers. The LSA packet describes the integrated states of the router interfaces belonging to a single area. These LSA packets are flooded only across a single area. The area has information about all the router IDs in that single area, including its own router information. All router link information within an area must be included within a single LSA.

Network Link State

This LSA packet format is also known as LSA Type 2. The DR generates network LSAs to update broadcast or multi-access networks. It contains the list of IP addresses of routers connected to the network and is flooded throughout a single area only. Network Link State LSA is generated only by the DR.

Network Summary Link State

This LSA packet format is also known as LSA Type 3. Summary Link States are generated by ABRs and are flooded throughout their associated areas. Each LSA describes a route to a destination network outside the area within the AS. Summary Link States describe routes to a network.

Summary ASBR Link State

This LSA packet format is also known as LSA Type 4. Summary ASBR Link States describe routes to ASBRs. Each Summary ASBR Link State describes a route to a destination ASBR.

AS External Link States

This LSA packet format is also known as LSA Type 5. AS External Link States are generated and flooded throughout the AS by ASBRs. Every LSA specifies a route to a destination in another AS. AS External Link States can also specify the default routes for the AS.

NSSA External LSA

This LSA packet format is also known as LSA Type 6. NSSA External LSA will not be propagated to any area other than NSSA. If any information is to be propagated throughout the AS, then LSA Type 6 is converted to LSA Type 5 by ABRs, and then it is propagated.

OSPF CONFIGURATION

Table 7.2 lists and explains some of the commonly used commands for configuring and debugging OSPF network configurations

TABLE 7.2   OSPF Configuration Commands

Command	Explanation
router(config)# router ospf process-id	Enables OSPF routing process
router(config-router)# network ip-address wildcard-mask area area-id	Specifies interface of the OSPF process along with the area ID
router(config-if)# ip ospf hello-interval seconds	Specifies Hello interval
router(config-if)# ip ospf dead-interval seconds	Specifies the dead interval
router(config-if)# ip ospf priority number	Elects the OSPF DR
router(config-if)# ip ospf cost number	Specifies the cost of sending a packet over an interface

(continued)

TABLE 7.2 *(continued)*

Command	Explanation
router(config-if)# ip ospf transmit-delay seconds	Specifies the time taken to send an update packet
router(config-if)# ip ospf retransmit-delay seconds	Specifies the time between retransmissions of LSAs
router(config-if)# ip ospf authentication-key key	Specifies authentication password in routing updates between neighboring routers
router(config-if)# ip ospf message-digest-key key-id md5 key	Specifies MD5 authentication through a key-id and key
router(config-if)# ip ospf authentication {message-digest | null}	Specifies the type of authentication
router(config-if)# ip ospf network {broadcast | non-broadcast | {point-to-multi-point [non-broadcast]}}	Specifies the network type used for the interface
router(config-if)# ip ospf network point-to-multipoint	Specifies point-to-multi-point network type
router(config-router)# neighbor ip-address cost number	Specifies a neighbor with an associated cost
router(config-if)# neighbor ip-address [priority number] [poll-interval seconds]	Defines a neighbor in a non-broadcast network
router(config-if)# ip ospf network point-to-multi-point non-broadcast	Configures an interface for point-to-multipoint non-broadcast network type
router(config-router)# neighbor ip-address [cost number]	Specifies a neighbor with an associated cost
router(config-if)# debug ip ospf events	Displays OSPF event information, such as selection of designated router and formation of router adjacencies
router(config-if)# debug ip ospf packet	Displays information contained in each OSPF packet, such as router ID and area ID

The syntax of the command to find the link state format information from the link state database is:

Router(config)# show ip ospf database

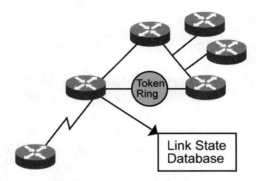

FIGURE 7.20   A link state database.

This command displays a summary of the topological link state database and describes the status of the different LSA formats. Figure 7.20 shows a link state database in a network.

The output for the show ip ospf database command is shown in Listing 7.1.

LISTING 7.1   Output for show ip ospf database Command

```
OSPF Router with ID (192.168.4.22) (Process ID 1)

Router Link States (Area 1)
Link ID        ADV Router      Age    Seq#          Checksum    Link Count
192.168.3.1    192.168.3.1     898    0x80000003    0xCE56      2
192.168.4.22   192.168.4.22    937    0x80000003    0xFD44      3

Summary Net Link States (Area 1)
Link ID        ADV Router     Age    Seq#          Checksum
172.16.1.0     192.168.3.1    848    0x80000005    0xD339
172.16.51.1    192.168.3.1    843    0x80000001    0xB329

Summary ASB Link States (Area 1)
Link ID        ADV Router     Age    Seq#          Checksum
192.168.1.1    192.168.3.1    912    0x80000003    0x93CC

Type-5 AS External Link States
Link ID     ADV Router      Age     Seq#          Checksum    Tag
11.0.0.0    192.168.1.1     1302    0x80000001    0x3FEA      0
12.0.0.0    192.168.1.1     1303    0x80000001    0x32F6      0
13.0.0.0    192.168.1.1     1303    0x80000001    0x2503      0
```

Link State Database is also called Topology Database.

Single Area Configuration

The steps to configure OSPF in a single area are:

1. Enable OSPF routing process using Router(config)# router ospf process-id command. The process ID can be within the range of 1 to 65535. The process-id allows you to run multiple OSPF routing processes on the same router. This process ID may not be the same in all routers. It is different from the process IDs used in IGRP and EIGRP routing protocols.
2. Find IP networks that are to be a part of the OSPF network using the Router(config)# network *address wildcard-mask* area area-id command. This command specifies the interfaces on which to send and receive updates, matching the address and the wild card mask.
3. Configure the loopback address using the Router(config)# interface loopback number command. This step is optional.
4. Change the OSPF priority on an interface using the Router(config)# ip ospf priority number(0-255) command. This step is optional.
5. Change the OSPF cost metric for a special interface using the Router(config)# ip ospf cost cost command. This step is optional.
6. Configure the authentication password using the Router(config)# ip ospf authentication-key password command. The password can be any valid text. This step is optional.

Consider the example of a single area router configuration shown in Figure 7.21. There are three routers, R1, R2, and R3, in the single OSPF area. These routers are on the Ethernet and are connected to each other using a switch.

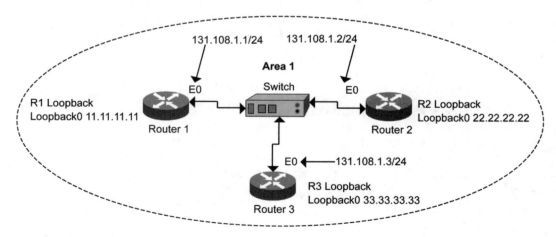

FIGURE 7.21 An OSPF single area router configuration.

The following section demonstrates the configuration of each router. The configuration of Router 1 is shown in Listing 7.2.

LISTING 7.2   Configuration of Router 1

```
!
version 12.1
!
hostname R1
!
no ip domain-lookup
!
ip subnet-zero
!
!
interface Loopback0
ip address 11.11.11.11 255.255.255.128
no shutdown
!
interface Ethernet0
ip address 131.108.1.1 255.255.255.0
no shutdown
!
.
.
router ospf 1
network 131.108.1.0 0.0.0.255 area 1
!
ip classless
ip http server
!
!
line con 0
line aux 0
line vty 0 4
!
end
```

In Listing 7.2, the hostname is configured to R1. Each interface is then assigned an IP address. R1 loopback is assigned to 11.11.11.11. Ethernet0 is assigned to the IP address 131.108.1.1. The OSPF routing protocol is enabled on the router, and the network address 131.108.1.0 is configured to be part of the OSPF network. The configuration of Router R2 is shown in Listing 7.3.

LISTING 7.3 Configuration of Router 2

```
!
hostname R2
!
no ip domain-lookup
!
ip subnet-zero
!
!
interface Loopback0
ip address 22.22.22.22 255.255.255.128
no shutdown
!
interface Ethernet0
ip address 131.108.1.2 255.255.255.0
no shutdown
!
interface Serial0
no ip address
shutdown
!
!
interface Serial1
no ip address
shutdown
!
.

.
router ospf 2
network 131.108.1.0 0.0.0.255 area 1
!
ip classless
ip http server
!
!
line con 0
line aux 0
line vty 0 4
!
end
```

In Listing 7.3, the hostname is specified as R2. For interface Loopback0, the IP address is assigned as 22.22.22.22. The IP address 131.108.1.2 is assigned to the

Ethernet0. The serial interfaces are not used for interconnection. As a result, no IP address is assigned to Serial0 and Serial1 interfaces. OSPF is enabled on R2, and the network address 131.108.1.0 is configured to be part of the OSPF network. The configuration of Router 3 is shown in Listing 7.4.

LISTING 7.4 Configuration of Router 3

```
!
hostname R3
!
no ip domain-lookup
!
ip subnet-zero
!
!
interface Loopback0
ip address 33.33.33.33 255.255.255.128
!
interface Ethernet0
ip address 131.108.1.3 255.255.255.0
no shutdown
!
.
.
router ospf 3
network 131.108.1.0 0.0.0.255 area 1
!
ip classless
ip http server
!
!
line con 0
line aux 0
line vty 0 4
login
!
end
```

In Listing 7.4, IP addresses are assigned to the OSPF Serial and Ethernet interfaces. OSPF is enabled using the router ospf command, and the network address 131.108.1.0 is configured to be part of the OSPF network.

Configuration steps for multiple areas can differ for each OSPF multiple area network, depending on the type of the networks that are interconnected. The steps to configure OSPF in multiple areas:

1. Enable the OSPF routing process using the Router(config)# router ospf `process-id` command. The process ID can be within the range of 1 to 65535. The process-id allows you to run multiple OSPF routing processes on the same router.
2. Find IP networks that are to be a part of the OSPF network using the Router(config)# network *address wildcard-mask* area `area-id` command. This command specifies the interfaces on which to send and receive updates, matching the address and the wild card mask.
3. Check if any interface is connected into a non-OSPF network using the Router(config)# area *area-id* range `address mask` command. This command consolidates non-OSPF routes on an ABR.
4. Configure route summarization in the ABR using the Router (config)# summary-address `address mask` command. This step is optional and consolidates non-OSPF routes on an ASBR.
5. Configure non-standard OSPF areas using the Router(config)# area `area-id stub` command.
6. Configure the cost for default routes propagated to non-standard areas using the Router(config)# area *area-id* default-cost `cost` command.
7. Configure virtual links if needed.

Virtual links are configured using the Router(config)# area *area-id* virtual-link *id-of-remote-router* command. The `areaid` command represents the ID of the OSPF area, which is not connected to Area 0. The `idremoterouter` command is the ID of the ABR in Area 0.

All areas in an OSPF AS should be physically connected to Area 0. If, in some cases, it is not possible to connect the areas to Area 0, then a virtual link between the non-connected areas to Area 0 is needed. The area through which the virtual link is configured is known as the transit area. The transit area cannot be an SA, and it must be configured between two ABRs.

The steps for multiple area configuration are valid for most cases of OSPF multiple area configuration.

Consider an example of a multiple area configuration in OSPF. In Figure 7.22, Area 1 is connected to Area 0 using the ABRs R2 and R3. A virtual link connects Area 1 to Area 0. R1 is connected to R2 using a switch.

The configuration of R1 is shown in Listing 7.5.

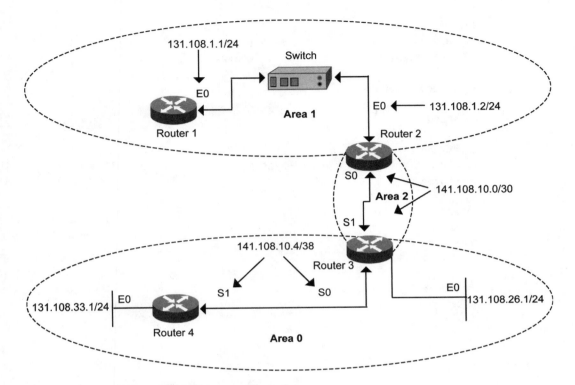

FIGURE 7.22 An OSPF multiple area configuration.

LISTING 7.5 Configuration of Router 1

```
!
version 12.1
no service single-slot-reload-enable
service timestamps debug uptime
service timestamps log uptime
no service password-encryption
!
hostname R1
!
!!
ip subnet-zero
no ip domain-lookup
!
!
interface LoopBack0
ip address 131.108.4.1 255.255.255.128
no shutdown
!
```

```
interface LoopBack1
ip address 131.108.4.129 255.255.255.128
no shutdown
!
interface LoopBack2
ip address 131.108.5.1 255.255.255.224
no shutdown
!
interface Ethernet0
ip address 131.108.1.1 255.255.255.0
no shutdown
!
interface Serial0
no ip address
shutdown
no fair-queue
!
interface Serial1
no ip address
shutdown
.
.
.
.
router ospf 1
log-adjacency changes
network 131.108.1.0 0.0.0.255 area 1
network 131.108.4.0 0.0.0.127 area 1
network 131.108.4.128 0.0.0.127 area 1
network 131.108.5.0 0.0.0.31 area 1
!
ip classless
ip http server
!
!
line con 0
line aux 0
line vty 0 4
!
end
```

In Listing 7.5, the hostname is configured as R1. Interfaces are assigned IP addresses. Networks with IP addresses 131.108.1.0, 131.108.4.0, 131.108.4.128, and
131.108.5.0 are configured to be part of the OSPF network. The configuration of R2
is shown in Listing 7.6.

LISTING 7.6   Configuration of Router 2

```
version 12.1
no service single-slot-reload-enable
service timestamps debug uptime
service timestamps log uptime
no service password-encryption
!
hostname R2
!
no ip domain-lookup
!
!
!
!
!
ip subnet-zero
!
!
!
!
interface LoopBack0
ip address 131.108.5.33 255.255.255.224
ip ospf network point-to-point
ip ospf cost 1000
no shut
!
interface LoopBack1
ip address 131.108.6.1 255.255.255.255
no shut
!
interface LoopBack2
ip address 131.108.6.2 255.255.255.255
no shut
!
interface Ethernet0
ip address 131.108.1.2 255.255.255.0
no shut
!
interface Serial0
ip address 141.108.10.1 255.255.255.252
no fair-queue
Clockrate 56000
no shut
```

```
!
interface Serial1
no ip address
shutdown
!
.
.
.
router ospf 2
log-adjacency-changes
area 2 virtual -link 141.108.12.1
network 131.108.1.0 0.0.0.255 area 1
network 131.108.5.32 0.0.0.31 area 1
network 131.108.6.1 0.0.0.0        area 1
network 131.108.6.2 0.0.0.0 area 1
network     141.108.10.0 0.0.0.3 area 2
!
ip classless
ip http server
!
!
line con 0
line aux 0
line vty 0 4
!
end
```

In Listing 7.6, the hostname is configured as R2. The Ethernet and Serial interfaces are assigned IP addresses. Virtual link is configured with IP address 141.108.12.1, which connects Area 2 to the backbone area. The network addresses are then configured and form the OSPF network. The configuration of Router 3 is shown in Listing 7.7.

LISTING 7.7   Configuration of Router 3

```
!
hostname R3
!
no ip domain-lookup
!
!
!
!
ip subnet-zero
```

```
!
!
!
!
interface Loopback0
ip address 141.108.9.1 255.255.255.128
ip ospf network point-to-point
no shutdown
!
interface Loopback1
ip address 141.108.9.129 255.255.255.128
ip ospf network point-to-point
no shutdown
!
interface Loopback2
ip address 141.108.12.1 255.255.255.0
ip ospf network point-to-point
no shutdown
!
interface Ethernet0
ip address 131.108.26.1 255.255.255.0
no shutdown
!
interface Serial0
ip address 141.108.10.6 255.255.255.252
clockrate 56000
no shutdown
!
interface Serial1
ip address 141.108.10.2 255.255.255.252
no shutdown
!
.
.
router ospf 3
log-adjacency-changes
area 1 virtual-link 131.108.6.2
network 131.108.26.0 0.0.0.255 area 0
network 141.108.9.0 0.0.0.127 area 0
network 141.108.9.128 0.0.0.127 area 0
network 141.108.10.0 0.0.0.3 area 2
network 141.108.10.4 0.0.0.3 area 0
network 141.108.12.1 0.0.0.255 area 0
!
```

```
ip classless
ip http server
!
!
line con 0
line aux 0
line vty 0 4
login
!
end
```

In Listing 7.7, the hostname is configured as R3. The Loopback, Serial, and Ethernet interfaces are assigned IP addresses. Virtual link is configured that connects Area 1 to the backbone area. The network addresses are added that need to be part of the OSPF network. The configuration of Router 4 is shown in Listing 7.8.

LISTING 7.8   Configuration of Router 4

```
!
hostname R4
!
no ip domain-lookup
!
ip subnet-zero
!
!
interface Loopback0
ip address 141.108.1.1 255.255.255.128
ip ospf network point-to-point
no shutdown
!
interface Loopback1
ip address 141.108.1.129 255.255.255.128
ip ospf network point-to-point
no shutdown
!
interface Loopback2
ip address 141.108.2.1 255.255.255.224
ip ospf network point-to-point
no shutdown
!
interface FastEthernet0/0
ip address 131.108.33.1 255.255.255.0
no shutdown
!
```

```
interface Serial0/0
ip address 141.108.10.5 255.255.255.252
clockrate 56000
no shutdown
!
.
.
router ospf 4
log-adjacency-changes
network 131.108.33.0 0.0.0.255 area 0
network 141.108.1.0 0.0.0.127 area 0
network 141.108.1.128 0.0.0.127 area 0
network 141.108.2.0 0.0.0.31 area 0
network 141.108.10.4 0.0.0.3 area 0
!
ip classless
no ip http server
!
!
line con 0
line aux 0
line vty 0 4
login
!
end
```

In Listing 7.8, the interfaces are assigned IP addresses. OSPF is then enabled on the router, followed by the configuration of network addresses that are a part of the OSPF network.

VERIFICATION OF OSPF OPERATIONS

This section discusses the various commands used to verify the OSPF routing process. The output for the show ip ospf command is shown in Listing 7.9.

LISTING 7.9 Output for show ip ospf Command

```
router# show ip ospf
OSPF is running, process id: 1, router id: 131.108.6.2
Number of areas: 1, normal: 1, stub: 0
Area: 11.22.33.44
Number of interfaces in this area is 1
Type of authentication none
SPF algorithm has run 3 times
SPF interval 6 seconds
```

The output for show ip ospf interface serial 0 command is shown in Listing 7.10.

LISTING 7.10   Output for show ip ospf interface serial 0 Command

```
Router# show ip ospf interface serial0
Serial0 is up, line protocol is up
Internet Address 10.10.10.1/24, Area 0
Process ID 1, Router ID 13.3.13.3, Network Type POINT_TO_POINT, Cost: 64
Transmit Delay is 1 sec, State POINT_TO_POINT,
Timer intervals configured, Hello 10, Dead 40, Wait 40, Retransmit 5
Hello due in 00:00:05
Index 1/1, flood queue length 0
Next 0x0(0)/0x0(0)
Last flood scan length is 1, maximum is 1
Last flood scan time is 0 msec, maximum is 0 msec
Neighbor Count is 1, Adjacent neighbor count is 1
Adjacent with neighbor 14.2.14.2
Suppress Hello for 0 neighbor(s)
```

The output for show ip ospf neighbor command is shown in Listing 7.11.

LISTING 7.11   Output for show ip ospf neighbor Command

```
Router# show ip ospf neighbor
Neighbor ID      Pri     State       Dead Time    Address       Interface
1.1.1.1           1      FULL/-      00:00:30     192.16.2.1    Serial0
```

The output for show ip ospf interface serial0 command is shown in Listing 7.12.

LISTING 7.12   Output for show ip ospf interface serial0 Command

```
Router# show ip ospf interface serial0
Serial0 is up, line protocol is up
Internet Address 192.16.2.1/24, Area 0
Process ID 1, Router ID 1.1.1.1, Network Type POINT_TO_POINT, Cost: 64
Transmit Delay is 1 sec, State POINT_TO_POINT,
Timer intervals configured, Hello 10, Dead 40, Wait 40, Retransmit 5
Hello due in 00:00:05
Index 1/1, flood queue length 0
Next 0x0(0)/0x0(0)
Last flood scan length is 1, maximum is 1
Last flood scan time is 0 msec, maximum is 0 msec
Neighbor Count is 1, Adjacent neighbor count is 1
Adjacent with neighbor 2.2.2.2
Suppress Hello for 0 neighbor(s)
```

The output for show ip ospf border-routers command is shown in Listing 7.13.

LISTING 7.13   Output for show ip ospf border-routers Command

```
Router# show ip ospf border-routers
Codes: i - Intra-area route, I-Inter-area route
Type  Dest Address  Cost  NextHop    Interface    ABR    ASBR   Area  SPF
I     131.108.1.2   10    0.0.0.31   Ethernet 2   TRUE   FALSE  0     3
```

The output for show ip ospf virtual-links command is shown in Listing 7.14.

LISTING 7.14   Output for show ip ospf virtual-links Command

```
Router# show ip ospf virtual-links
Interface address: 131.108.6.2 (POS 1/1/1) cost: 1, state: P To P,
transit area: 11.22.32.44 hello: 10, dead: 40, retrans: 5 nbr id:
10.1.2.103, nbr address: 10.1.2.103 nbr state: Full, nbr mode: Master,
last hello: 38
```

The output for show ip ospf database command is shown in Listing 7.15.

LISTING 7.15   Output for show ip ospf database Command

```
Router# show ip ospf database
OSPF RouterOSPF Router with ID () (Process ID 0)
Router Link States (Area) Link ID ADV Router Age Seq# Checksum Link
count Net Link States (Area) Link ID ADV Router Age Seq# Checksum
```

The output for show ip route summary command is shown in Listing 7.16.

LISTING 7.16   Output for show ip route summary Command

```
Router# show ip route summary
Route source Networks
connected 5
static 2
ospf intra-area 1585
total 7024
```

SUMMARY

In this chapter, we looked at the routing process using OSPF routing protocol. We also discussed the various packet format and configuration commands in OSPF. In the next chapter, we will discuss the IS_IS routing protocol.

POINTS TO REMEMBER

- OSPF routers gather link state information and use shortest path algorithm for calculating the optimal path for routing data packets.
- OSPF routers calculate loop-free paths to different destinations using Djikstra's algorithm and represent these paths in the form of a graph called the Shortest Path First (SPF) tree.
- Single Area (SA) is a logical subdivision of the greater OSPF domain, grouping routers that run OSPF with identical topological databases.
- Totally Stubby Area (TSA) is Cisco proprietary and is used when there are few networks with limited connectivity with the remaining network.
- Not So Stubby Area (NSSA) is mainly used in redistribution of routing information.
- Designated Router acts as the hub and manages the flooding process in a multiaccess network.
- Backup Designated Router (BDR) acts as the new DR in case the DR fails.
- Routers participating in the election apart from the elected DR and BDR are known as DRothers.
- BDR is elected in a multi-access OSPF network so that the network does not collapse in the event of a DR failure.
- The four types of OSPF routers are Internal Router (IR), Area Border Router (ABR), Backbone Router (BR), and Autonomous System Boundary Router (ASBR).
- All OSPF packets consist of eight fields for the header and one field for data.
- The LSA packet describes the integrated states of the router interfaces belonging to a single area.
- The DR generates network LSAs in order to update broadcast or multi-access networks.
- There are seven types of LSA formats in OSPF routing protocol.

8 Intermediate System to Intermediate System

INTRODUCTION TO IS_IS

Intermediate System to Intermediate System (IS_IS) routing protocol is an Interior Gateway Protocol (IGP) that supports true Internet Protocol (IP) environments and true Open Systems Interconnection (OSI) environments as well as dual environments. IS_IS protocol can be used for routing both IP and OSI network traffic. As a result, this protocol is also known as Integrated IS_IS protocol. An IS_IS router that supports only IP traffic is known as an IP-only router, and the IS_IS router that supports only OSI traffic is called an OSI-only router. If an IS_IS router supports both IP and OSI environments, it is known as a dual IS_IS router.

Two types of network topologies are supported by the IS_IS routing protocol:

Broadcast Networks: Local Area Networks (LANs)

Point-To-Point Networks: Circuit Switched Networks, such as twisted cable network

The IS_IS protocol uses Connectionless Network Service (CLNS) addresses for supporting dual environments. CLNS addresses include Area ID, System ID, and Network Selector (NSEL) field. The IS_IS addresses are not based on IP subnetwork addresses. As a result, a separate partial route recalculation is performed if routing information for an IP network is updated.

Large Internet Service Providers (ISPs) use IS_IS to leverage interconnection support to large networks. IS_IS is appropriate for ISPs and business services similar to ISPs because of the need to connect both OSI- and IP-based networks.

IS_IS packets are encapsulated in the data link layer, so IS to IS routing can operate in both OSI- and IP-based environments. IS_IS is more efficient than the OSPF protocol because of its flexibility. OSPF does not provide dual environment support and is only used in IP-based networks.

IS_IS TERMINOLOGY

This section discusses the terminology used in IS_IS routing protocol. The terms are:

OSI: An international standard protocol architecture.

CLNP: An OSI protocol that is similar to IP.

Protocol Data Unit or Packet Data Unit (PDU):   A single data unit of the corresponding protocol used across the network.

Link State Protocol Data Unit (LSP):   A data unit of the IS_IS protocol. LSP is analogous to the Link State Advertisement (LSA) in the OSPF protocol. LSP refers to the packet containing routing information.

End System (ES):   A host system in the IS_IS network. ES is an OSI terminology for host.

Intermediate System (IS):   A router in IS_IS terminology.

Designated Intermediate System (DIS):   The Designated Router (DR) in IS_IS network.

Pseudo Node (PN):   The broadcast link emulated as a virtual link by DIS.

Subnetwork Point of Attachment (SNPA):   The data link interface of the IS_IS network. This is the point at which a device is connected to the network.

Network Service Access Point (NSAP):   The OSI address. It consists of area address, system Identifier, and NSEL selector byte. If NSEL for an NSAP is equal to 0, it is known as Network Entity Title (NET). NSEL for a router is always set to 0.

Network Layer Protocol ID (NLPID):   The octet field that identifies a network layer protocol.

IS_IS Hello (IIH) Packets: Maintain neighbor information and adjacencies in IS_IS.

Adjacency: Nearby systems in IS_IS network.

End System to Intermediate System (ES-IS):   A protocol similar to IS_IS that delivers the ISO-defined CLNS.

The SNPA for a LAN connection is the address of the interface itself. The virtual circuit identifier is the SNPA for a Wide Area Network (WAN).

IS_IS OPERATION

In an IS_IS network, each IS establishes adjacencies by sending IIH packets, and it generates an LSP. These LSPs that contain information about the neighbors are sent to each router in the IS_IS routing domain. Figure 8.1 shows the working of the IS_IS routing protocol. In Figure 8.1, Routers A, B, C, D, and E represent the different IS_IS routers. The metric of each of the interfaces is also given.

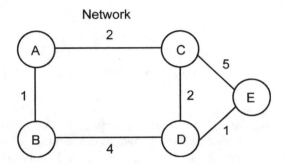

FIGURE 8.1   Routers in an IS_IS network.

Each IS_IS router generates and maintains its own link state database, as shown in Figure 8.2. Every router in the IS_IS network must have updated accurate LSPs from every other router in the routing domain. If the LSPs are not updated, the

Link state database

A		B		C		D		E	
C	2	A	1	A	2	B	4	C	5
B	1	D	4	D	2	C	2	D	1
				E	5	E	1		

FIGURE 8.2   Link state databases in an IS_IS network.

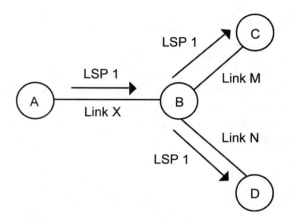

FIGURE 8.3 LSP flooding in an IS_IS network.

IS_IS routing algorithm may fail or may be less efficient. LSPs are propagated to each router through a process known as flooding. Figure 8.3 illustrates the process of flooding. Routers A, B, C, and D form an IS_IS network. In Figure 8.3, if a router receives an LSP in link X, then it propagates the received LSP to every link other than link X. If B receives LSP 1, the router forwards it to all links other than the link from which the LSP 1 was received. Similarly, C and D follow the same flooding process by which LSP 1 reaches all routers in the IS_IS network.

NOTE

Each router uses the timestamps, sequence number, and checksum in LSPs to decide which LSP is the latest and whether flooding is essential or not.

TIP

Refer to the section "Link State PDU Formats" in this chapter to see the LSP formats.

A router can calculate the nearest path to its neighbors, using the shortest path algorithm once the router has all the LSPs in its link state database. This database, containing the entire shortest path to the neighbors, constitutes the IS_IS routing table for that specific router. Accordingly, each router in the routing domain calculates and keeps the updated routing table. The major steps for execution of the IS_IS routing process are:

- The IS sends IIH packets to all IS_IS enabled interfaces to find and establish adjacencies with nearby systems.

- Each IS maintains neighborhood relationships, using the authentication information available in the IIH packet.
- The ISs build an LSP database and flood LSPs to all adjacent systems except the system from which the LSP was received.
- The routing table for IS is generated and updated, using the shortest path algorithm.
- The IS uses the routing table to find the route of a new data packet and forwards the packet along the same route as in the routing table.

The shortest path algorithm is executed when the IS_IS network topology is to be calculated. When IP routing information alone is calculated, the shortest path algorithm is not executed; instead, only Partial Route Calculation (PRC) is executed so that less CPU time is used.

Neighbor Discovery and Adjacencies

An IS_IS router discovers neighbors by flooding IIHs to all the routers within the routing domain. Once the updated routing information is available from IIHs, it is recorded in the link state database for that particular router.

Areas in IS_IS

In IS_IS, an area identifies the functionality of each IS within the routing domain. These areas define the sub-boundary within the IS_IS routing domain. These sub-boundaries, in turn, define the functionality of each IS available within that area. The IS_IS routing domain can be subdivided into Level 1 (L1) and Level 2 (L2) areas.

Level 1 Area

Typically, an IS falls under only one area. The boundary of the area is on the link that connects two separate areas. The IS_IS L1 area maintains its own link state database. Per OSI standard, an L1 router should know the topology of its own area. When OSI routing configuration is needed, L1 area is used. All ISs within the L1 perform routing between adjacent ISs within the same area.

Level 2 Area

This is the backbone area, which connects multiple L1 areas. The L2 area connects two different L1 areas within the same IS_IS routing domain. All routers can be configured as L2 routers in an IP-based network. The ISs in the

L2 area fall in both L1 and L2 areas. As a result, they are referred to as L1-L2 routers.

L1 routers are also referred to as non-backbone area routers, and L2 routers are referred to as backbone area routers.

Figure 8.4 depicts the areas in the IS_IS routing domain. Consider four routers, R1, R2, R3, and R4, in an IS_IS routing domain. In Figure 8.4, R3 and R4 are in two separate IS_IS areas, Area 1 and Area 2. R3 and R4 perform routing within the same area only. These routers will not forward any packets to a router outside their area. R1 and R2 are L1-L2 routers that connect two different IS_IS L1 areas. These routers forward packets from Area 1 to Area 2.

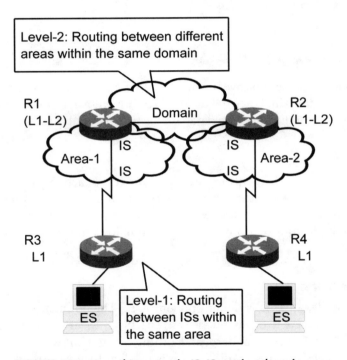

FIGURE 8.4 L1 and L2 areas in IS_IS routing domain.

If an L1 router gets an LSP from an external area router, the LSP is forwarded to the nearest L2 router, which, in turn, forwards the LSP to the System ID specified in the CLNS address System ID field. If an L1 router gets an LSP from an internal area router, then it forwards that LSP to the correct System ID specified in the CLNS address.

Consider the example shown in Figure 8.5. There are three different L1 areas, Area 89.001, Area 89.0002, and Area 89.003, and an L2 area. In Figure 8.5, an IS_IS LSP is generated by Area 89.001 L1-L2 router. The destination of this L1-L2 router is Area 89.003. The routing table of Area 89.001 L1-L2 router contains the shortest route to the L1-L2 router in Area 89.003. Using this routing table information, Area 89.001 L1-L2 router forwards the LSP to the nearest L1-L2 router of Area 89.003.

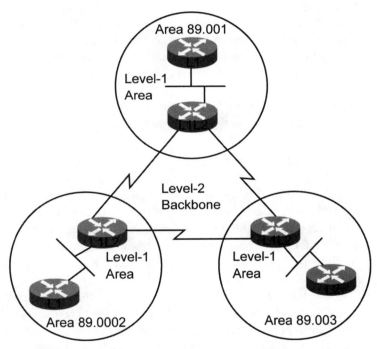

FIGURE 8.5   The different areas and L1, L2, and L1-L2 routers in an IS_IS routing domain.

Hierarchical Structure

The IS_IS routing protocol follows a two-level routing hierarchy with two different levels of routers. This hierarchical structure minimizes routing traffic and increases scalability. The two levels of routing hierarchy for IS_IS protocol are discussed.

Level 1 Routing

L1 routing is also called intra-area routing and occurs within an IS_IS area. L1 routers know the topology of their area, including information about all ISs and ESs within the area. However, L1 routers are unaware of any router or destination information outside their area. These L1 routers have routes not only for the systems within their area but also a default route. This default route will point to a L2 router used for external routing.

Level 2 Routing

L2 routing is also called inter-area routing and occurs between multiple IS_IS areas. L2 routing is based on the Area ID. Djikstra's shortest path algorithm is executed for finding the shortest path to other areas. L2 routing is based on the System ID after the IS_IS packet reaches the destination area. L2 routers know the destinations that are reachable using each L2 router, but they do not have information about L1 routing. These L2 routers can exchange information directly with an external router that is located outside the routing domain.

There is also one more level of routing named Level 3 (L3) routing that deals with routing between different routing domains.

Figure 8.6 illustrates the IS_IS hierarchical routing structure. In Figure 8.6, the routers in Level 1 IS_IS areas are non-backbone routers. There are two routers in the backbone area. The backbone area routers are L2 routers that connect Area 89.001 and Area 89.002. The routers within the L1 area act as L1 routers. The router that connects Area ID: 89.001 and Area ID: 89.0002 acts as both L1 and L2 routers. When an LSP is forwarded from the non-backbone router to another area, it passes through the backbone area routers to the destination area.

The shortest path algorithm is used to calculate the shortest paths to IS_IS areas and ESs, not the shortest paths to IP networks. As a result, even if the IP network configuration changes, recalculation of shortest paths is not required.

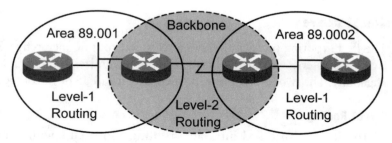

FIGURE 8.6 The IS_IS hierarchical routing structure.

Intermediate System Types

This section discusses the types of routers present in an IS_IS routing domain. Please note that the term "router" can interchangeably be used for "IS" in IS_IS terminology.

L1 Intermediate Systems

L1 routers are not directly connected to any other devices residing in another area. An L1 router does not have any knowledge about destinations outside its own area. L1 routers are analogous to Internal Routers (IRs) in OSPF terminology.

L1 routers perform routing within an area based on the ID portion of the IS_IS packet. If the destination address specified in the IS_IS packet is not within the same area as the L1 router, then this router handles those IS_IS packets to the nearest L2 router.

An L1 router has neighbors only in the same area and has an L1 link state database that contains the whole routing information for that specific area. The L1 uses the nearest L2 router to exit the area.

L2 Intermediate Systems

L2 routers connect multiple IS_IS areas to each other. One L2 router is always connected to another L2 router or to an L1-L2 router in a different area. L2 routers are similar to backbone routers in OSPF.

L2 routers perform routing based on the area address in the IS_IS packet. The L2 router does routing toward the area ID without considering the internal structure of the area. This router can also take up the role of an L1 router for its area. L2 routers may have neighbors in other areas and an L2 link state database with all routing information about inter-area routing.

L1-L2 Intermediate Systems

L1-L2 routers interconnect an L1 router and an L2 router and maintain different link states for these connections. The L1-L2 router is analogous to an ABR in the OSPF protocol. The IS_IS LSP maintains a bit that shows that the L1-L2 router is connected to other networks. The L1-L2 routers can have neighbors in any area and two link state databases—one for intra-area routing and the other for inter-area routing.

Figure 8.7 shows the three types of IS_IS routers, L1, L2, and L1-L2. In Figure 8.7, the LSPs from the L1 router and the L2 router are forwarded to L1 routers in other areas through the L1-L2 router. When an LSP is generated by the L1 router and is passed to the L1 router in another area, the process of forwarding the LSP to the destination is:

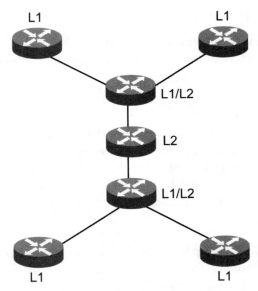

FIGURE 8.7 The L1, L2, and L1-L2 routers in the IS_IS routing domain.

1. L1 router computes the nearest L1-L2 router in the network, depending on the metric of the interfaces interconnected.
2. LSP is passed to the L1-L2 router identified in step 1.
3. L1-L2 router computes the shortest path to destination L1-L2 router.
4. LSP is passed to destination L1 router by the destination L1-L2 router.

Table 8.1 lists the features of the three types of routers present in an IS_IS routing domain.

TABLE 8.1 IS_IS Router Features

L1 Router	L2 Router	L1-L2 Router
Non-backbone router	Backbone router	Similar to OSPF ABR
Knows about intra-area routes only	Knows about inter-area routes only	Knows about both intra-area and inter-area routes
Maintains L1 link state state database	Maintains L2 link state database	Maintains two link databases

Network Entity Title

Each IS in an IS_IS area has a unique address within the area. This unique address is known as Network Entity Title (NET). The NET contains two main fields, System ID and Area ID. The System ID uniquely identifies a system within an IS_IS area, whereas the Area ID uniquely identifies an area in the IS_IS network.

The System ID is analogous to the router ID in OSPF and is used by L1 routers for locating a specific system within an area. L2 routers use the Area ID to locate a specific area within an IS_IS network. There are thee types of NETs for the IS_IS routing protocol. They are:

- Octet Format NET
- OSI NSAP Format NET
- Globally Unique NSAP Format NET

Octet Format NET

Figure 8.8 shows an example of an octet format NET. The fields are:

Area ID: Identifies the IS_IS area where the router is situated. In Figure 8.8, the Area ID is 05.

System ID: Identifies the router located within the IS_IS domain. In Figure 8.8, the System ID is 0000.3456.a9ab.

NSEL: Refers to the network-level services offered to the IS_IS router. The Network Selector is always 00 in the octet format NET.

Area ID	System ID	NSEL
05	0000.3456.a9ab	00

FIGURE 8.8   The octet format NET.

OSI NSAP Format NET

Figure 8.9 shows an example of an OSI NSAP format NET.

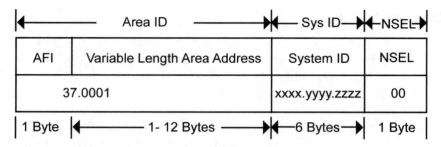

FIGURE 8.9 The OSI NSAP format NET.

The fields are:

Area ID: This field is divided into Authority and Format Identifier (AFI) and variable length area address. AFI represents the first octet of all NSAP addresses. In Figure 8.9, the Area ID is 37.0001. Table 8.2 maps the network address domain with its AFI value.

TABLE 8.2 Address Domain AFI Value Mapping

Address Domain	AFI Value	Explanation
X.121	37	International plan for public data networks
ISO DCC	39	ISO Data Country Code
ISO 6523	47	Telex
Local	49	Local use within the network domain

System ID: Identifies the router located within the IS_IS domain. In Figure 8.9, the System ID is xxxx.yyyy.zzzz.

NSEL: Value of NSEL is 00 on an IS_IS router.

Globally Unique NSAP Format NET

The Globally Unique NSAP (GUNSAP) NET format is used for large interconnected telecommunications systems such as ATM switches. Figure 8.10 shows the GUNSAP NET format.

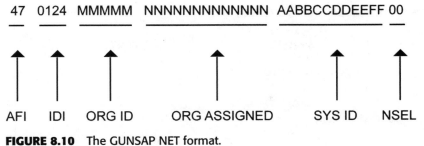

FIGURE 8.10   The GUNSAP NET format.

The fields are:

AFI: Represents the first octet of the GUNSAP address; used to identify the format of rest of the GUNSAP address. In Figure 8.10, the AFI is 47.

IDI: In Figure 8.10, the IDI is 0124.

ORG ID: In Figure 8.10, the ORG ID is MMMMM.

ORG Assigned: In Figure 8.10, the ORG Assigned is NNNNNNNNNNNNNN.

System ID: Identifies the router located within the IS_IS domain. In Figure 8.10, the system ID is AABBCCDDEEFF.

NSEL: Value of NSEL is 00 on an IS_IS router.

IS_IS FUNCTIONS

An IS_IS routing network has two types of functions:

Subnetwork independent functions: Provide full network PDU traffic between any set of adjacent ISs. These functions are independent of the specific subnetwork or data link service executing below them. Routing and congestion control are examples of subnetwork independent functions.

Subnetwork dependent functions: Conceals attributes of the subnetwork or data link service from the subnetwork independent functions. Reception and transmission of IS_IS PDUs over a particular subnetwork, exchanging IIHs to establish router and host adjacencies, flooding, and link state database synchronization, are examples of subnetwork dependent functions.

Subnetwork Independent Functions

There are two main IS_IS subnetwork independent functions. They are:

- Congestion Control
- Routing

Congestion Control

Congestion control is the management of local resources by an IS_IS router. To keep the IS_IS network traffic at a minimum, the IS_IS router uses timers and queues for controlling the IS_IS PDUs.

Routing

The routing function routes IS_IS networks PDU along a pre-defined path that has the lowest metric. The routing function is further divided into four major processes:

- Receiving process
- Updating process
- Decision-making process
- Forwarding process

Receiving Process

This process is the starting point where all data, including routing information, user data, error information, and control packets, are received. This process passes routing information and control packets, such as IIHs, LSPs, and SNPs, to the updating process and passes user data and error information to the forwarding process.

Updating Process

This process manages L1 and L2 link state databases and floods LSPs throughout the area. An IS_IS router updates link and adjacencies information in the updating process.

Every LSP records the remaining lifetime for the LSP. If the remaining lifetime for the LSP reaches 0, the LSP is kept in the link state database for an additional 60 seconds. This additional lifetime is known as Zero Age Life Time. If the corresponding router does not update the LSP even after the Zero Age Life Time, the LSP is deleted from the link state database.

Decision-making Process

This process executes the shortest path algorithm on the link state database and generates the forwarding database. It is this forwarding database that is referred for routing packets over the IS_IS network.

Forwarding Process

This process uses the forwarding database for propagating the IS_IS packets over the network toward the destination of the data packet. Figure 8.11 depicts the subnetwork independent routing functions.

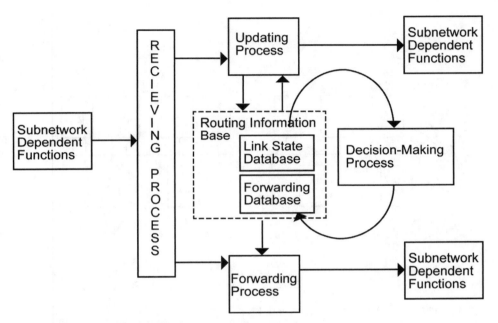

FIGURE 8.11   The subnetwork independent routing functions in an IS_IS routing domain.

Subnetwork Dependent Functions

The subnetwork dependent functions conceal the attributes of different subnetworks from the independent functions. The subnetwork dependent functions are:

- Reception and transmission of IS_IS PDUs over subnetwork
- Exchange of IIHs to establish router and host adjacencies
- Flooding and link state database synchronization

Reception and Transmission of IS_IS PDUs over Subnetwork

Each subnetwork is represented as a Pseudo Node (PN) to minimize the amount of routing traffic. The ISs in the subnetwork elect the DIS to act as a PN. The election is based on the value in the priority field in IIH PDUs. Figure 8.12 shows an IS_IS subnetwork PN.

Each IS on the subnetwork reports only a single link to the PN instead of reporting a link to every other IS in the broadcast subnetwork. The DIS then constructs an LSP database for the whole subnetwork.

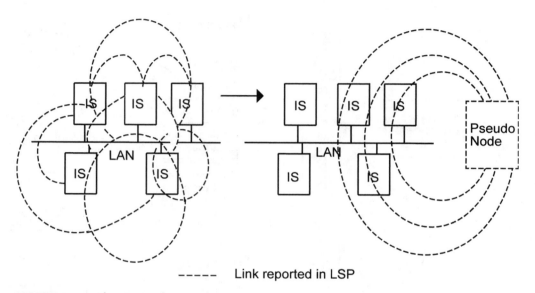

----- Link reported in LSP

FIGURE 8.12 The IS_IS subnetwork PN.

Exchange of IIHs to Establish Router and Host Adjacencies

Every IS exchanges IIH PDUs to obtain neighbor information. Point-to-point IIH PDUs are transmitted in non-broadcast subnetworks that do not belong to an external routing domain. On broadcast subnetworks, L1 and L2 LAN IIHs are transmitted to generate L1 and L2 adjacencies, respectively. There are three states of broadcast subnetwork adjacencies, INIT, UP, and DOWN.

L1 adjacencies are accepted only if a match in one or more area addresses is found. If the LAN IIH PDU passes the adjacency requirement, then the particular adjacency is set to INIT state. An adjacency is set to the UP state to ensure two-way connectivity.

IIHs are exchanged every 10 seconds on a point-to-point link by default. In a broadcast network, IIHs are exchanged every 10 seconds except for the DIS. IIHs are exchanged every 3.3 seconds for DIS. Figure 8.13 shows the different broadcast subnetwork adjacency states.

The commands and outputs used to establish adjacencies and connection with the IS_IS neighbors are shown in Listing 8.1 and Listing 8.2.

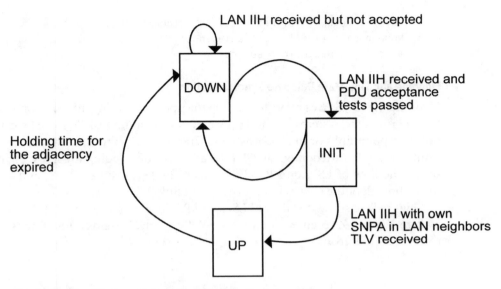

LAN IIH received but not accepted

DOWN

LAN IIH received and
PDU acceptance
tests passed

INIT

Holding time for
the adjacency
expired

LAN IIH with own
SNPA in LAN neighbors
TLV received

UP

FIGURE 8.13   The broadcast subnetwork adjacency states.

LISTING 8.1   Output for show clns neighbors Command

```
Router# show clns neighbors

System Id Interface SNPA      State Holdtime   Type   Protocol
R2        Se0      *HDLC*       Up 18          L1     IS_IS
R4        Et0      0000.0b92.de4c Up 10        L2     IS_IS
```

LISTING 8.2   Output for show clns interface serial 0 Command

```
Router# show clns interface serial 0
  Serial0 is up, line protocol is up
  Checksums enabled, MTU 1200, Encapsulation HDLC
  ERPDUs enabled, min. interval 12 msec.
  RDPDUs enabled, min. interval 110 msec., Addr Mask enabled

Congestion Experienced bit set at 4 packets
  CLNS fast switching disabled
  CLNS SSE switching disabled
  DEC compatibility mode OFF for this interface
  Next ESH/ISH in 14 seconds
  Routing Protocol: IS_IS
  Circuit Type: level-1
  Interface number 0x1, local circuit ID 0x101
```

```
Level-1 Metric: 10, Priority: 64, Circuit ID: R2.00
Number of active level-1 adjacencies: 1
Next IS_IS Hello in 6 seconds
```

Flooding and Link State Database Synchronization

Flooding the IS_IS LSPs refers to sending the updated routing information to each of the intermediate systems in the IS_IS network. Flooding of IS_IS LSPs can be in point-to-point links or in broadcast links. In point-to-point-links, a reliable mechanism using acknowledgments (PSNP) is adopted for flooding the LSPs. Figure 8.14 shows the flow of LSPs and PSNPs between the IS_IS routers X, Y, and Z; it also shows how flooding occurs in point-to-point links.

Figure 8.14 shows the flow of LSPs and PSNPs from X to Y to Z. PSNPs are sent by Z to Y when Z receives an LSP from Y. Similarly, Y sends a PSNP to X when Y receives an LSP from X.

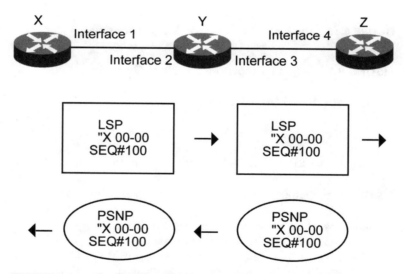

FIGURE 8.14 The flooding in point-to-point links.

In broadcast links, the flooding of IS_IS LSPs depends on periodic advertisements of CSNPs. Figure 8.15 shows flooding in broadcast links of IS_IS routers, Z, M, N, and O.

IS_IS PACKET FORMATS

This section discusses the different IS_IS packets and their general formats.

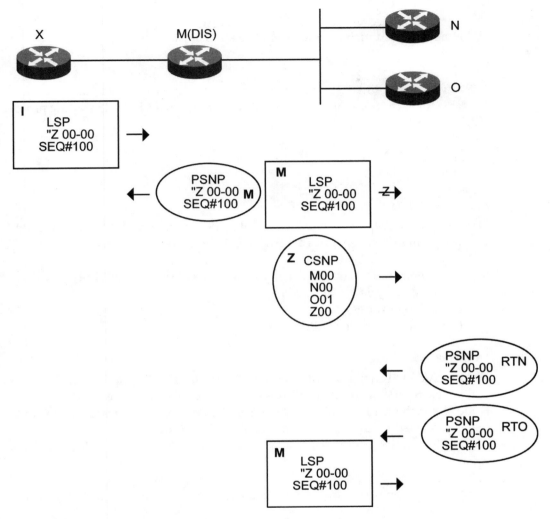

FIGURE 8.15   The flooding of LSPs in broadcast links.

IS_IS PDU Format

IS_IS PDUs are directly encapsulated within a data link frame. The IS_IS PDU formats available within an IS_IS network are:

- IIH
- LSP
- CSNP
- PSNP

IIH

IIH establishes and maintains adjacency relationships.

LSP

LSP advertises the link state information in the IS_IS network to update the routing table for each IS_IS router. There are two types of LSPs in IS_IS routing protocol. They are:

- Non Psuedo Node LSP
- Pseudo Node LSP

Non Psuedo Node LSP

Non Psuedo Node LSPs are LSPs generated by ISs. Every IS in the IS_IS routing domain generates and propagates a fresh Non Psuedo Node LSP when:

- A new neighbor IS comes alive in the IS_IS network and then shuts down
- A new IP prefix is added or eliminated
- The link cost varies
- The update period for LSPs expires

Psuedo Node LSP

Psuedo Node LSPs refer to LSPs generated by a DIS. IS_IS includes only one DIS that assists ISs to synchronize the link state databases. The DIS generates one Psuedo Node LSP each for L1 and L2. A PN refers to the LAN, and a separate Psuedo Node LSP is generated for each LAN connected to the IS_IS routing domain.

The DIS in the IS_IS network generates and propagates the Psuedo Node LSPs when:

- A new neighbor IS comes alive into the IS_IS network and then shuts down
- The update period for LSPs expires
- A new LAN is connected to the IS_IS routing domain

CSNP

CSNP is an update that includes a whole list of LSPs known to the IS_IS router. CSNPs are generated and flooded by DIS every 10 seconds and explain every LSP in the link state database.

PSNP

PSNP requests missing route information after receiving CSNPs and acknowledges a routing update on point-to-point links.

IS_IS Packet Header

The format of a common IS_IS packet header is shown in Figure 8.16. The IS_IS packet header consists of different fields. The packet header fields and their functions are:

Intra-domain Routing Protocol Discriminator: Identifies the network layer assigned to an IS_IS network. The binary value is 10000011.

Length Indicator: Indicates length of fixed header in octets.

Version/Protocol ID Extension: Indicates current value of 1 for version and identifies the value for protocol extension, if any.

ID Length: Indicates the total length of the System ID. The ID Length should be the same for all nodes in the domain.

PDU Type: Indicates the L1 or L2 LSP.

Version: Set value to 1.

Reserved: Set to 0 if the LSP is on transmission. An IS ignores this field value on reception of the LSP.

Maximum Area Addresses: Specifies maximum number of area addresses allowed for this IS_IS area. The value for this field ranges from 1 to 254.

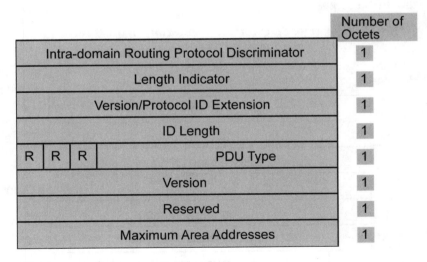

	Number of Octets
Intra-domain Routing Protocol Discriminator	1
Length Indicator	1
Version/Protocol ID Extension	1
ID Length	1
R R R PDU Type	1
Version	1
Reserved	1
Maximum Area Addresses	1

FIGURE 8.16 A common IS_IS packet header.

IS_IS Link State Database General Packet Format

The IS_IS link state database general packet format is shown in Figure 8.17.

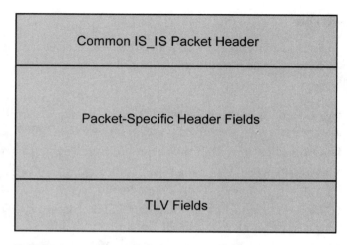

FIGURE 8.17 The IS_IS link state database general packet format.

The IS_IS link state database general packet format contains the common IS_IS packet header, packet-specific header fields, and the Type Length Value (TLV) field. This TLV field contains routing-related information such as type, length, and value.

Link State PDU Format

The IS_IS link state PDU format is shown in Figure 8.18.

				Octets
Common IS_IS Packet Header				8
PDU Length				2
Remaining Lifetime				2
LSP ID				ID Length + 2
Sequence Number				4
Checksum				2
P	ATT	LSPDBOL	IS Type	1
TYPE LENGTH VALUE FIELDS				Variable

FIGURE 8.18 The link state PDU packet format.

The link state PDU packet format contains:

Common IS_IS Packet Header: Refers to the common IS_IS packet.

PDU Length: Refers to the total length of the IS_IS PDU in octets. This length includes the length of the common header and TLVs.

Remaining Lifetime: Indicates the time in seconds before the LSP is purged. If the value of this field is 0, the IS waits for another 60 seconds for an updated LSP. If after 60 seconds an updated LSP is not received from the IS, the LSP is purged.

LSP ID: Discussed later in this section.

Sequence Number: Contains the sequence number of LSP sent from a particular IS.

Checksum: Contains a numeric value that checks for the accuracy of the whole IS_IS LSP.

Partition (P): Provides information about the partition repairing capability of the originator of the LSP.

Attached (ATT): Indicates that the LSP originator is attached to another area if this bit is set to "high." ATT represents bits 4 to 7 of the octet.

LSPDBOL: Indicates that the originator's link state database is overloaded and should be avoided or alternate routing calculation methods should be used for reducing the load of the LSP originator. This is indicated if the bit is set to "high." Bit 3 of the octet represents LSPDBOL.

IS Type: Represented by bit 1 and bit 2 of the octet. If only bit 1 is set, it is an L1 router. If both bits are set, it is an L2 router.

Type Length Value (TLV) Fields: Contains routing-related information such as type, length, and value.

The LSPID format is shown in Figure 8.19.
The LSPID format includes:

System ID: Refers to the identifier of the IS in Non-Psuedo Node or the DIS in PN in the IS_IS network. In Figure 8.19, 00b0.0090.abcd is the system ID.

00b0.0090.abcd.03-01

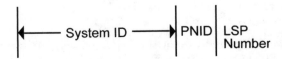

FIGURE 8.19 The link state packet identifier format.

PN ID: Remains zero for LSPs generated by an IS in Non-Psuedo Node and will be non-zero for LSPs generated by DIS in PN. In Figure 8.19, the PN ID is 03, which is non-zero. As a result, it indicates that this LSP is generated by a DIS.

LSP Number: Represents the fragmentation number for each LSP. In Figure 8.19, the LSP number is 01.

COMPARING IS_IS AND OSPF PROTOCOLS

IS_IS and OSPF routing protocols differ from each other. The comparison is shown in Table 8.3.

TABLE 8.3 Comparing IS_IS and OSPF Protocols

IS_IS	OSPF
Area borders are on links. The IS_IS network is divided into routing domains and the area boundary links are specified as exterior links. No IS_IS routing messages are passed to an exterior link.	Area borders are easily distinguished due to the presence of ABRs.
Areas are divided into Level 1 and Level 2.	Areas are divided into single area and multiple areas.
Supports both IP and OSI network traffic.	Supports only IP traffic.
Supports routing for an IP-based and OSI-based interconnected network.	Supports routing only for IP-based networks.
Backbone routers can reside in any area. The only requirement for the backbone to function is an unbroken chain of L2 and L1-L2 routers.	Backbone area is mandatory and it is connected to each of the OSPF areas through an ABR or a virtual link.
IS_IS LSPs contain a checksum field. If a router receives an LSP with an invalid checksum, the IS deletes the LSP and re-floods the LSP so that it can be resent.	Only the originating router has the right to delete the LSP.
IS_IS has no Backup Designated Router (BDR) because the IS_IS borders are on links.	OSPF elects a BDR in case the DR fails.

DESIGN CONSIDERATIONS

The parameters to be considered while designing ISP networks using IS_IS protocol for routing are discussed in this section.

Sub-Optimal Level 1 Routing

When a packet is forwarded outside the L1 area, by default the L1 router forwards the packet to the nearest L1-L2 router based on the Attach bit set in the L1 LSP. In Figure 8.20, Routers A, B, C, D, E, and F form an IS_IS network. Router A points to C because it is nearer. However, the nearest L1-L2 router may not be the most optimal path. This process of the router pointing to the nearest L1-L2 router for forwarding an L1 LSP is known as the sub-optimal routing process.

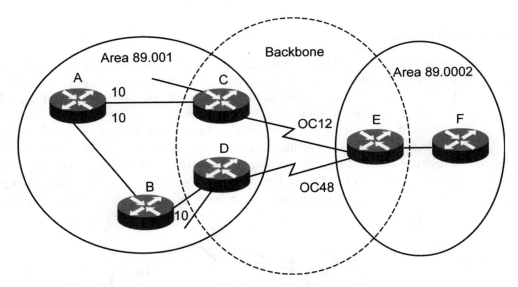

FIGURE 8.20   The initial IS_IS routing configuration in the ISP.

Figure 8.21 shows the solution for reducing the sub-optimal routing. All the routers are placed within a single IS_IS L1 area or L2 area. As a result, each router will be performing only L1 or L2 routing, eliminating sub-optimal routing. In Figure 8.21, A, B, C, D, E, and F are within the same IS_IS area with Area ID 89.001. There is no need for sub-optimal routing in this scenario.

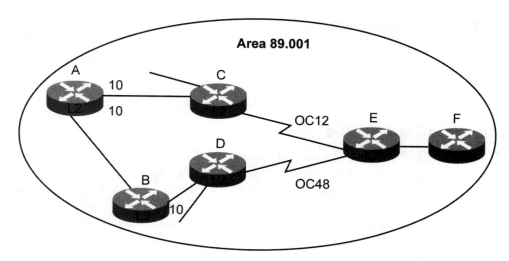

FIGURE 8.21: The optimized IS_IS routing configuration.

Mesh Configuration

Consider the example depicted in Figure 8.22. Routers M, O, Q, N, and P are in mesh configuration of the IS_IS network. This network uses minimal flooding of LSPs between different routers in the network. The routing traffic overhead is kept at a minimum level in such a mesh configuration.

The IS_IS protocol can execute in broadcast networks, using a mesh design, as shown in Figure 8.22. If one interface does not work, then IS_IS fails in the mesh network. It is best to use point-to-point sub-interfaces where IS_IS reliability is more crucial in cases such as Frame-Relay broadcast networks. However, the mesh design is advantageous in providing minimal routing overhead traffic between the routers.

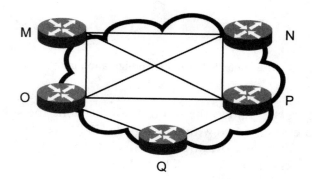

FIGURE 8.22 A mesh router IS_IS configuration.

IS_IS CONFIGURATION COMMANDS

To configure an IS_IS network:

1. Find the IS_IS Area in which routing is performed.
2. Identify where the IS_IS router will be located within the area, and identify the interfaces upon which IS_IS is to be enabled.
3. Enable IS_IS on the router with the Router# router isis command. Configure the NETs using the Router# net command.
4. Enable IS_IS on interfaces using the Router# ip router isis command.

Router# router isis tag Command

This command enables or disables the IS_IS protocol on the IS. The tag is an optional parameter that specifies a string identifier for router identification. For example, Router(config)# router isis myrouter enables the IS_IS routing protocol in the router and specifies the tag as myrouter for router identification.

Router# net netw-ent-title Command

This command configures a NET for the routing process. This command takes one parameter netw-ent-title as input and specifies the IS_IS area address and the System ID of the IS_IS router. All routers within the same IS_IS area use the same length NET.

Router# max-areas number Command

This command sets the maximum number of areas for the IS_IS routing process. It takes an integer value between 1 and 255 as a parameter. The default value for max-areas in IS_IS is 3. For example,

```
Router# max-areas 2
Router# net 43.001.0000.0000.0001.00
Router# net 44.001.0000.0000.0001.00
Router# show isis
```

In this example, the maximum number of IS_IS areas is set to 2. The two different areas are configured using the net command. The show isis command will show the details of all IS_IS areas configured, as shown in Listing 8.3.

LISTING 8.3   Output of show isis Command

```
Router# show isis
Global ISIS information
```

```
ISIS process tag: Router1
System ID: 1900.6500.4001
NET: 43.0001.0000.0000.0001.00
NET: 44.0001.0000.0000.0001.00
Maximum number of areas: 2
```

Router# isis hello-interval seconds {level-1 | level-2} Command

This command configures the time interval between two IIHs that the ISs send to establish adjacencies. This command takes the time interval between the hello packets in seconds as a parameter. For example,

```
Router# isis hello-interval 10 Level 1
Router# isis hello-interval 5 Level 2
```

In the example, the IS_IS Level 1 hello-interval goes up to 10 seconds, and Level 2 hello-interval goes up to 5 seconds.

Router# isis hello-multiplier number Command

This command configures the hello-multiplier for the current interface. The command accepts one numeric value between 3 and 1000. This value decides how long to hold onto an IIH before deciding an adjacency has failed. The default multiplier value is 10 times the IIH hello interval. For example,

```
Router#isis hello-interval 2
Router#ihello-mutiplier 8
```

In the example, the hello-interval is configured as 2 seconds and the hello-multiplier as 8 seconds.

Router# isis hello-padding {all | first-packet-only} Command

This command enables or disables the hello padding for IS_IS IIHs. This feature is disabled for IS_IS, by default. It takes one optional parameter to specify whether all the IS_IS packets need to be padded or only the first packet needs to be padded. Padding is done for all IS_IS IIHs in the Router# isis hello-padding all command. Padding is done only for the first IIH in the Router# isis hello-padding first packet-only command.

Router# isis retransmit-interval seconds Command

This command is used in a point-to-point network to set the time interval between successive retransmissions of IS_IS link state PDUs. The command takes parameter seconds as input, which give the transmit-interval time period. The parameter seconds can take any value between 0 and 65535. The default transmit-interval time for IS_IS is 5 seconds. For example, the command Router#isis retransmit-interval 7 sets the retransmit-interval to 7 seconds.

Router# isis circuit-type {level-1 | level-1-2 | level-2-only} Command

This command configures the adjacency type of an interface Level 1, Level 1-2, or Level 2. The default setting for circuit-type is Level 1-2. The command Router# isis circuit-type level-1 sets the circuit-type as Level 1. The command Router# isis circuit-type level-1-2 sets the circuit-type as Level 1-2.

Router# isis csnp-interval seconds {level-1 | level-2} Command

This command configures the IS_IS CSNP interval. It takes the CSNP time interval in seconds as the parameter. The CSNP intervals for Level 1 and Level 2 can be configured separately by optionally specifying Level 1 or Level 2 in the command. For example,

```
Router# isis csnp-interval 50 Level 1
Router# isis csnp-interval 30 Level 2
```

CSNP intervals are specified for Level 1 and Level 2 routing separately as 50 seconds and 30 seconds, respectively.

Router# isis max-broadcast-pkts number Command

This command configures the maximum number of LSPs that can be transmitted over the interface at any point of time. Valid values for the parameter number are between 0 and 65535. The default value for the maximum packets that can be broadcast over an interface is 10. For example, the command Router# isis max-broadcast-pkts 15 sets the maximum number of LSPs that can be transmitted over an interface as 15.

Router# isis metric number {level-1 | level-2} Command

Metric is the relative cost for transmission of information over the IS_IS interface. The default metric for an interface in an IS_IS network is 10. This command configures the metric for a specific interface. The levels can be specified if the metric differs for Level 1 and Level 2 information transmission. The parameter number can take any value from 0 to 63. For example, the command Router# isis metric 40 level-1 sets the metric as 40 for Level 1 routing. The command Router# isis metric 10 level-2 sets the metric as 10 for Level 2 routing.

Router# isis min-broadcast-interval msecs Command

This command configures the minimum time in milliseconds that an LSP is to be kept before it is transmitted over an interface. It takes one parameter in msecs, which can range from 0 to 65535. The default minimum broadcast interval for an LSP over an interface is 250 milliseconds. For example, Router# isis min-broadcast-interval 300 sets the minimum broadcast interval for an LSP to 300 milliseconds.

Router# isis network-type {broadcast | point-to-point} Command

This command sets the type of the interface in an IS_IS routing environment as broadcast or point-to-point. The command Router# isis network-type broadcast sets the interface as broadcast interface. The command Router# isis network-type point-to-point sets the interface as point-to-point interface.

Router# isis password passwd {level-1 | level-2} Command

This command is used for authentication for usage of a specific interface. Two levels of password can be set for the same interface. By default, this authentication mechanism for interfaces is disabled in an IS_IS network. The command Router# isis password mypwd level-1 sets the Level 1 password for the current interface as mypwd.

Router# isis priority value {level-1 | level-2} Command

This command is used to set the value for setting the priority. The values can be from 0 to 255. The default priority of a router in an IS_IS network is 64. The IS_IS priority value is used to elect a DR in a broadcast network. For example, the command Router# isis priority 200 level-1 sets the priority of the router as 200 for Level 1 routing.

Router# isis psnp-interval Command

This command configures the IS_IS PSNP interval. It takes the PSNP time interval in seconds as the parameter. The PSNP intervals for Level 1 and Level 2 are configured separately by optionally specifying Level 1 or Level 2 in the command. The default PSNP interval for IS_IS is 2 seconds. For example,

```
Router# isis psnp-interval 10 Level 1
Router# isis psnp-interval 40 Level 2
```

The PSNP intervals are specified for Level 1 and Level 2 routing separately as 10 seconds and 40 seconds, respectively.

Router# isis add area area-id Command

This command is used to add an area to the IS_IS network. For example, the command Router#isis add 49.0003 adds the area with Area ID 49.0003 to the IS_IS network.

Router# isis add interface interface-id Command

This command adds a new interface to the IS_IS routing area. For example, the command Router# isis add interface interface2 adds an interface named interface2 to the IS_IS network.

Router# isis set system-id sysid Command

This command is used to configure the system ID of the IS. For example, the command can be written as:

```
Router# isis set system-id 1900.6502.8004 sets the System ID of the
router to 1900.6500.8004.
```

Router# isis isp-interval msecs Command

This command configures the time interval between two successive LSP transmissions by an IS. It takes only one parameter, the time interval between two successive LSPs in milliseconds. For example, the command Router# isis lsp-interval 200 sets the time interval between two successive LSP transmissions as 200 milliseconds.

IS_IS VERIFICATION COMMANDS

Consider the IS_IS configuration shown in Figure 8.23. Router 1 with System ID 190.65.4.1 is connected to Router 2 with System ID 190.65.14.2 through Ethernet interface e1. Router 2 is connected to Router 3 with System ID 190.65.28.4 through serial interface s1.

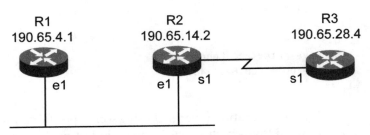

FIGURE 8.23 An network configured with IS_IS routing protocol.

The configuration for Router 1 is shown in Listing 8.4.

LISTING 8.4   Configuration for Router 1

```
Version 11.2
!
hostname R1
!
interface loopback0
  ip address 190.65.4.1   255.255.255.255
!
interface Ethernet1
  ip address 190.65.24.1 255.255.255.0
  ip router isis
!
router isis
  passive-interface Loopback0
  net 49.001.1900.6500.4001.00
!
```

The configuration for Router 2 is shown in Listing 8.5.

LISTING 8.5   Configuration for Router 2

```
Version 11.2
!
hostname R2
!
interface Loopback0
  ip address 190.65.14.2 255.255.255.255
!
interface Ethernet1
  ip address 190.65.24.2 255.255.255.0
  ip router isis
!
interface Serial1
  ip address 190.65.28.4 255.255.255.253
  ip router isis
!
router isis
  passive-interface Loopback0
  net 49.001.1900.6501.4002.00
!
```

The IS_IS configuration is verified using some specific commands. The output for the command Router# show isis is shown in Listing 8.6.

LISTING 8.6   Output for Router# show isis Command

```
Router# show isis
Global ISIS information
ISIS process tag: Router1
System ID: 1900.6500.4001
NET: 47.001.1900.6500.4001.00
Maximum number of areas: 3
There is 1 manual area address
47
There is 1 active area address
47
ISIS level-1
ISIS is enabled on 4 interfaces
Distance : 250
Maximum of 1 path per route
Number of SPF runs, L1: 14, L2: 5
```

Listing 8.7 shows the output for the Router# show clns neighbor command.

LISTING 8.7   Output for Router# show clns neighbor Command

```
Router# show clns neighbor
      System Id Interface    SNPA      State Holdtime Type    Protocol
        R2      Et1    0000.0b54.b778   Up 24        L1L2     IS_IS
```

Listing 8.8 shows the output for the Router# show clns interface Ethernet1 command.

LISTING 8.8   Output for the Router# show clns interface Ethernet1 Command

```
Router# show clns interface Ethernet1
Ethernet 1 is up, line protocol is up
Checksums enabled, MTU 1497, Encapsulation SAP
Routing Protocol : IS_IS
Circuit Type: Level-1-2
Interface number: 0x1, local circuit ID:0x1
Level-1 Metric: 10, Priority: 64, circuit ID:R2.01
Number of active level-1 adjacencies: 1
Level-2 Metric: 10, Priority: 64, Circuit ID: R2.01
Number of active level-2 adjacencies:1
Next IS_IS LAN Level-1 Hello in 8 seconds
Next IS_IS LAN Level-2 Hello in 2 seconds
```

The Router# show clns interface Ethernet1 command provides the interface details. Listing 8.9 shows the output for the Router# show isis database command.

LISTING 8.9   Output for Router# show isis database Command

```
Router# show isis database
IS_IS Level-1 Link State Database
   LSPID        LSP Seq Num     LSP Checksum    LSP Holdtime    ATT/P/OL
   R1.00-00     0x0000008C      0x3243          67              0/0/0
   R2.00-00 *   0x00000089      0x524A          27              0/0/0
   R2.01-00 *   0x0000006A      0x824C          07              0/0/0
   R3.00-00     0x00000091      0x1223          12              0/0/0

   IS_IS Level-2 Link State Database

   LSPID     LSP Seq Num    LSP Checksum      LSP Holdtime    ATT/P/OL
   R1.00-00     0x00000010     0x323C         17              0/0/0
   R2.00-00 *   0x00000019     0x222A         125             0/0/0
   R2.01-00 *   0x0000003A     0x221B         07              0/0/0
   R3.00-00     0x00000071     0x822A         132             0/0/0
```

The Router# show isis database command shows the contents of the IS_IS database. These contents vary dynamically as the network configuration changes. Listing 8.10 shows the output for the Router1# show ip route isis command.

LISTING 8.10   Output for Router1# show ip route isis Command

```
Router1# show ip route isis

i L1    190.65.14.2 /32    [125/10] via 190.65.14.2 Ethernet1
i L1    190.65.24.1 /32    [125/10] via 190.65.24.2 Ethernet1
```

The Router1# show ip route isis command shows the route followed by an LSP from a particular router. This can be verified against the IP addresses and the cost of the interfaces that are available in the IS_IS network.

TROUBLESHOOTING COMMANDS IN IS_IS

This section discusses the commands used for troubleshooting in the IS_IS routing domain.

A sample output for the Router# show isis command is shown in Listing 8.11. This command displays the global IS_IS configuration information. The command is used to find out whether the IS_IS routing protocol is enabled on a specified router or not. The system-level information about the IS_IS execution is also displayed using this command.

LISTING 8.11   Output for the Router# show isis Command

```
Router#show isis
Global ISIS information
ISIS process tag: Router1
System ID: 1900.6500.4001
NET: 47.001.1900.6500.4001.00
Maximum number of areas: 3
There is 1 manual area address
47
There is 1 active area address
47
ISIS level-1
ISIS is enabled on 4 interfaces
Distance: 250
Maximum of 1 path per route
Number of SPF runs, L1: 14, L2: 5
```

The Router# show isis database {level-1 | | 1 | level-2 | | 2 |detail | lspid } command is used to display the current IS_IS link state database. The various options can be specified as:

```
Router# show isis database level-1
Router# show isis database l1
Router# show isis database level-2
Router# show isis database l2
```

The command router# show isis database detail displays each LSP content in detail. Otherwise, only a summary of LSP contents is displayed.

To display the LSP content, the command Router# show isis database lspid is used, where lpsid is the identifier of the LSP that is to be displayed.

The command Router# show ip route isis shows the route followed by an LSP through the IS_IS network.

The command Router# show isis spf-log displays the time and reason for the shortest path algorithm execution.

The command Router# show clns protocol provides the system ID, IS Type, Area ID, and interface information along with their metrics.

The command Router# show clns neighbor displays the System ID, Interface ID, SNPA, Holdtime, IS Type, and protocol used. The output for this command is shown in Listing 8.12.

LISTING 8.12   Output for Router# show clns neighbor Command

```
Router# show clns neighbor
System ID Interface SNPA              State

R1                 E1              0000.0b19.331a  UP
Holdtime           Type            Protocol
20                 L1              IS_IS
```

The command Router# show isis topology displays all the routers within an area for an L1 router. For an L2 router, this command shows both local routers as well as routers outside the L1 area.

SUMMARY

In this chapter, you looked at the operation and functioning of the IS_IS routing protocol. The chapter also discussed the various configuration commands of IS_IS. In the next chapter, you will learn about BGP routing protocol.

POINTS TO REMEMBER

- IS_IS protocol is an IGP that supports true IP environments and true OSI environments, as well as dual environments.
- IS_IS uses CLNS addresses that include Area ID, System ID, and NSEL field.
- In IS_IS, each IS establishes adjacencies by sending IIH packets; each IS also generates an LSP.
- IS_IS routing domain is subdivided into Level 1 and Level 2 areas.
- All ISs within the L1 perform routing between adjacent ISs within the same area.
- The L2 area connects two different L1 areas within the same IS_IS routing domain.
- The two levels of routing hierarchy for IS_IS protocol are L1 and L2 routing.
- The three types of ISs in IS_IS are L1, L2, and L1-L2 routers.
- Each IS in IS_IS areas has a unique address within the area known as NET.
- IS_IS has two types of functions, subnetwork independent and subnetwork dependent functions.

- The two major design considerations in an IS_IS network are sub-optimal L1 routing and mesh configuration.
- Router# router isis tag command enables or disables the IS_IS protocol.
- Router# net netw-ent-title command configures a NET.
- Router# max-areas number command sets the maximum number of areas for IS_IS.
- Router# isis hello-interval seconds {level 1 | level 2} command configures the time interval between two IIHs that the ISs send to establish adjacencies.
- Router# isis hello-multiplier number command configures the hello-multiplier for the current interface.
- Router# isis hello-padding {all | first-packet-only} command enables or disables the hello padding for IS_IS IIHs.
- Router# isis retransmit-interval seconds command is used in a point-to-point network to set the time interval between successive retransmissions of IS_IS link state PDUs.
- Router# isis circuit-type {level-1 | level-1-2 | level-2-only} command configures the adjacency type of an interface Level 1, Level 1-2 or Level 2 only.
- Router# isis csnp-interval seconds {Level 1 | Level 2} command configures the IS_IS CSNP interval.
- Router# isis max-broadcast-pkts number command configures the maximum number of LSPs that can be transmitted over the interface.
- Router# isis metric number {level-1 | level-2} command configures the metric for a specific interface.
- Router# isis min-broadcast-interval msecs command.
- Router# isis network-type {broadcast | point-to-point} command.
- Router # isis password passwd {level-1 | level-2} command.
- Router# isis priority value {level-1 | level-2} command sets the value for setting the priority.
- Router# isis lsp-interval msecs command configures the time interval between two successive LSP transmissions.
- Router# show isis command displays the global IS_IS configuration information.
- Router# show isis database {level-1 | | 1 | level-2 | | 2 |detail | lspid } command is used to display the current IS_IS link state database.
- Router# show isis database detail command displays each LSP content.
- Router# show isis spf-log displays the time and reason for the shortest path algorithm execution.
- Router# show ip route isis shows the route followed by an LSP through the IS_IS network.

9 Border Gateway Protocol

INTRODUCTION TO BGP

Border Gateway Protocol (BGP) enables the exchange of routing information between different Autonomous Systems (ASs) across networks. This routing protocol provides a loop-free interdomain for ASs to communicate with each other. BGP is used to connect networks of different Internet Service Providers (ISPs).

BGP OPERATION

The routing information in BGP setup is a list of Autonomous System Numbers (ASNs) to reach a particular destination IP network. This list of ASNs is known as the AS path. BGP routers exchange information by setting up a TCP connection at port 179. When BGP is configured between two routers and the TCP connection is set up, they are known as BGP peers or BGP neighbors.

BGP devices exchange the entire routing table during initial data exchange. Routing updates are exchanged only when there is some change in routing information. Other types of messages, such as periodic keep-alive messages, and notifications, are also exchanged between BGP peers. Figure 9.1 shows the basic BGP configuration in routers.

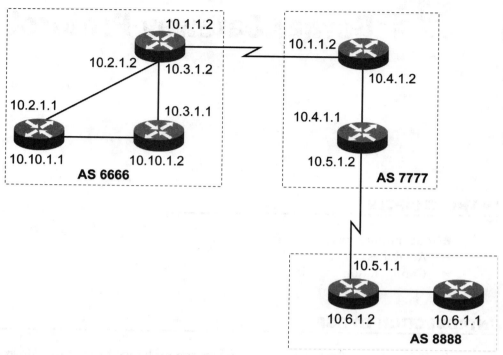

FIGURE 9.1 Routers in BGP configuration.

The configuration of Router A1 is:

```
router bgp 6666
neighbor 10.1.1.1 remote-as 7777
neighbor 10.2.1.1 remote-as 6666
neighbor 10.3.1.1 remote-as 6666
```

The configuration of Router A2 is:

```
router bgp 6666
neighbor 10.2.1.2 remote-as 6666
neighbor 10.10.1.2 remote-as 6666
```

The configuration of Router A3 is:

```
router bgp 6666
neighbor 10.3.1.2 remote-as 6666
neighbor 10.10.1.1 remote-as 6666
```

The configuration of Router B1 is:

```
router bgp 7777
neighbor 10.1.1.2 remote-as 6666
neighbor 10.4.1.1 remote-as 7777
```

The configuration of Router B2 is:

```
router bgp 7777
neighbor 10.4.1.2 remote-as 7777
neighbor 10.5.1.1 remote-as 8888
```

The configuration of Router C1 is:

```
router bgp 8888
neighbor 10.5.1.2 remote-as 7777
neighbor 10.6.1.1 remote-as 8888
```

The configuration of Router C2 is:

```
router bgp 8888
neighbor 10.6.1.2 remote-as 8888
```

The commands that are configured on the routers ensure basic BGP settings. The functions of the commands are:

router bgp: Enables a router to communicate with other routers in the BGP setup. This command also introduces a router to the ASN corresponding to the ISP or the corporate entity.

neighbor: Informs the router about the peering router running BGP. This address is the destination IP address with which a TCP connection at port 179 needs to be set up.

CAUTION

Ensure IP level reachability between any two sets of potential BGP peers so that a BGP session is established. BGP peers need not even be directly connected.

BGP operates in two main modes, internal and external BGP.

Internal BGP

When BGP is used to exchange routing updates between routers belonging to the same AS, it is known as Internal Border Gateway Protocol (iBGP). All routers running BGP with external ASs need to run iBGP with each other to form a full mesh. This is required because a BGP router does not forward routes learned from one iBGP peer to another. Figure 9.2 shows iBGP routers.

In Figure 9.2, there are four ASs with ASNs, 6666, 7777, 8888, and 9999. AS 7777 acts as the transit AS for traffic between AS 6666-8888, 6666-9999, and 8888-9999. In AS 7777, there are three BGP routers, B1, B2, and B3, among which only B1-B2 and B1-B3 communicate with each other.

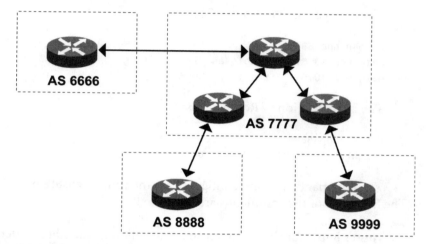

FIGURE 9.2   Working of iBGP.

For transit traffic between AS 6666-8888, B1 forwards routes via external BGP to B2. Router B2 advertises these routes to its external BGP peer, C1. Similarly, in the reverse direction, routes traverse from C1 to A1.

In case of transit traffic between AS 6666-9999, B1 forwards routes via external BGP to B3. Router B3 advertises these routes to its external BGP peer, D1. Similarly, in the reverse direction, routes traverse from D1 to A1.

However, consider the case of transit traffic from AS 8888-9999. B2 learns routing information from C1 and forwards it to B1. This information is not forwarded to B3 because it was learned by B2 from an iBGP peer. B3 is another iBGP peer and is not aware of the routes from AS 8888. It cannot forward the routing information to D1.

Similarly, C1 and D1 have no knowledge of each other's routes and are unable to exchange traffic using AS 7777 as the transit AS. This problem is overcome if there is a full iBGP mesh in AS 7777. A full mesh in AS 7777 can be achieved by running iBGP between Routers B2 and B3.

You can also use an IGP within routers in an AS and redistribute routes to and from BGP. However, iBGP is more customizable and ensures that all external neighbors get a consistent view of the AS.

TIP

External BGP

When BGP is used between routers belonging to different ASs, it is termed as External Border Gateway Protocol (eBGP). The eBGP neighbors are directly connected by WANs. In Figure 9.3, eBGP is configured between A1 in AS 6666 and B1 in AS 7777. In Figure 9.3, A1 and B1 are connected via two equal bandwidth WAN links with WAN IP addresses, 10.1.1.0/30 and 10.1.1.4/30.

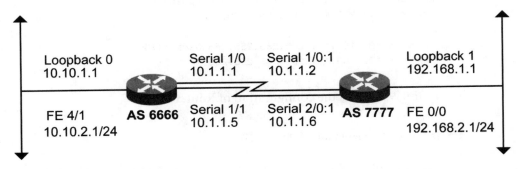

FIGURE 9.3 Working of eBGP.

There are two options to configure eBGP sessions between the directly connected WAN IP addresses. However, load balancing of traffic between AS 6666 and AS 7777 will not be ensured in this process.

The ebgp-multihop command ensures load balancing. This command indicates that the configured neighbors are not directly connected. It also indicates that there may be more than a single WAN link between the two ASs. The loopback interfaces can be used for configuring BGP in such cases.

Loopback interface is not physical; it is always accessible until any one interface of the router is alive and functional. The use of loopback interface ensures maximum uptime and stability of the BGP session. Loopback interfaces are used with both iBGP and eBGP. In the case of eBGP, it also allows load balancing among equal bandwidth links. Listing 9.1 shows the configuration of Router A1.

LISTING 9.1 Configuration of Router A1

```
interface Loopback0
ip address 10.10.1.1 255.255.255.255
```

```
!
interface FastEthernet4/1
ip address 10.2.2.1 255.255.255.0
!
interface Serial1/0
ip address 10.1.1.1 255.255.255.252
!
interface Serial1/1
ip address 10.1.1.5 255.255.255.252
!
router bgp 6666
neighbor 192.168.1.1 remote-as 7777
neighbor 192.168.1.1 ebgp-multihop
neighbor 192.168.1.1 update-source Loopback 0
network 10.2.2.0 mask 255.255.255.0
!
ip route 192.168.1.1 255.255.255.255 Serial1/0
ip route 192.168.1.1 255.255.255.255 Serial1/1
```

Listing 9.2 shows the configuration of Router B1.

LISTING 9.2 Configuration of Router B1

```
interface Loopback1
ip address 192.168.1.1 255.255.255.255
!
interface FastEthernet0/0
ip address 192.168.2.1 255.255.255.255
!
interface Serial1/0:1
ip address 10.1.1.2 255.255.255.252
!
interface Serial2/0:1
ip address 10.1.1.6 255.255.255.252
!
router bgp 7777
neighbor 10.10.1.1 remote-as 6666
neighbor 10.10.1.1 ebgp-multihop
neighbor 10.10.1.1 update-source Loopback 1
network 192.168.2.0
!
ip route 10.10.1.1 255.255.255.255 Serial1/0:1
ip route 10.10.1.1 255.255.255.255 Serial2/0:1
```

eBGP neighbors need not be directly connected. You may need an extra configuration command to indicate this. However, this will not cause any problems in setting up the BGP session and exchange of BGP routing updates.

Advertising the Networks

In a BGP session between two peering routers, you need to ensure that the routes, which are being advertised to the session's peer, reach the external BGP peers effectively. This process is known as advertising of routes. There are two ways in which an AS can advertise its networks:

- Use network command wherein the networks to be advertised are explicitly specified.
- Redistribute routes present in the routing table. These routes may be either static routes or dynamic routes, learned by any other routing protocol.

Advertisement methods are required for routes not learned by BGP. All routes learned by BGP routers are installed in the BGP table and are automatically passed on to BGP peers.

Network Command

The network command is used to mention routes, which are to be injected to the BGP from the routing table. This command will only work for those routes that are present in the router's routing table, whether it is a directly connected route, a static route, or a route known to the router via a dynamic routing protocol. Figure 9.4 shows the use of the network command to advertise networks.

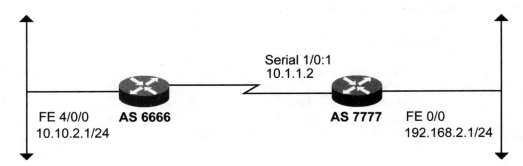

FIGURE 9.4   Advertising networks using the network command.

The configuration of Router A1 is:

```
router bgp 6666
neighbor 10.1.1.2 remote-as 7777
network 10.2.2.0 mask 255.255.255.0
```

The configuration of Router B1 is:

```
router bgp 7777
neighbor 10.1.1.1 remote-as 6666
network 192.168.2.0 mask 255.255.255.0
```

In Figure 9.4, in A1, 10.2.2.0/24 is a directly connected network that is being injected in the BGP table for advertisement using the network command. Similarly in B1, 192.168.2.0/24 is a directly connected network that is being advertised using the network command.

Listing 9.3 shows the routing table for A1.

LISTING 9.3   Routing Table for A1

```
RouterA1# show ip bgp
  BGP table version is 805985, local router ID is 10.10.2.1
  Status codes: s suppressed, d damped, h history, * valid, > best, i -
     internal
  Origin codes: i - IGP, e - EGP, ? - incomplete
  Network      Next Hop    Metric LocPrf Weight    Path
  *> 10.10.2.0/24   0.0.0.0     0 32768 i
```

Listing 9.4 shows the routing table for B1.

LISTING 9.4   Routing Table for B1

```
RouterB1# show ip bgp
  BGP table version is 805985, local router ID is 192.168.2.1
  Status codes: s suppressed, d damped, h history, * valid, > best, i -
     internal
  Origin codes: i - IGP, e - EGP, ? - incomplete
  Network      Next Hop    Metric LocPrf Weight    Path
  *> 192.168.2.0/24   0.0.0.0     0 32768 i
```

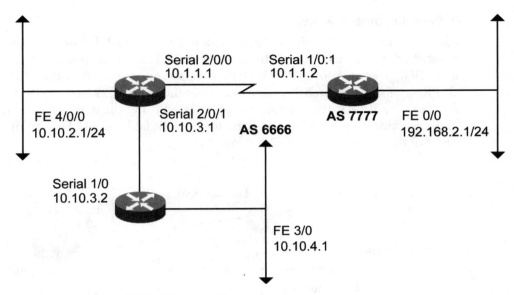

FIGURE 9.5 Redistribution of static routes.

Redistributing Static Routes

Redistribution of static routes is another method of injecting networks into the routing BGP table. Figure 9.5 shows the redistribution of static routes. In Figure 9.5, three routers, A1, A2, and B1, are connected. A1 and A2 belong to AS 6666, while B1 belongs to AS 7777. eBGP is running between A1 and B1. A1 is directly connected to A2, and the static route entry has been made for the network 10.10.4.0. The redistribute static command is used to redistribute the network 10.10.4.0, whereas the redistribute connected command is used to redistribute the directly connected networks 10.10.3.0/30 and 10.10.2.0/24. The configuration of Router A1 is:

```
router bgp 6666
neighbor 10.1.1.2 remote-as 7777
distribute connected
distribute static metric 2
!
ip route 10.10.4.0 255.255.255.0 10.10.3.2
!
```

Redistributing Dynamic Routes

IGP routes can also be redistributed to BGP to be advertised. Figure 9.6 shows the redistribution of dynamic routes from an IGP to BGP. In Figure 9.6, A1 of AS 7777 is the BGP peer of B1 of AS 8888. OSPF is being run as the IGP within AS 7777. OSPF is redistributed in BGP. A1 learns about the network 192.168.10.0/24 from A2 via OSPF. In A1, the network is not declared using the network command. However, this network is still injected to the BGP table of A1. This occurs because of the redistribution of OSPF into BGP.

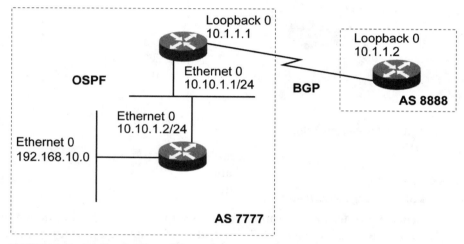

FIGURE 9.6   Redistribution of dynamic routes.

The configuration of Router A1 is:

```
router bgp 7777
neighbor 10.1.1.2 remote-as 8888
redistribute ospf 1
!
router ospf 1
network 10.1.1.0 0.0.0.255 area 0
network 10.10.1.0 0.0.0.255 area 0
!
```

The configuration of Router B1 is:

```
router bgp 8888
neighbor 10.1.1.1 remote-as 7777
```

IGP routes tend to be redistributed back to BGP, when you use the method of redistribution of dynamic routes. As a result, the network command and redistribution of static routes are the preferred methods of advertising networks.

Careful use of access lists ensures the safe redistribution of IGP routes to BGP routes.

TIP

It is an industry practice to not redistribute IGP routes to BGP, because the global routing tables modify every time there is a change in the network.

NOTE

Aggregation

Aggregation combines multiple routes so that a single route is advertised. The more it is possible to aggregate at the edge of a network, the more it reduces the size of routing tables, thereby reducing the processing load on routers.

In Figure 9.7, BGP runs between the ASs 6666-9999, 7777-9999, and 8888-9999. B2 receives updates about networks 192.168.96.0/23 from AS 6666, 192.168.119.0/24 from AS 7777, and 192.168.112.0/23 from AS 8888. The figure shows the networks being aggregated to 192.168.96.0/19 at B1.

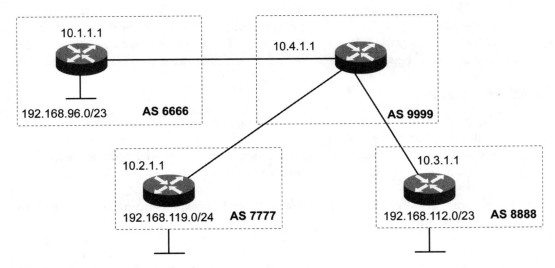

FIGURE 9.7   Route aggregation in BGP.

Configuration of Router A1 is:

```
router bgp 6666
neighbor  10.4.1.1 remote-as 9999
network 192.168.96.0 mask 255.255.254.0
```

Configuration of Router B1 is:

```
router bgp 7777
neighbor  10.4.1.1 remote-as 9999
network 192.168.119.0
```

Configuration of Router C1 is:

```
router bgp 8888
neighbor  10.4.1.1 remote-as 9999
network 192.168.112.0  mask 255.255.254.0
```

Configuration of Router D1 is:

```
router bgp  9999
neighbor  10.1.1.1 remote-as 6666
neighbor  10.2.1.1 remote-as 7777
neighbor  10.3.1.1 remote-as 8888
aggregate-address 192.168.96.0 255.255.224.0
```

The aggregate address advertises the supernet 192.168.96.0/19 from D1. Each of the ASs 6666, 7777, and 8888 are using smaller networks of prefix lengths /23 or /24, allocated by the ISP, AS 9999, to them. These networks are aggregated at the ISP D1, while being advertised to the Internet, into the supernet 192.168.96.0/19.

Scalability Problems

All BGP routers in an AS need to communicate with each other using iBGP to maintain consistency in the routing tables. This is feasible in small ASs with five or six BGP routers. However, as the network grows, the number of BGP routers also increases. The growth may occur in terms of:

Smaller ISPs: Higher number of upstream providers from whom bandwidth is provisioned.

Mid-sized ISPs: New relationships with higher-level ISPs. Growth will also be in terms of enhancement in number of customers who would like to run BGP with mid-sized ISPs.

Corporates: New bandwidth provisioned from multiple service providers.

Setting up a full iBGP mesh becomes more tedious with the growth of the network. A separate iBGP session needs to be set up in each BGP router corresponding to other BGP routers in the network. There will be a burden on router resources, such as memory and CPU. BGP does not scale well in case of larger networks. Route reflectors and confederations help to overcome scalability problems.

Growth in both ISP and corporates will also come from the addition of new locations in the network.

Route Reflectors

Route reflectors exchange the entire BGP routing table among themselves, to maintain uniformity. They reflect all the routes that the clients learned from the peers. They also reflect the routes that the other route-reflector servers learned from the client.

You need to identify certain primary BGP routers in the network. These routers should have maximum connectivity with external AS. The remaining routers are configured as router-reflector clients. A full iBGP mesh is established only between those routers configured as servers, each of which will have a number of clients attached to it. Figure 9.8 shows the configuration and working of a route-reflector scenario.

In Figure 9.8, there are seven routers, B1, B2, B3, B4, B5, B6, and B7, in AS 7777. B1 and B5 are running eBGP, B1 with AS 6666, and B5 with ASs 9999 and 8888. B1 is configured as a route reflector. B2, B3, and B4 are the route-reflector clients. B5 is the other route reflector in AS 7777 with B6 and B7 as clients. B1 and B5 are iBGP peers and reflect all external routes learned by the clients, who, in turn, reflect all the routes to the route reflector. As a result, all routers in AS 7777 know of all BGP routes and maintain consistency, without a full iBGP mesh.

The configuration of Router B1 is:

```
router bgp 7777
neighbor 10.1.1.1 remote-as 6666
neighbor 192.168.2.1 remote-as 7777
```

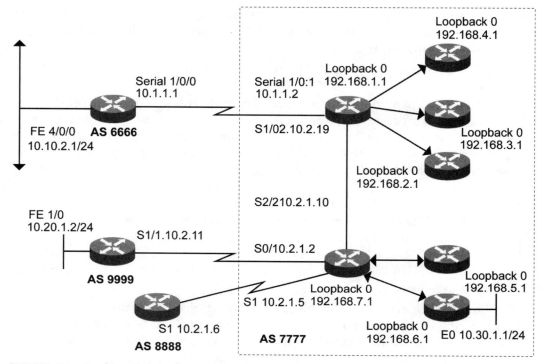

FIGURE 9.8 Working and configuration of a route reflector.

```
neighbor 192.168.2.1 route-reflector client
neighbor 192.168.3.1 remote-as 7777
neighbor 192.168.3.1 route-reflector client
neighbor 192.168.4.1 remote-as 7777
neighbor 192.168.4.1 route-reflector client
neighbor 10.2.1.9 remote-as 7777
```

The configuration of Router B2 is:

```
router bgp 7777
neighbor 192.168.1.1 remote-as 7777
```

The configuration of Router B3 is:

```
router bgp 7777
neighbor 192.168.1.1 remote-as 7777
```

The configuration of Router B4 is:

```
router bgp 7777
neighbor 192.168.1.1 remote-as 7777
```

The configuration of Router B5 is:

```
router bgp 7777
neighbor 10.2.1.1 remote-as 9999
neighbor 10.2.1.6 remote-as 8888
neighbor 192.168.4.1 remote-as 7777
neighbor 192.168.4.1 route-reflector client
neighbor 192.168.5.1 remote-as 7777
neighbor 192.168.6.1 route-reflector client
neighbor 10.2.1.9 remote-as 7777
```

The configuration of Router B6 is:

```
router bgp 7777
neighbor 192.168.7.1 remote-as 7777
```

The configuration of Router B7 is:

```
router bgp 7777
neighbor 192.168.7.1 remote-as 7777
```

Consider the networks 10.10.2.0/24 and 10.30.1.0/24. The route 10.10.2.0/24 is learned by B1 via iBGP and is reflected to B2, B3, and B4. This route is available in the BGP table of all these three routers, being reflected by the route reflector B1. B5 learns the route 10.30.1.0/24 from client B7 and reflects it to all the iBGP and eBGP peers.

Confederations

A confederation is the second method to avoid a full mesh iBGP. In this approach, the AS is subdivided into smaller or mini ASs. A mini AS functions as an AS in itself. The smaller ASs are allotted a private ASN that varies from 64512 to 65535. The ASs need to be configured in full iBGP mesh and have a single connection with each other. Figure 9.9 shows BGP running between AS 6666 and AS 7777. eBGP is running between B1 of AS 7777 and A3 of AS 6666.

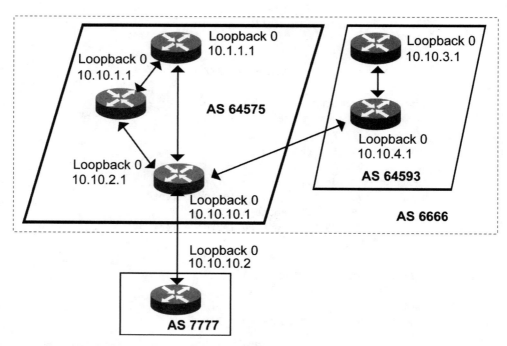

FIGURE 9.9 Working and configuration of a confederation.

In Figure 9.9, AS 6666 is divided into two mini ASs, AS 64575 and AS 64593. A1, A2, and A3 in mini AS 64575 are configured in full iBGP mesh. A4 and A5 in mini AS 64593 are also configured in full iBGP mesh. A single iBGP session is configured between the two mini ASs, which run between A3 and A4.

The configuration of Router A3 is shown in Listing 9.5.

LISTING 9.5 Configuration of Router A3

```
router bgp 64575
bgp confederation identifier 6666
bgp confederation peers 64593
neighbor 10.10.1.1 remote-as 64575
neighbor 10.1.1.1 remote-as 64575
neighbor 10.10.3.1 remote-as 64593
neighbor 10.10.10.2 remote-as 7777
```

The configuration of Router A4 is shown in Listing 9.6.

LISTING 9.6 Configuration of Router A4

```
router bgp 64593
bgp confederation identifier 6666
```

```
bgp confederation peers 64575
neighbor 10.10.10.1 remote-as 64575
neighbor 10.10.4.1 remote-as 64593
```

Route Flap Dampening

The stability of a route is measured in terms of the number of times it has flapped over a given period of time. The higher the number of flaps, the more unstable the route is. A penalty value is assigned to a route whenever it flaps. When the penalty value crosses a threshold value, advertisement of the route is suppressed. BGP uses the method of route flap dampening to take care of this problem. The route is re-advertised after the flaps are reduced and the route is stabilized again. Some of the terms used in route flap dampening are:

Penalty: Attached to a route with each flap.

Half-life: Configured to a value between 1 and 45 minutes. Half-life is the time interval after which the penalty value is reduced by half. The default half-life is 15 minutes.

Suppress Limit: Configured to a value between 1 and 20000, with 2000 being the default. A route is suppressed if the penalty is higher than the suppress limit.

Suppressed Route: Suppression of a route occurs when the penalty is higher than the suppress limit. As a result, the route is not advertised to BGP peers, even if it is reachable.

Reuse Limit: Configured to a value between 1 and 20000, with 750 being the default value. When the penalty of a suppressed route falls below the reuse limit, it is advertised again.

History Entry: Stores information about the recent flaps of a route.

A route that is flapping receives a penalty of 1000 corresponding to each flap. The penalty value increases with each new flapping. When the penalty value reaches the preconfigured suppress limit, the advertisement of the route is suppressed. The route continues to accumulate penalty with each new flap even in this suppressed condition. If, on the other hand, the route shows no flaps once it is suppressed, the penalty value reduces by half with each half-life period. Once the penalty value falls below the reuse limit, the route is brought out of its suppressed state and re-advertised.

Aggregates are always stable as they are always advertised.

NOTE

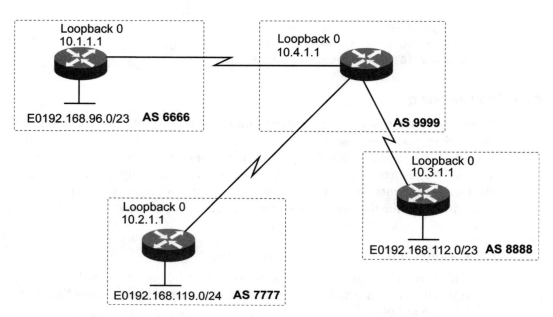

FIGURE 9.10 Route flap dampening in a BGP network.

In Figure 9.10, there are three ASs, AS 6666, AS 7777, and AS 8888. BGP runs between Routers A1-B1 and B1-C1. AS 6666 has a route 192.168.10.0/24, which is unstable. In Figure 9.10, consider a situation where B1 is configured for route flap dampening.

The configuration of A1 is:

```
Router bgp 6666
neighbor 10.2.1.1 remote-as 7777
network 192.168.10.0
```

The configuration of B1 is:

```
Router bgp 7777
bgp dampening
neighbor 10.1.1.1 remote-as 6666
neighbor 10.3.1.1 remote-as 8888
network 172.16.11.0
```

The configuration of C1 is:

```
router bgp 8888
neighbor 10.2.1.1 remote-as 7777
```

Listing 9.7 shows the routing table for B1 when all routes are stable and there is no flapping.

LISTING 9.7   Routing Table for B1

```
RouterA1# show ip bgp
  BGP table version is 75, local router ID is 172.16.11.1
  Status codes: s suppressed, d damped, h history, * valid, > best, i -
    internal
  Origin codes: i - IGP, e - EGP, ? - incomplete
  Network      Next Hop      Metric LocPrf Weight Path
  *> 192.168.10.0/2    10.1.1.1      0        6666i
  *> 172.16.11.0/24    0.0.0.0       0        32768i
```

Route 192.168.10.0/24 is not stable. Consider a flap in this route. The BGP entry of the network in the BGP table of B1 is shown in Listing 9.8.

LISTING 9.8   BGP Entry of the Network in BGP Table of B1

```
RouterA1# show ip bgp 192.168.10.0/24
BGP routing table entry for 192.168.10.0/24, version 76
Paths: (1 available, no best path)
6666(history entry)
from 10.10.1.1(192.168.10.5)
Origin IGP, metric 0, external
Dampinfo: penalty 1000, flapped 1 times in 0:03:23
```

The penalty value is 1000 after the first flapping, but the route is still not suppressed, as the value is lower than the default suppress limit of 2000. The route has flapped once in the last 3 minutes 23 seconds, per the command output shown in Listing 9.8. For each subsequent flap of the route, a penalty of 1000 will be added after every 15 minutes.

Suppose the penalty value increases to 2500 and exceeds the suppress limit after some more flaps. Listing 9.9 shows the BGP table entry corresponding to route 192.168.10.0/24 in B1.

LISTING 9.9 BGP Table Entry in B1

```
RouterB1# show ip bgp 192.168.10.0/24
BGP routing table entry for 192.168.10.0/24, version 82
Paths: (1 available, no best  path)
6666(suppressed due to dampening)
from 10.10.1.1(192.168.10.5)
Origin IGP, metric 0, external
Dampinfo: penalty 2500, flapped 3 times in 15:20:02, reuse in 00:40:30
```

Now the penalty value exceeds the suppress limit. As a result, the route is suppressed in any advertisements. The reuse time is also shown, provided that there are no further flaps.

BGP ATTRIBUTES

BGP uses a large number of parameters to decide the best path to an AS from another AS, while providing a loop-free interdomain for routing. These parameters are known as BGP attributes. The different BGP attributes that are discussed in this section are:

- AS_path Attribute
- Origin Attribute
- Next-Hop Attribute
- Weight Attribute
- Local Preference Attribute
- Multi-exit Discriminator Attribute
- Community Attribute

AS_path Attribute

A packet traverses many ASs to reach a destination AS. The measure of the path is in terms of the ASs the packet traversed on the way. This path, typically a list of ASs separated by a space, is known as an AS path.

When an update is sent out by a BGP speaker to its eBGP peer, it appends its ASN to the AS_path attribute of that route as existing in its BGP table. If it is a path originated at the BGP speaker, that is, a route internal to the AS, the attribute will only be a single AS long. However, in the case of a transit route, the AS_path attribute could be a string of ASs the packet has already traversed. The current ASN will be appended to the attribute while sending the updates.

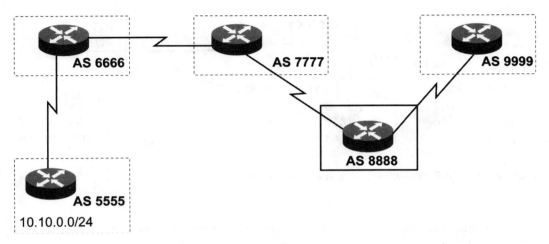

FIGURE 9.11 BGP network with AS_path attribute values.

In Figure 9.11, the updates are traveling through a string of ASs starting from Router Z1 in AS 5555 to Router D1 in AS 9999. Consider the route 10.10.0.0/24 from AS 5555 in Figure 9.11. The route is internal to AS 5555. In BGP updates from Router Z1 to Router A1, this route will have an AS path of 5555. The AS path becomes 5555-6666 when the same route is passed to Router B1 in AS 7777.

Similarly at Router C1, the AS_path attribute of the route is 5555 6666 7777, and at Router D1 it is 5555 6666 7777 8888. If Router D1 sends updates about this route to any external BGP peer, it will advertise route 10.10.0.0/24 with AS_path attribute of 5555-6666-7777-8888-9999.

Origin Attribute

The origin attribute of a BGP route provides information about the birth of the route and how it became a part of the BGP routing table. A route can become a part of the BGP table in many ways. It may be either an external route or an internal route. Based on these factors, three values of the origin attribute are defined:

EGP: Assigned to routes, which are learned by the EGP

IGP: Assigned to routes, which are internal to an AS

Incomplete: Assigned to routes that are redistributed into BGP

The incomplete routes are also internal to the AS, but their origin is unknown or incomplete. These routes are neither learned from eBGP nor injected

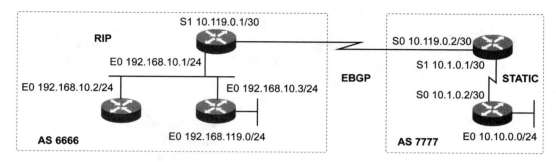

FIGURE 9.12 BGP network with origin attribute values.

using network command. The redistribution can be from connected, static, or any IGP.

Figure 9.12 depicts a BGP network where the routes are associated with various values of the origin attribute. In Figure 9.12, A1 belongs to AS 6666, and B1 belongs to AS 7777. Within AS 6666, an IGP, RIP is running among the Routers A1, A2, and A3. Static routing is defined between the Routers B1 and B2 in AS 7777. In this example, the network will be configured in two different ways, to demonstrate the variation in the origin attributes of two specific routes, 192.168.119.0/24 and 10.10.0.0/24.

Look at the first configuration option. The configuration of Router A1 is:

```
router bgp 6666
neighbor 10.119.0.2 remote-as 7777
redistribute rip
!
router rip
network 10.119.0.0
network 192.168.10.0
!
```

Configuration of Router B1 is:

```
router bgp 7777
neighbor 10.119.0.1 remote-as 6666
network 10.1.0.0 mask 255.255.255.0
network 10.10.0.0 mask 255.255.255.0
!
ip route 10.10.0.0 255.255.255.0 10.1.0.2
```

In this case, the origin attribute of the different routes will be as listed in Table 9.1.

TABLE 9.1   Routes and Origin Attributes

Route	Origin Attribute
192.168.119.0/24 as available in BGP table of B1	Incomplete as it is injected in BGP table by a redistribute command
10.10.0.0 as available in BGP table of A1	IGP as it is declared using network command

Let us look at the second configuration option. The configuration of Router A1 is:

```
router bgp 6666
neighbor 10.119.0.2 remote-as 7777
network 192.168.119.0
!
router rip
```

Configuration of Router B1 is:

```
router bgp 7777
neighbor 10.119.0.1 remote-as 6666
network 10.1.0.0 mask 255
redistribute static
!
ip route 10.1.0.0 255.255.255.0 10.1.0.2
```

In this case, the origin attribute of the different routes will be as listed in Table 9.2.

TABLE 9.2   Routes and Origin Attributes

Route	Origin
192.168.119.0/24 as available in BGP table of B1	IGP as it is injected in BGP table by a network command
10.10.0.0 as available in BGP table of A1	Incomplete as it is declared using redistribute command

Next-Hop Attribute

Next-hop attribute of a route is the IP address to which the packets are forwarded. If the BGP routers belong to different ASs that are running eBGP, the next-hop attribute of the routes is the IP address specified in neighbor IP-address remote-as command. If a route originates within the AS, the BGP routers set the next-hop attribute to its own IP address. When routes are learned from some other ASs, the BGP router changes the next-hop attribute to itself before sending updates.

However, two routers exchanging iBGP routing updates between themselves do not alter the next-hop attribute of any route. This attribute is passed on without any alteration, containing the same value as learned from an eBGP peer. Figure 9.13 shows the next-hop attribute in each route and how it is altered as the updates traverse various ASs.

In Figure 9.13, Routers A1 and A2 are running iBGP among themselves in AS 6666. Routers B1 and B2 are running iBGP in AS 7777. A1 and B1 are running eBGP between AS 6666 and AS 7777, and B2 and C1 are running BGP between AS 7777 and AS 8888.

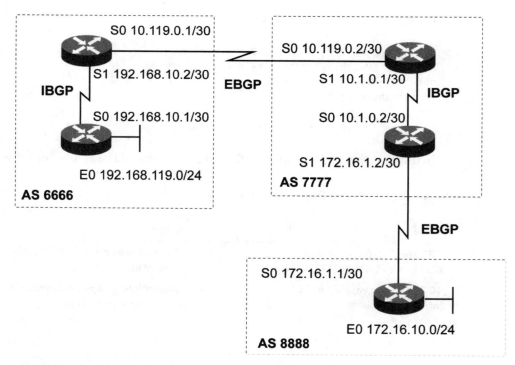

FIGURE 9.13 BGP network with next-hop attributes.

The configuration of Router A1 is:

```
router bgp 6666
neighbor 10.119.0.2 remote-as 7777
neighbor 192.168.10.1 remote-as 6666
!
```

The configuration of Router A2 is:

```
router bgp 6666
neighbor 192.168.10.1 remote-as 6666
```

The configuration of Router B1 is:

```
router bgp 7777
neighbor 10.119.0.1 remote-as 6666
neighbor 10.1.0.2 remote-as 7777
!
```

The configuration of Router B2 is:

```
router bgp 7777
neighbor 10.1.0.1 remote-as 7777
neighbor 172.16.1.1 remote-as 8888
!
```

The configuration of Router C1 is:

```
router bgp 8888
neighbor 172.16.1.1 remote-as 7777
```

Table 9.3 lists the next-hop attribute of the available routes. You will also see how they change as the packets pass from one end of the network to another. Consider two routes: 192.168.119.0/24 and 172.16.10.0/24.

TABLE 9.3   Next-hop Attributes of Routes

Route	Router	Next-hop Attribute	Explanation
192.168.119.0/24	A1	192.168.10.1	This route originated at A2
192.168.119.0/24	B1	10.119.0.1	Next-hop attribute is changed by A1 to its own value while sending BGP updates

(continued)

TABLE 9.3 *(continued)*

Route	Router	Next-hop Attribute	Explanation
192.168.119.0 /24	B2	10.119.0.1	Next-hop attribute remains unchanged as B1 and B2 are running iBGP
192.168.119.0/24	C1	172.16.1.2	Next-hop attribute is changed by B2 to its own value while sending BGP updates
172.16.10.0/24	B2	172.16.1.1	Next-hop attribute is set by C1 while sending EBGP updates
172.16.10.0/24	B1	172.16.1.1	Next-hop attribute remains unchanged as B1 and B2 are running iBGP
172.16.10.0/24	A1	10.119.0.2	Next-hop attribute is changed by B1 to its own value while sending BGP updates
172.16.10.0/24	A2	10.119.0.2	Next-hop attribute remains unchanged as B1 and B2 are running iBGP

In Table 9.3:

- B2 should have a route to 10.119.0.1
- B1 should have a route to 172.16.1.1
- A2 should have a route to 10.119.0.2

These routes can either be given statically or should be available in the IGP routing table. The next-hop attribute remains unchanged by iBGP routers. Reachability of next-hop IP addresses must be ensured so that packets destined for iBGP routes are not dropped.

Next-hop Attribute and Multi-access Media

The next-hop attribute does not apply to multi-access media. In this case, the behavior is slightly different; this is explained using an example. In Figure 9.14,

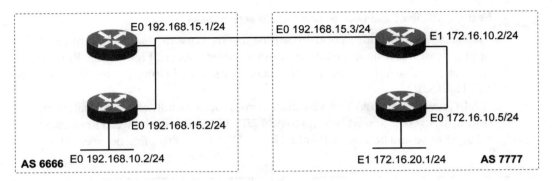

FIGURE 9.14 Next-hop attribute with multi-access media.

Routers A1, A2, and B1 share the same multi-access media and belong to the same subnet 192.168.15.0/24. In Figure 9.14, A1 and A2 belong to AS 6666, and B1 belongs to AS 7777. IGP runs between A1 and A2 in AS 6666, and it runs between B1 and B2 in AS 7777. A1 and B1 use BGP.

The configuration of Router A1 is:

```
router bgp 6666
neighbor 192.168.15.3 remote-as 7777
neighbor 192.168.15.2 remote-as 6666
network 192.168.10.0
!
```

The configuration of Router B1 is:

```
router bgp 7777
neighbor 192.168.15.1 remote-as 6666
```

In Figure 9.14, A1 advertises the network 192.168.10.0/24 to its eBGP peer B1. This network is connected to A2 and learned via IGP by A1. The next-hop attribute set should ideally be 192.168.15.1. But in this case, the next-hop attribute set by A1 while advertising the route to B1 is 192.168.15.2. This is because A1, A2, and B1 share the same multi-access media. A packet destined for network 192.168.10.0/24 from B1 need not first reach A1 and then be forwarded to A2. The packet can be directly sent to A2 over the multi-access media. This property of BGP eliminates the extra hop from the packet path.

Next-hop Attribute and Non-Broadcast Multi-access Media

The next-hop behavior in non-broadcast multi-access media is different from that of broadcast multi-access media. Consider the example in Figure 9.15. Routers A1, A2, and B1 share the same multi-access media and belong to the same subnet, 192.168.15.0/24.

In Figure 9.15, A1, A2, and B1 share non-broadcast multi-access media. A1 and A2 belong to AS 6666, and B1 belongs to AS 7777. IGP is running between Routers A1 and A2 in AS 6666 and between B1 and B2 in AS 7777. BGP is running between A1 and B1.

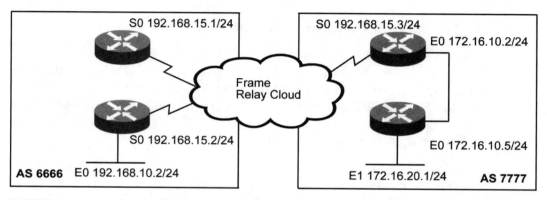

FIGURE 9.15 Next-hop attribute with non-broadcast multi-access media.

The configuration of Router A1 is:

```
router bgp 6666
neighbor 192.168.15.3 remote-as 7777
neighbor 192.168.15.2 remote-as 6666
network 192.168.10.0
!
```

The configuration of Router B1 is:

```
router bgp 7777
neighbor 192.168.15.1 remote-as 6666
```

In Figure 9.15, A1 will advertise network 192.168.10.0/24 to B1 with a next-hop not of itself but of A2. However, there is no Permanent Virtual Circuit (PVC) between A2 and B1. As a result, B1 fails to route a packet destined for network 192.168.10.0/24 to A2. The packet needs to be routed to A1 with whom PVC exists. This routing is enabled with the help of the next-hop-self command.

The configuration of Router A1 is:

```
router bgp 6666
neighbor 192.168.15.3 remote-as 7777
neighbor 192.168.15.2 remote-as 6666
neighbor 192.168.15.2 next-hop-self
network 192.168.10.0
!
```

The configuration of Router B1 is:

```
router bgp 7777
neighbor 192.168.15.1 remote-as 6666
```

With this configuration, A1 will advertise 192.168.10.0 with a next-hop of itself, that is, 192.168.15.1.

PVC is a logical connection across a packet switched network, such as Frame-Relay. It is a permanently virtual circuit for which call setup and call termination are not required.

Weight Attribute

Weight attribute enables the selection of the best path to a destination. This attribute is totally local to a router and is not carried in BGP updates between peers.

The value of weight falls in the range of 0-65535. The default value of weight is 32768 for routes originating within the router and 0 for other routes. It can be set to any value within the prescribed upper and lower limits. The higher the value of weight attribute associated with a route, the more preferred it is.

Figure 9.16 shows four ASs—AS 6666, AS 7777, AS 8888, and AS 9999. BGP is running between A1-B1, A1-D1, B1-C1, and C1-D1.

In Figure 9.16, the IP addresses are the loopback addresses of routers between which BGP is being run. Consider route 192.168.10.0/24 belonging to AS 8888. This AS is not directly connected to AS 6666. However, it receives updates of route 192.168.10.0/24 from two quarters—AS 7777 and AS 9999. Various weight attributes can be set for both of these paths.

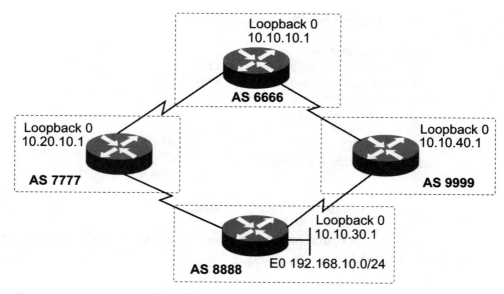

FIGURE 9.16 BGP network with the weight attributes.

Consider a scenario where the path via AS 9999 is preferred over the path via AS 7777. There are two methods to set the weight attribute:

- Using route map
- Using neighbor weight command

Using Route Map

The configuration for Router A1 using the route map is shown in Listing 9.10.

LISTING 9.10 Configuration for Router A1 Using Route map

```
router bgp 6666
neighbor 10.10.20.1 remote-as 7777
neighbor 10.10.20.1 route map WEIGHTB in
neighbor 10.10.40.1 remote-as 9999
neighbor 10.10.20.1 route map WEIGHTD in
!
route map WEIGHTB permit 10
match ip address 1
set weight 100
!
```

```
route map WEIGHTB permit 20
match ip address 2
!
route map WEIGHTD permit 10
match ip address 1
set weight 500
!
route map WEIGHTD permit 20
match ip address 2
!
access-list 1 permit 192.168.10.0 0.0.0.255
access-list 2 permit any
!
```

In Listing 9.10, the weight of 100 is assigned to updates about network 192.168.10.0/24 received from AS 7777, and the weight of 500 is assigned to updates about network 192.168.10.0/24 received from AS 9999. The route from AS 9999 has a higher value of weight attribute. As a result, AS 9999 will be preferred over AS 7777 for routing packets destined to network 192.168.10.0/24.

Using neighbor weight Command

The configuration for Router A1 using the neighbor weight command is shown in Listing 9.11.

LISTING 9.11 Configuration of Router A1 Using neighbor weight Command

```
router bgp 6666
neighbor 10.10.20.1 remote-as 7777
neighbor 10.10.20.1 weight 100
neighbor 10.10.40.1 remote-as 9999
neighbor 10.10.20.1 weight 500
!
```

In Listing 9.11, all updates received from AS 7777 are assigned a weight of 100, and those received from AS 9999 are assigned the weight of 500. The update about route 192.168.10.0/24 is received from both AS 7777 and AS 9999. However, AS 9999 will be the preferred path due to the higher weight value.

The weight attribute is specific to Cisco IOS and is not part of standard BGP implementation across all platforms.

Local Preference Attribute

The local preference attribute is used for path selection in BGP, if there are multiple paths for the same destination available. The default value of local preference is 100. The higher the local preference value, the more preferred a path is. Unlike the weight attribute, local preference is not local to a router and is sent out with BGP updates, but only within the AS.

Figure 9.17 shows four ASs—AS 6666, AS 7777, AS 8888, and AS 9999. eBGP is run between A1-B1, A1-D1, B1-C1, and C1-D1. iBGP is run between A1-A2 in AS 6666. The IP addresses are the loopback addresses of the different routers that run BGP. Consider the route 192.168.10.0/24 belonging to AS 8888.

In Figure 9.17, AS 8888 is not directly connected via BGP session with AS 6666. However, AS 6666 receives updates of the route 192.168.10.0/24 from B1 of AS 7777 and D1 of AS 9999.

Consider a situation where the path via AS 9999 will be preferred. There are two methods to set the local preference attribute:

- Using route maps
- Using bgp default-local-preference command

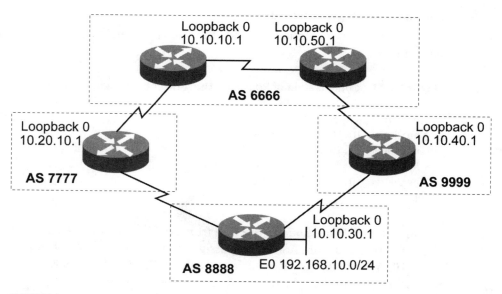

FIGURE 9.17   BGP network with local preference attribute.

Setting Local Preference Using Route Maps

Consider the example illustrated in Figure 9.17. In the figure, in A1, the local preference attribute value for route 192.168.10.0/24 is received from B1 of AS 7777.

This value is set to 500. In A2, the local preference attribute value for route 192.168.10.0/24 is received from D1 of AS 9999. This value is set to 1000.

Both A1 and A2 will come to the decision that the path via AS 9999 will be the preferred path to reach network 192.168.10.0/24 in AS 8888. As a result, all traffic originating from or passing via AS 6666 will be forwarded to A2. The configuration of Router A1 is shown in Listing 9.12.

LISTING 9.12   Configuration of Router A1

```
router bgp 6666
neighbor 10.10.20.1 remote-as 7777
neighbor 10.10.20.1 route map LOCALPREFA1 in
neighbor 10.10.50.1 remote-as 6666
!
route map LOCALPREFA1 permit 10
match ip address 1
set local-preference 500
!
access-list 1 permit 192.168.10.0 0.0.0.255
!
```

The configuration of Router A2 is shown in Listing 9.13.

LISTING 9.13   Configuration of Router A2

```
router bgp 6666
neighbor 10.10.40.1 remote-as 9999
neighbor 10.10.40.1 route map LOCALPREFA2 in
neighbor 10.10.10.1 remote-as 6666
!
route map LOCALPREFA2 permit 10
match ip address 1
set local-preference 1000
!
access-list 1 permit 192.168.10.0 0.0.0.255
!
```

Using bgp default-local-preference Command

The configuration of Router A1 using bgp default-local-preference command is:

```
Router bgp 6666
neighbor 10.10.20.1 remote-as 7777
neighbor 10.10.50.1 remote-as 6666
```

```
bgp default-local-preference 500
!
```

The configuration of Router A2 using bgp default-local-preference command is:

```
router bgp 6666
neighbor 10.10.40.1 remote-as 9999
neighbor 10.10.10.1 remote-as 6666
bgp default-local-preference 1000
!
```

Look at the example in Figure 9.17. In A1, all updates received are set to a local preference value of 500. Updates received in A2 are set to a local preference value of 1000. The local preference value is higher for A2. As a result, the preferred path will be A2 for all routes whose updates are being received from both AS 7777 and AS 9999.

eBGP updates do not contain a local preference attribute.

Multi-Exit Discriminator Attribute

The Multi-exit Discriminator (MED) attribute is used to decide the preferred entry point to an AS that has multiple entry points with its peering ASs. The entry points are routers running eBGP with other ASs. The lower the MED attribute is, the more preferred a route is.

While the weight attribute is local to a router and the local preference attribute is local to an AS, the MED attribute is exchanged among ASs. A BGP router sends the MED value of its routes to its eBGP peer. This value of routes is compared only within the same AS to decide the best path to another AS. The MED value of routes received from an AS is reset to 0 while sending out the update to another peer.

The MED value indicates the internal condition of an AS and is specific to the AS. As a result, it is not normal practice to compare MED values of routes from different ASs. However, one can compare MED values from different ASs, using the bgp always-compare-med command.

Consider Figure 9.18 where BGP is running between ASs 6666-7777, 6666-8888, and 7777-8888.

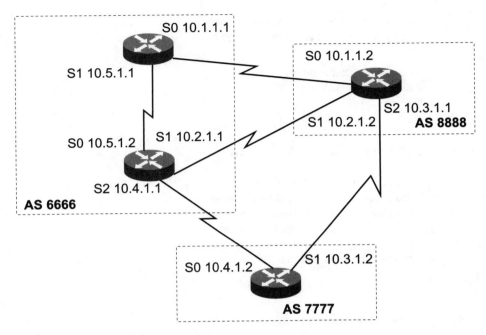

FIGURE 9.18 BGP network with MED values.

The routers shown in Figure 9.18 are configured such that the MED attribute is used for deciding the entry point to an AS. The configuration of A1 is:

```
router bgp 6666
neighbor 10.1.1.1 remote-as 8888
neighbor 10.1.1.1 route map MED1
neighbor 10.5.1.2 remote-as 6666
!
route map MED1 permit 20
set metric 100
!
```

The configuration of Router A2 is:

```
router bgp 6666
neighbor 10.2.1.2 remote-as 8888
neighbor 10.2.1.2 route map MED2
```

```
neighbor 10.5.1.1 remote-as 6666
neighbor 10.4.1.2 remote-as 7777
!
route map MED2 permit 20
set metric 150
!
!
```

The configuration of Router B1 is:

```
router bgp 7777
neighbor 10.4.1.1 remote-as 6666
neighbor 10.3.1.1 remote-as 8888
neighbor 10.3.1.1 route map MED3
!
route map MED3 permit 20
set metric 75
!
```

The configuration of Router C1 is:

```
router bgp 8888
neighbor 10.1.1.2 remote-as 6666
neighbor 10.2.1.1 remote-as 6666
neighbor 10.3.1.2 remote-as 7777
```

In the example depicted in Figure 9.18, A1 and A2 are advertising routes to C1 with different MED values of 100 and 150, respectively. B1 also advertises routes of AS 6666 to C1, but with a metric value equal to 75. This value is lower than the metric values of the routes received from A1 and A2. The MED value of routes received by different ASs is not compared by default. As a result, A1 remains the preferred outbound point for traffic of AS 6666 originating from AS 8888.

However, it is possible to compare MED values of routes received from different ASs by configuration. In this case, all traffic from AS 8888 to AS 6666 will take the path via B1.

Configuration of Router C1 is:

```
router bgp 8888
neighbor 10.1.1.1 remote-as 6666
neighbor 10.2.1.1 remote-as 6666
neighbor 10.3.1.2 remote-as 7777
bgp always-compare-med
```

Community Attribute

The community attribute is used to group destinations into entities known as communities, to which some uniform policies can be applied. There are some pre-defined communities in BGP, such as no-export, no-advertise, local-AS, and Internet. Routes having:

No-export: are not advertised to eBGP peers

No-advertise: are not advertised to any peers

Local-AS: are not advertised outside local AS

Internet: are advertised to all routers

Consider Figure 9.19, where BGP is running between AS 6666-7777 and AS 7777-8888. In the example depicted in Figure 9.19, pre-defined communities are used to restrict the flow of BGP updates. The configuration of Router A1 is:

```
router bgp 6666
neighbor 10.1.1.2 remote-as 7777
neighbor 10.1.1.2 route map comm1 out
neighbor 10.1.1.2 send-community
network 192.168.10.0
!
route map COMM1 permit 10
```

FIGURE 9.19 BGP network to illustrate community attribute.

```
match ip address 1
set community no-export
!
access-list 1 permit 192.168.10.0 0.0.0.255
```

The configuration of Router B1 is:

```
router bgp 7777
neighbor 10.1.1.1 remote-as 6666
neighbor 10.2.1.1 remote-as 8888
!
```

The configuration of Router C1 is:

```
router b1
router bgp 8888
neighbor 10.2.1.2 remote-as 7777
```

In the configurations, A1 declares network 192.168.10.0/24 to B1 with a community attribute of no-export. This means that the route can only be further advertised to iBGP peers and not to any eBGP peer. As a result, this route will not be present in the routing table of C1, as it will not be advertised.

COMMON BGP COMMANDS

This section introduces some commonly used BGP troubleshooting and debugging commands. Table 9.4 lists these commands.

TABLE 9.4 Troubleshooting and Debugging Commands

Command	Purpose
show ip bgp summary	Shows brief summary of BGP peers and BGP activity on the router
show ip bgp neighbors	Shows detailed information of BGP peers
show ip bgp nei "neighbor-IP" adv	Shows the list of routes that are getting advertised to the BGP peer identified by the IP neighbor-IP

TABLE 9.4   *(continued)*

Command	Purpose
clear ip bgp "neighbor-IP"	Resets BGP session corresponding to BGP peer identified by neighbor-IP
clear ip bgp *	Reset all BGP sessions on the router
show ip bgp	Shows the BGP table of the router
show ip bgp "network"	Shows all BGP attributes associated with the route identified by the IP address network
show ip route bgp	Shows the BGP routes installed in the routing table of the router
show ip bgp community "comm"	Shows routes matching the community comm
debug ip bgp "neighbor-IP"	Generates debugging information corresponding to the BGP peer with IP address neighbor-IP
debug ip bgp dampening	Generates debugging information for route-flap dampening
debug ip bgp events	Generates debugging information for various BGP events
debug ip bgp in	Generates debugging data for BGP inbound information
debug ip bgp keepalives	Generates debugging data for periodic BGP keepalives
debug ip bgp out	Generates debugging data for BGP outbound information
debug ip bgp updates	Generates debugging data for BGP updates

Consider the example depicted in Figure 9.20. We will consider this network to show sample outputs of some of the commands discussed. In Figure 9.20, the configuration of Router A1 is:

```
router bgp 6666
neighbor  10.10.0.2 remote-as 7777
neighbor  10.10.0.2 description REMOTE
network 10.1.1.0 mask 255.255.255.0
```

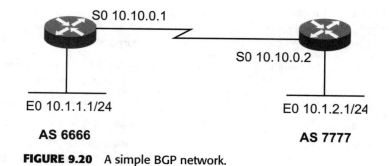

FIGURE 9.20 A simple BGP network.

The configuration Router B1 is:

```
router bgp 7777
neighbor  10.10.0.1 remote-as 6666
network 10.1.2.0 mask 255.255.255.0
```
The output for show ip bgp sum command is shown in Listing 9.14.

LISTING 9.14 Output of show ip bgp sum Command

```
RouterA1# show ip bgp sum
BGP router identifier 10.10.0.1, local AS number 6666
BGP table version is 805985, main routing table version 805985
10 network entries and 12 paths using 578 bytes of memory
30 BGP path attribute entries using 50 bytes of memory
66 BGP rrinfo entries using 2368 bytes of memory
8 BGP AS-PATH entries using 82 bytes of memory
0 BGP community entries using 0 bytes of memory
0 BGP route map cache entries using 0 bytes of memory
0 BGP filter-list cache entries using 0 bytes of memory
0 received paths for inbound soft reconfiguration
BGP activity 12/13 prefixes, 45/50 paths, scan interval 60 secs
Neighbor    V    AS MsgRcvd MsgSent   TblVer   InQ OutQ Up/Down
   State/PfxRcd
10.10.0.2   4 6666    53324    53270    805985     0    0 06:46:03  5
```

The output for show ip bgp neighbors command is shown in Listing 9.15.

LISTING 9.15 Output of show ip bgp neighbors Command

```
RouterA1# show ip bgp neighbors
sh ip bgp neighbors 10.10.0.2
```

```
BGP neighbor is 10.10.0.2,  remote AS 6666, external link
Description: REMOTE
BGP version 4, remote router ID 10.10.0.2
BGP state = Established, up for 5d11h
Last read 00:00:00, holdtime is 180, keepalive interval is 60 seconds
Neighbor capabilities:
Route refresh: advertised and received(new)
Address family IPv4 Unicast: advertised and received
Received 100359 messages, 0 notifications, 0 in queue
Sent 100337 messages, 0 notifications, 0 in queue
Route refresh request: received 0, sent 0
Default minimum time between advertisement runs is 30 seconds
For address family: IPv4 Unicast
BGP table version 8800548, neighbor version 8800542
Index 1, Offset 0, Mask 0x2
Inbound soft reconfiguration allowed
Inbound path policy configured
Outgoing update network filter list is
Route map for incoming advertisements is
0 accepted prefixes consume 0 bytes
Prefix advertised 150, suppressed 0, withdrawn 114
Number of NLRIs in the update sent: max 1, min 0
Connections established 16; dropped 15
Last reset 5d11h, due to Peer closed the session
External BGP neighbor may be up to 0 hops away.
Connection state is ESTAB, I/O status: 1, unread input bytes: 0
Local host: 10.10.0.1, Local port: 179
Foreign host: 10.10.0.2, Foreign port: 11003
Enqueued packets for retransmit: 0, input: 0  mis-ordered: 0 (0 bytes)


Event Timers (current time is 0x165D8D050):
Timer          Starts       Wakeups        Next
Retrans        7966         1              0x0
TimeWait       0            0              0x0
AckHold        7910         7247           0x0
SendWnd        0            0              0x0
KeepAlive      0            0              0x0
GiveUp         0            0              0x0
PmtuAger       0            0              0x0
DeadWait       0            0              0x0
iss: 1197207348  snduna: 1197360030  sndnxt: 1197360030  sndwnd:  16289
irs: 3480772276   rcvnxt: 3480923155  rcvwnd:  15966 delrcvwnd:   418
SRTT: 300 ms, RTTO: 303 ms, RTV: 3 ms, KRTT: 0 ms
minRTT: 0 ms, maxRTT: 500 ms, ACK hold: 200 ms
```

```
Flags: passive open, nagle, gen tcbs
Datagrams (max data segment is 536 bytes):
Rcvd: 15077 (out of order: 0), with data: 7910, total data bytes: 150878
Sent: 15519 (retransmit: 1), with data: 7964, total data bytes: 152681
```

The output for show ip bgp neighbors 10.10.0.2 adv command is shown in Listing 9.16.

LISTING 9.16   Output of show ip bgp neighbors 10.10.0.2 adv Command

```
RouterA1# show ip bgp neighbors 10.10.0.2 adv
BGP table version is 805985, local router ID is 10.10.0.1
Status codes: s suppressed, d damped, h history, * valid, > best, i -
    internal
Origin codes: i - IGP, e - EGP, ? - incomplete
Network          Next Hop            Metric LocPrf Weight
Path
*> i10.1.1.0/24             0.0.0.0                   0          32768 i
```

The output for show ip bgp neighbors 10.10.0.2 routes command is shown in Listing 9.17.

LISTING 9.17   Output of show ip bgp neighbors 10.10.0.2 routes Command

```
RouterA1# show ip bgp neighbors 10.10.0.2 routes
BGP table version is 805985, local router ID is 10.10.0.1
Status codes: s suppressed, d damped, h history, * valid, > best, i -
internal
 Origin codes: i - IGP, e - EGP, ? - incomplete
Network          Next Hop            Metric LocPrf Weight
Path
*> 10.1.2.0/24    10.0.0.2           0          0 7777 i
```

The output for show ip bgp command is shown in Listing 9.18.

LISTING 9.18   Output of show ip bgp Command

```
RouterA1# show ip bgp
BGP table version is 805985, local router ID is 10.10.0.1
Status codes: s suppressed, d damped, h history, * valid, > best, i -
    internal
Origin codes: i - IGP, e - EGP, ? - incomplete
Network          Next Hop            Metric LocPrf Weight
Path
```

```
*> i10.1.1.0/24    0.0.0.0          0         32768 i *>
10.1.2.0/24        10.0.0.2         0         0 7777 i 151 i
ISP-Router# show ip route bgp
B    10.1.1.0/24 [200/0] via 10.10.0.2, 04:12:12
```

BGP PATH SELECTION CRITERIA

In a BGP session, the best path to a destination is selected based on the values of the different BGP attributes. BGP selects only one path out of the routes available and installs it in the routing table for each destination. The BGP peer advertises only this selected path to the other peers. BGP follows a step-by-step method, known as the BGP decision algorithm, to select the best path. The BGP Decision Criteria flow-chart in Figure 9.21 depicts the steps involved in selecting the best possible path.

The lowest origin attribute type is IGP, followed by EGP and incomplete.

Router ID is the loopback address of the IP address, if one is defined. Otherwise, it is the highest IP address among those configured on the different interfaces of the router.

BGP POLICY-BASED ROUTING

This section discusses the different methods of route filtering enabled in BGP. The various BGP policy-based route-filtering methods are:

- Route maps Filtering
- Prefix-based Filtering
- AS Path Access List
- Community-based Filtering

Route maps Filtering

Route maps are used extensively in BGP configuration. They are a Cisco IOS feature and are not BGP-specific. Route maps filter incoming and outgoing routes in BGP updates. There is total control on the routes a BGP router wants to advertise or accept from its peer and install in the routing table. The syntax of a route map command is:

Route map "name of the route map" permit/deny instance-number

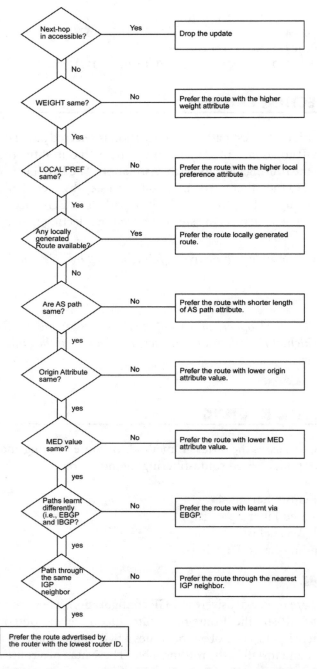

FIGURE 9.21   A BGP decision criteria flowchart depicting the steps involved in selecting the best possible path.

A route map can have multiple instances. A route is matched against some criteria in these instances, and the required action is undertaken per the criteria. The instances are sequenced in a particular order and processed in the same order. The sequence number falls in the range of 1-65535.

Consider an example defining a route map called MYROUTEMAP in Listing 9.19.

LISTING 9.19 Defining Route map MYROUTEMAP

```
Route map MYROUTEMAP permit 10
Match "Condition 1"
Match condition 2
...
...
match condition n
Set "Action 1"
Route map MYROUTEMAP permit 20
Match "Condition 1"
...
...
...
Match condition n
Set "Action 2"
```

When a route map is applied to any routing update, it is processed starting from the first instance. If the update does not meet the match specified in the first instance, it moves onto the next instance and so on, until a match is found or until there are no more instances to process. The various match criteria to be set in a route map are listed in Table 9.5.

TABLE 9.5 Match Criteria Available in Cisco IOS

Criterion	Explanation
as_path	BGP AS path list
community	BGP community list
extcommunity	BGP/VPN extended community list
interface	First-hop interface of route
ip	IP specific information
length	Packet length
metric	Metric of route
route-type	Route-type of route
tag	Tag of route

The various criteria to be set in a route map are listed in Table 9.6.

TABLE 9.6 Set Criteria Available in Cisco IOS

Criterion	Explanation
as_path	Prepend string for BGP AS-path attribute
automatic-tag	Automatically computed TAG value
comm-list	BGP community list for deletion
community	BGP community attribute
dampening	BGP route flap dampening parameters
default	Default information
extcommunity	BGP extended community attribute
interface	Output interface
ip	IP-specific information
level	Location to import route
local-preference	BGP local preference path attribute
metric	Metric value for destination routing protocol
metric-type	Type of metric for destination routing protocol
origin	BGP origin code
tag	Tag value for destination routing protocol
traffic-index	BGP traffic classification number for accounting
weight	BGP weight for routing table

Consider Figure 9.22 that depicts a route map in BGP. In Figure 9.22, we have used a route map MYROUTEMAPAOUT in Router A1 to filter outgoing updates. Two route maps are used to filter outgoing and incoming updates in Router B1. They are MYROUTEMAPBOUT and MYROUTEMAPBIN, respectively.

The configuration of Router A1 is shown in Listing 9.20.

LISTING 9.20 Configuration of Router A1

```
route map MYROUTEMAPAOUT permit 10
match ip address 1
set metric 5
!
route map MYROUTEMAPAOUT permit 20
match ip address 2
set community 300
```

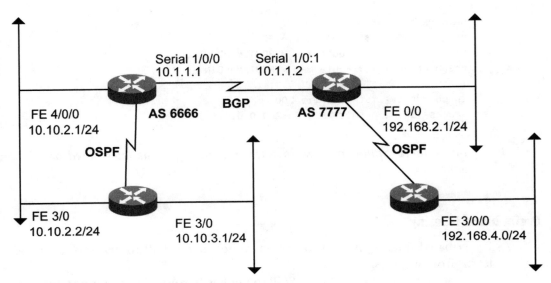

FIGURE 9.22   A route map in the BGP network.

```
!
router bgp 6666
network 10.10.2.0 mask 255.255.255.0
network 10.10.3.0 mask 255.255.255.0
neighbor 10.1.1.2 remote-as 7777
neighbor 10.1.1.2 route map MYROUTEMAPAOUT out
!
access-list 1 permit 10.10.2.0 0.0.0.255
access-list 2 permit 10.10.3.0 0.0.0.255
```

The configuration of Router B1 is shown in Listing 9.21.

LISTING 9.21 Configuration of Router B1

```
route map MYROUTEMAPBIN permit 20
match community 1
set weight 200
!
route map MYROUTEMAPBOUT permit 10
match ip address 2
set as-path prepend 6666 6666 6666
!
router bgp 7777
network 192.168.2.0
```

```
network 192.168.3.0
neighbor 10.1.1.1 remote-as 6666
neighbor 10.1.1.1 route map MYROUTEMAPBOUT out
neighbor 10.1.1.1 route map MYROUTEMAPBIN in
!
ip community-list 1 permit 300
access-list 2 permit 192.168.2.0 0.0.0.255
```

Route maps cannot be used to filter incoming BGP updates based on IP addresses.

Prefix-based Filtering

Prefix-based filtering is used to restrict routes that a BGP router advertises and learns from its peers.

Consider an example where a BGP router has 10 routes in its routing table. The BGP router can choose any select number of routes less than 10 and only advertise them to its peer. It may also receive 100 routes from its peer, but the router can decide to install only 50 of them selectively in the BGP table. These customizations can be done using prefix-based filtering.

Figure 9.23 shows a BGP session between Routers A1 of AS 6666 and B1 of AS 7777. It also shows the networks belonging to both the ASs. Prefix-based filtering can be used to control the networks to be advertised and installed in the routing table.

Prefix-based filtering can be performed in two ways:

- Using access-list
- Using prefix-list

Prefix Filtering Using Access-list

Filtering can be preformed using the access-list 1 in configurations. Listing 9.22 shows the configuration of Router A1.

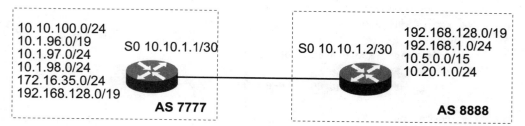

FIGURE 9.23   Prefixed-based filtering in a BGP network.

LISTING 9.22 Configuration of Router A1

```
router bgp 7777
neighbor 10.10.1.2 remote-as 8888
neighbor 10.10.1.2 distribute-list 1 out
network 10.1.96.0 mask 255.255.224.0
network 10.1.97.0 mask 255.255.255.0
network 10.1.98.0 mask 255.255.255.0
network 172.16.35.0 mask 255.255.255.0
network 172.16.22.0 mask 255.255.254.0
!
access-list 1 deny 10.1.97.0 0.0.0.255
access-list 1 deny 172.16.22.0 0.0.1.255
access-list 1 permit 0.0.0.0 255.255.255.255
```

In Listing 9.22, among the networks available in the routing table, only a few networks are filtered out while sending updates to Router B1. Here, supernet 10.1.96.0/19, and all but one higher prefix network from the same supernet, 10.1.97.0/24, is allowed to be advertised to B1. Among the networks available to A1, 10.1.96.0/19, 10.1.98.0/24, and 172.16.35.0/24 are advertised. Network 10.1.97.0/24 is not advertised.

Listing 9.23 shows the configuration of Router B1.

LISTING 9.23 Configuration of Router B1

```
router bgp 8888
neighbor 10.10.1.1 remote-as 7777
neighbor 10.10.1.1 distribute-list 2 in
network 192.168.128.0 mask 255.255.224.0
network 192.168.1.0 mask 255.255.255.0
network 10.0.0.0 mask 255.254.0.0
network 10.20.1.0 mask 255.255.255.0
!
access-list 2 permit 10.1.97.0 0.0.0.255
```

In B1, all networks advertised by A1 are not to be a part of the routing table. An inbound filtering is performed, thereby allowing only network 10.1.97.0/24 from AS 6666 to become a part of the routing table of B1.

Prefix Filtering Using Prefix-list

Filtering can be performed using the prefix-list 1 in configurations. Listing 9.24 shows the configuration for router A1.

LISTING 9.24   Configuration of Router A1

```
router bgp 7777
neighbor 10.10.1.2 remote-as 8888
neighbor 10.10.1.2 prefix-list 1 out
network 10.1.96.0 mask 255.255.224.0
network 10.1.97.0 mask  255.255.255.0
network 10.1.98.0 mask 255.255.255.0
network 172.16.35.0 mask 255.255.255.0
network 172.16.22.0 mask 255.255.254.0
!
prefix-list 1 seq 10  deny 10.1.97.0/24 le 24
prefix-list 1 seq 20 permit 10.1.96.0/19 le 24
prefix-list 1 seq 30 permit 172.16.22.0/24 le 24
!
```

In Listing 9.24, all but one higher prefix networks from the same supernet, which is 10.1.97.0/24, is filtered out. Out of all the networks available to A1, 10.1.96.0/19, 10.1.98.0/24, and 172.16.35.0/24 are advertised. The more specific 10.1.97.0/24 is not passed on.

Listing 9.25 shows the configuration of Router B1.

LISTING 9.25   Configuration of Router B1

```
router bgp 8888
neighbor 10.10.1.1 remote-as 7777
neighbor 10.10.1.1 prefix-list 2 in
network 192.168.129.0 mask 255.255.224.0
network 192.168.1.0 mask 255.255.255.0
network 10.0.0.0 mask 255.254.0.0
network 10.20.1.0 mask 255.255.255.0
!
prefix-list 2 seq 10 permit 10.1.97.0/24 le 24
```

In Listing 9.25, all networks advertised by A1 are not to be a part of the routing table. An inbound filtering is performed to allow only network 10.1.97.0/24 from AS 7777 to become a part of the routing table of B1.

The same result is obtained using access-lists or prefix-lists. However, prefix-list is more flexible and easy to configure, especially in case of supernets.

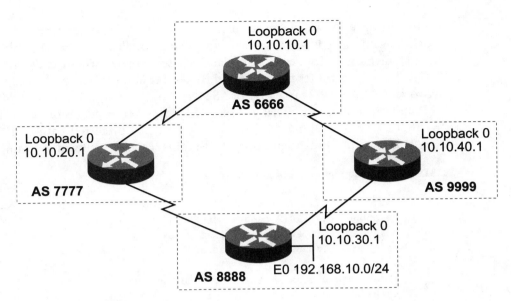

FIGURE 9.24   Filtering using AS path lists.

AS Path Access Lists

BGP incoming and outgoing updates can be filtered on the basis of the AS path. Consider the example in Figure 9.24. There are four ASs, namely AS 6666, AS 7777, AS 8888, and AS 9999. BGP is running between Routers A1-B1, A1-D1, B1-C1, and C1-D1.

In Figure 9.24, the IP addresses shown are the loopback addresses of the various routers among whom BGP is being run. Look at route 192.168.10.0/24 belonging to AS 8888. The AS path attribute is used to prefer one path to another. AS 8888 receives updates of the route 192.168.10.0/24 from both AS 7777 and AS 9999. The AS path access-lists are used to filter updates.

Listing 9.26 shows the configuration of Router A1, where the path via AS 9999 will be preferred.

LISTING 9.26   Configuration of Router A1

```
router bgp 6666
neighbor 10.10.20.1 remote-as 7777
neighbor 10.10.20.1 filter-list 1 in
neighbor 10.10.40.1 remote-as 9999
!
ip as-path-access-list 1 deny 7777
ip as-path-access-list 1 permit.*
```

In Listing 9.26, an incoming list is set at A1. All updates received will be matched against as-path-access-list 1 before being installed in the routing table. The first statement in the list denies all updates whose AS path attribute ends with 7777. This will deny any routes received from B1. As a result, the only path available for network 192.168.10.0/24 is via D1 of AS 9999.

Similarly, outbound filtering is set at B1, to prevent updates of network 192. 168.10.0/24 from reaching A1. This ensures that the preferred path to 192.168. 10.0/24 is via D1. The configuration of B1 in this scenario is shown in Listing 9.27.

LISTING 9.27   Configuration of Router B1

```
router bgp 7777
neighbor 10.10.20.2 remote-as 6666
neighbor 10.10.20.2 filter-list 2 out
neighbor 10.10.30.1 remote-as 8888
!
ip as-path-access-list 2 deny 8888
ip as-path-access-list 2 permit.*
```

All updates originating from AS 8888 will be denied by as-path-access-list 2.

Community-based Filtering

Route filtering can also be performed using the community attribute. Figure 9.25 shows BGP running between the ASs 6666, 7777, and 8888. The Routers A1, B1, and C1 are connected by WAN links.

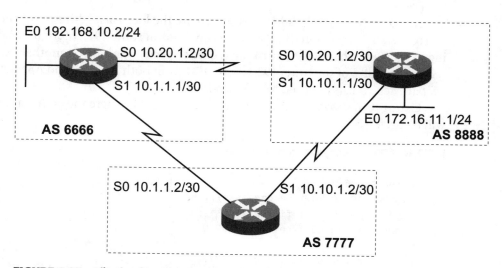

FIGURE 9.25   Filtering based on community attribute.

In Figure 9.25, the configuration for Router A1 is:

```
Router bgp 6666
neighbor 10.1.1.2 remote-as 7777
neighbor 10.20.1.1 remote-as 8888
network 192.168.10.0
```

The configuration for Router B1 is shown in Listing 9.28.

LISTING 9.28 Configuration of Router B1

```
router bgp 7777
neighbor 10.1.1.1 remote-as 6666
neighbor 10.10.1.1 remote-as 8888
neighbor 10.10.1.1 route map comm1 out
neighbor 10.10.1.1 send-community
!
route map COMM1 permit 10
match ip address 9
set community 100
!
access-list 9 permit 192.168.10.0  0.0.0.255
!
```

The configuration for Router C1 is shown in Listing 9.29.

LISTING 9.29 Configuration of Router C1

```
router bgp 8888
neighbor 10.10.1.2 remote-as 7777
neighbor 10.10.1.2 route map COMM2 in
neighbor 10.20.1.2 remote-as 6666
neighbor 10.20.1.2 weight 200
network 172.16.11.0 mask 255.255.255.0
!
route map COMM2 permit 10
match community 5
set weight 500
!
ip community-list 5 permit 100
!
```

In the configurations, B1 learns route 192.168.10.0/24 from A1. An outgoing route map is set between B1 and C1 to modify some attributes of the outgoing update. The community attribute for route 192.168.10.0/24 to 100 is set using route map COMM1, while sending it out to C1. The command send-community is used to send this attribute to C1, as it is not sent by default. At C1, the update of 192.168.10.0/24 is received with a community attribute value of 100. The route map COMM2 is used to modify updates matching this community value to weight equal to 500.

C1 also receives the update of route 192.168.10.0/24 from A1 directly. The weight value of this update is set to 200, using the neighbor weight command in C1. As a result, traffic from AS 8888 to network 192.168.10.0/24 will take the path via AS 7777 due to the higher weight attribute.

SUMMARY

In this chapter, you have learned about the different concepts of BGP routing protocol. You have also looked at the operation and configuration of the BGP protocol. The next chapter will discuss the concept of redistribution.

POINTS TO REMEMBER

- BGP enables the exchange of routing information between different ASs.
- When BGP exchanges routing updates between routers belonging to the same AS, it is known as iBGP.
- When BGP exchanges between routers belonging to different ASs, it is termed as eBGP.
- Advertising of routes is performed using the network command and by redistributing routes present in the routing table.
- Aggregation combines multiple routes so that a single route is advertised.
- Route reflectors and confederations help to overcome scalability problems in BGP.
- AS is subdivided into mini ASs in the confederation method to avoid a full mesh iBGP.
- The origin attribute of a BGP route provides information about the birth of the route and how it became a part of the BGP routing table.
- The next-hop attribute of a route is the IP address to which the packets are forwarded.
- The weight attribute enables the selection of the best path to a destination. This attribute is local to a router and is not carried in BGP updates between peers.
- The local preference attribute is used for path selection in BGP, if there are multiple paths for the same destination available.

- The MED attribute is used to decide the preferred entry point to an AS that has multiple entry points with its peering ASs.
- The community attribute is used to group destinations into entities known as communities, to which some uniform policies can be applied.
- The various BGP policy-based route-filtering methods are route map filtering, prefix-based filtering, AS path access list, and community-based filtering.

10 Redistribution

REDISTRIBUTION CONCEPTS

Redistribution of routes is the sharing of routing information among different routing protocols. By default, any two routing protocols do not share routing information among themselves in a router. Consider two routers running two different routing protocols, say, Open Shortest Path First (OSPF) and Interior Gateway Routing Protocol (IGRP). The OSPF router is reachable only via OSPF learned routes, and the IGRP router is reachable only via IGRP learned routes. The OSPF router can never reach the IGRP router and vice versa. Implementation of redistribution in OSPF and IGRP routers will ensure their reachability, irrespective of whether they belong to the OSPF or IGRP domain.

In the preceding chapters, you have looked at the routing protocols in isolation. Each routing protocol is a self-contained domain. However, redistribution among routing protocols is necessary if, in an organization:

- Network is spread over a vast geographical expanse
- Network is to be expanded
- Network should be scalable
- Network should effectively support Cisco and non-Cisco routers
- Network should support old and new versions of hardware

Command Syntax

The redistribution of one routing protocol into another is not a default process. The redistribute command is used for enabling redistribution. The syntax of the command is:

```
redistribute protocol [process-id] {level-1 | level-1-2 | level-2} [as-
number] [metric metric-value] [metric-type type-value] [match {internal
| external 1 | external 2}]
[tag tag-value] [route map map-tag] [subnets]
```

Table 10.1 explains the options in the redistribute command.

TABLE 10.1 Options in redistribute Command

Option	Explanation
protocol	Protocol from which routes are redistributed into another protocol. The values include bgp, connected, egp, igrp, isis, mobile, ospf, static [ip], or rip. The static and connected keywords are used when static and connected routes are being redistributed.
process-id	Process ID is: ASN value in case of BGP, IGRP or EIGRP. OSPF process ID in case of OSPF. Optional in case of IS_IS. If a name is given for the IS_IS routing process, this is to be used while configuring.
level-1	Value specific to IS_IS. It indicates that IS_IS level 1 routes are redistributed into other routing protocols.
level-1-2	Value specific to IS_IS. It indicates that both IS_IS level 1 and level 2 routes are redistributed into other routing protocols.
level-2	Value specific to IS_IS. It indicates that IS_IS level 2 routes are redistributed into other routing protocols.
as-number	Value is optional. This indicates ASN for the redistributed route.
metric	Value is associated with the redistributed route.
metric-type	Value specific to few of the routing protocols being redistributed. In case of OSPF, there two metric values: Type 1 external route Type 2 external route The default value is 2. In case of IS_IS, there are two metric values: Internal—IS_IS metric that is < 63. External—IS_IS metric that is > 64 < 128 The default value is internal.
match {internal |	Value specific to OSPF routes. It specifies the criteria to redistribute OSPF routes into other routing protocols.

TABLE 10.1 *(continued)*

Option	Explanation
external 1 | external 2	The values are: internal—Routes that are internal to an AS external 1—Routes that are external to AS but are imported into OSPF as type 1 external route external 2—Routes that are external to AS but are imported into OSPF as type 2 external route
tag	Value used by Autonomous System Boundary Routers (ASBR) to communicate. This 32-bit value is associated with OSPF external routes. The remote ASN is used for routes redistributed from BGP and EGP, if no tag is specified. The value remains 0 for all other routes.
route map	Value used to control import of route from a different routing domain. If a route map is configured, only specific routes are denied or allowed to redistribute. If this keyword is not specified, all routes from a routing protocol are imported. However, if this keyword is specified and no route maps are configured or listed, no route redistribution would take place.
map-tag	Name of the route map being configured.
subnets	Value is specific to redistribution into OSPF. The subnets of a particular network to be redistributed into OSPF are also specified.

There is no process-ID for RIP, because it is not defined using a process-ID or ASN.

METRICS AND REDISTRIBUTION

Consider two protocols, X and Y. When a route from X is redistributed into Y, it has to be assigned a metric value of Y. This is required because X and Y protocols have different metrics. The route of X will originally have a metric, which would be typical of the X protocol. When this route is redistributed into Y, it would appear as a Y protocol route to other routers in the routing domain. As a result, a metric typical of the Y protocol has to be assigned to the route.

For example, a default metric value of 20 is taken for routes redistributed into OSPF from any protocol except BGP, for which the value is 1. The metric value can be assigned to a route in two ways:

Using redistribute command: Metric is added per each routing protocol. It can vary from one redistributed routing protocol to another.

Using default-metric command: Metric is set for all redistributed routes.

It is mandatory for some routing protocols to set the metric. For other protocols, the default metric value is taken if no value is specified.

PROTOCOL-SPECIFIC REDISTRIBUTION

The process of redistribution varies from one protocol to another. We will discuss the concept of redistribution specific to each protocol in the following sections.

Redistribution of Static and Connected Routes

The connected routes are configured in the various interfaces of a router, while static routes are defined in a router and not learned by any routing protocol. Connected routes are directly declared in most routing protocols, but you can also redistribute them.

Consider a connected network as shown in Figure 10.1, in which an Ethernet interface of each router is configured. Routers A1, A2, and A3 are part of a RIP

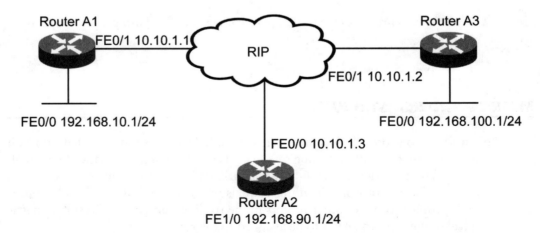

FIGURE 10.1 A connected network with a RIP domain.

domain. A1 and A2 are declared in the RIP domain, using the redistribute command. The network 192.168.100.0/24 is declared conventionally in A3.

The configuration of Router A1 is:

```
interface FastEthernet0/0
ip address 192.168.10.1 255.255.255.0
!
interface FastEthernet0/1
ip address 10.10.1.1 255.255.255.0
!
router rip
redistribute connected
default-metric 1
!
```

The configuration of Router A2 is:

```
interface FastEthernet0/0
ip address 10.10.1.3 255.255.255.0
!
interface FastEthernet1/0
ip address 192.168.10.1 255.255.255.0
!
router rip
redistribute connected
default-metric 1
!
```

The configuration of Router A3 is:

```
interface FastEthernet0/0
ip address 192.168.100.1 255.255.255.0
!
interface FastEthernet0/1
ip address 10.10.1.2 255.255.255.0
!
router rip
network 192.168.100.0
!
```

The routing table of A2 is shown in Listing 10.1.

LISTING 10.1 Routing Table of A2

```
RouterA2# show ip route
Codes: C - connected, S - static, I - IGRP, R - RIP, M - mobile, B - BGP
```

```
D - EIGRP, EX - EIGRP external, O - OSPF, IA - OSPF inter area
N1 - OSPF NSSA external type 1, N2 - OSPF NSSA external type 2
E1 - OSPF external type 1, E2 - OSPF external type 2, E - EGP
i - IS_IS, L1 - IS_IS level-1, L2 - IS_IS level-2, * - candidate
  default
U - per-user static route, o - ODR

Gateway of last resort is not set
 R 192.168.10.0/24 [120/1] via 10.10.1.1, 00:00:16, FastEthernet0/0
 R 192.168.100.0/24 [120/1] via 10.10.1.2, 00:00:16, FastEthernet0/0
 C 192.168.10.1/24 is directly connected, FastEthernet1/0
 C 10.10.1.0/24 is directly connected, FastEthernet0/0
```

The networks 192.168.10.0/24 and 192.168.100.0/24 are learned as RIP networks irrespective of the fact that they are announced, using the network command or redistribute connected.

Consider a scenario where we add Router A4, which is not a part of the RIP domain. It is connected to the router via a serial link, and a static route is defined in A1 for networks attached to A4, as shown in Figure 10.2.

Network 172.16.25.0/24 is statically routed from A1 to A4. This static route, in turn, is redistributed into RIP at A1.

Configuration for Router A1 is:

```
interface FastEthernet0/0
ip address 192.168.10.1 255.255.255.0
!
```

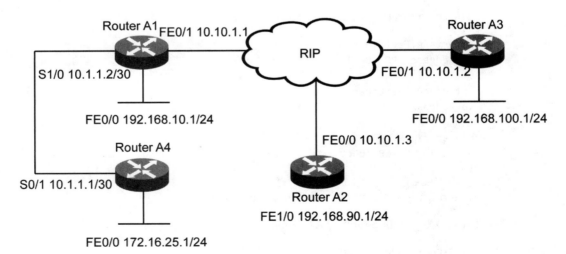

FIGURE 10.2 Adding a router that is not part of the RIP domain.

```
interface FastEthernet0/1
ip address 10.10.1.1 255.255.255.0
!
interface Serial1/0
ip address 10.1.1.2 255.255.255.252
!
router rip
redistribute connected
redistribute static
default-metric 1
!
ip route 172.16.25.0 255.255.255.0 10.1.1.1
!
```

Configuration for Router A4 is:

```
interface FastEthernet0/0
ip address 172.16.25.1 255.255.255.0
!
interface Serial0/1
ip address 10.1.1.1 255.255.255.252
!
ip route 0.0.0.0 0.0.0.0 10.1.1.2
!
```

Configuration for Router A2 is:

```
interface FastEthernet0/0
ip address 10.10.1.3 255.255.255.0
!
interface FastEthernet1/0
ip address 192.168.90.1 255.255.255.0
!
router rip
redistribute connected
!
```

Configuration for Router A3 is:

```
interface FastEthernet0/0
ip address 192.168.100.1 255.255.255.0
!
interface FastEthernet0/1
ip address 10.10.1.2 255.255.255.0
```

```
!
router rip
network 192.168.100.0
!
```

The routing table for A2 is shown in Listing 10.2.

LISTING 10.2   Routing Table of A2

```
RouterA2# show ip route
Codes: C - connected, S - static, I - IGRP, R - RIP, M - mobile, B - BGP
   D - EIGRP, EX - EIGRP external, O - OSPF, IA - OSPF inter area
   N1 - OSPF NSSA external type 1, N2 - OSPF NSSA external type 2
   E1 - OSPF external type 1, E2 - OSPF external type 2, E - EGP
   i - IS_IS, L1 - IS_IS level-1, L2 - IS_IS level-2, * - candidate
     default
   U - per-user static route, o - ODR

Gateway of last resort is not set
   R 192.168.10.0/24 [120/1] via 10.10.1.1, 00:00:16, FastEthernet0/0
   R 192.168.100.0/24 [120/1] via 10.10.1.2, 00:00:16, FastEthernet0/0
   R 172.16.25.0/24 [120/1] via 10.10.1.2, 00:00:16, FastEthernet0/0
   C 192.168.90.1/24 is directly connected, FastEthernet1/0
   C 10.10.1.0/24 is directly connected, FastEthernet0/0
```

The network 172.16.25.0/24, which was redistributed from static Router A1, is available in the routing table of A2 as a RIP route. Similarly, static and connected routes can be redistributed into all other routing protocols like IGRP, EIGRP, OSPF, and BGP.

Redistribution in Dynamic Routing Protocols

Static, connected, as well as dynamically learned routes can be redistributed into any dynamic routing protocol. The redistribution of routes into different dynamic routing protocols is discussed in the following sections.

Redistribution into RIP (Version 1 and Version 2)

Listing 10.3 shows the different protocols that can be redistributed into RIP.

LISTING 10.3   Redistribution of Protocols into RIP

```
router rip
network 192.168.10.0
redistribute static
```

```
redistribute igrp 100
redistribute eigrp 100
redistribute ospf 1
redistribute isis
redistribute bgp 6666
default-metric 4
```

Consider the scenario where static routes, IGRP and EIGRP with ASN 100, OSPF with process-ID 1, IS_IS, and BGP with ASN 6666 are being redistributed into RIP. The redistribution occurs with a metric value of 4 for all redistributed routes. You can set different metrics for the different redistributed routes into RIP, as shown in Listing 10.4.

LISTING 10.4   Metrics for Redistributed Routes into RIP

```
router rip
network 192.168.10.0
redistribute static metric 8
redistribute igrp 100 metric 6
redistribute eigrp 100 metric 4
redistribute ospf 1 metric 1
redistribute isis metric 3
redistribute bgp 6666 metric 2
```

Redistribution into IGRP or EIGRP

Listing 10.5 shows the protocols that can be redistributed into IGRP or EIGRP.

LISTING 10.5   Redistribution of Protocols into IGRP or EIGRP

```
router igrp/eigrp 100
network 192.168.10.0
redistribute static
redistribute ospf 1
redistribute rip
redistribute isis
redistribute bgp 6666
default-metric 20000 100 250 5 1500
```

IGRP and EIGRP have the same metrics. As a result, the configuration for redistribution into IGRP and EIGRP would be similar. Consider a scenario where static routes, RIP, OSPF with process ID 1, IS_IS, and BGP with ASN 6666 are being redistributed into IGRP or EIGRP. The redistribution occurs with a metric value of 20000, 100, 250, 5, and 1500, in the order of bandwidth delay reliability load MTU, respectively, for all redistributed routes. You can set different metrics for the different redistributed routes into IGRP or EIGRP, as shown in Listing 10.6.

LISTING 10.6 Metrics for Redistributed Routes into IGRP or EIGRP

```
router igrp/eigrp 100
network 192.168.10.0
redistribute static metric 10000 100 250 3 1500
redistribute ospf 1 metric 20000 100 255 5 1500
redistribute rip metric 10000 60 250 5 1500
redistribute isis metric 20000 90 255 5 1500
redistribute bgp 6666 metric 15000 200 250 2 1500
```

Redistribution into OSPF

Listing 10.7 shows the different protocols that can be redistributed into OSPF.

LISTING 10.7 Redistribution of Protocols into OSPF

```
router ospf 1
network 192.168.10 0.0.255.255 area 0
redistribute static
redistribute rip
redistribute igrp 100
redistribute eigrp 100
redistribute isis
redistribute bgp 6666
default-metric 10
```

Consider a scenario where static routes IGRP and EIGRP with ASN 100, RIP, IS_IS, and BGP with ASN 6666, are redistributed into OSPF. The redistribution occurs with a metric value of 10 for all redistributed routes. You can set different metrics for the different redistributed routes into RIP, as shown in Listing 10.8.

LISTING 10.8 Metrics for Redistributed Routes into RIP

```
router ospf 1
network 192.168.10 0.0.255.255 area 0
redistribute static metric 200 subnets
redistribute rip metric 200 subnets
redistribute igrp 100 metric 100 subnets
redistribute eigrp 100 metric 100 subnets
redistribute isis metric 10 subnets
redistribute bgp 6666 metric 10 subnets
```

Different metrics are set for the different protocols being redistributed. The keyword subnets is used to redistribute subnets of major networks, which are otherwise not redistributed. The default metric value of OSPF is 20 for non-BGP routes and 1 for BGP routes.

Redistribution of Routes among IGRP and EIGRP

EIGRP is the enhanced version of IGRP. As a result, the basic features remain the same. If IGRP and EIGRP are run on a router with the same ASN, redistribution occurs automatically. You need not configure them explicitly. However, redistribution between IGRP and EIGRP can be disabled.

The redistribution between IGRP and EIGRP differs from normal redistribution because the metrics of the IGRP route are directly compared with those of EIGRP routes. The difference in administrative distances is not taken into account.

Figure 10.3 shows IGRP and EIGRP domains. Routers B1 and B2 speak only IGRP, and Routers B4 and B5 speak only EIGRP, whereas Router B3 speaks both IGRP and EIGRP.

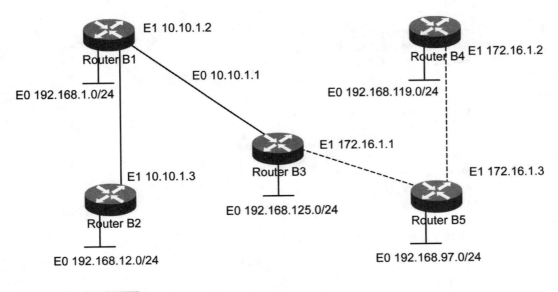

FIGURE 10.3 A connected network with IGRP and EIGRP domains.

Configuration of Router B1 is:

```
!
router igrp 100
network 192.168.1.0
network 10.10.1.0
!
```

Configuration of Router B2 is:

```
!
router igrp 100
network 192.168.12.0
network 10.10.1.0
!
```

Configuration of Router B3 is:

```
!
router igrp 100
network 192.168.125.0
network 10.10.1.0
!
router eigrp 100
network 192.168.125.0
network 172.16.1.0
!
```

Configuration of Router B4 is:

```
!
router eigrp 100
network 192.168.119.0
network 172.16.1.0
!
```

Configuration of Router B5 is:

```
!
router igrp 100
network 192.168.97.0
network 172.16.1.0
!
```

Both EIGRP and IGRP are configured with the same ASN on B3. Router B3 can reach both the IGRP and EIGRP learned networks. The routing table for B1 is shown in Listing 10.9.

LISTING 10.9   Routing Table of B1

```
RouterB1# show ip route
Codes: C - connected, S - static, I - IGRP, R - RIP, M - mobile, B - BGP
   D - EIGRP, EX - EIGRP external, O - OSPF, IA - OSPF inter area
   N1 - OSPF NSSA external type 1, N2 - OSPF NSSA external type 2
   E1 - OSPF external type 1, E2 - OSPF external type 2, E - EGP
   i - IS_IS, L1 - IS_IS level-1, L2 - IS_IS level-2, * - candidate
      default
   U - per-user static route, o - ODR

Gateway of last resort is not set
   C 192.168.1.0/24 is directly connected, Ethernet0
   C 10.10.1.0/24 is directly connected, Ethernet1
   I 192.168.12.0/24 [100/1100] via 10.10.1.3, 00:00:33,
   I 192.168.125.0/24 [100/1100] via 10.10.1.1, 00:00:33,
   I 192.168.119.0/24 [100/1100] via 10.10.1.1, 00:00:33,
   I 192.168.97.0/24 [100/1100] via 10.10.1.1, 00:00:33,
```

The routing table for B2 is shown in Listing 10.10.

LISTING 10.10   Routing Table of B2

```
RouterB2# show ip route
Codes: C - connected, S - static, I - IGRP, R - RIP, M - mobile, B - BGP
   D - EIGRP, EX - EIGRP external, O - OSPF, IA - OSPF inter area
   N1 - OSPF NSSA external type 1, N2 - OSPF NSSA external type 2
   E1 - OSPF external type 1, E2 - OSPF external type 2, E - EGP
   i - IS_IS, L1 - IS_IS level-1, L2 - IS_IS level-2, * - candidate
      default
   U - per-user static route, o - ODR

Gateway of last resort is not set
   C 192.168.12.0/24 is directly connected, Ethernet0
   C 10.10.1.0/24 is directly connected, Ethernet1
   I 192.168.1.0/24 [100/1100] via 10.10.1.2, 00:00:33,
   I 192.168.125.0/24 [100/1100] via 10.10.1.1, 00:00:33,
   I 192.168.119.0/24 [100/1100] via 10.10.1.1, 00:00:33,
   I 192.168.97.0/24 [100/1100] via 10.10.1.1, 00:00:33,
```

The routing table for B4 is shown in Listing 10.11.

LISTING 10.11   Routing Table of B4

```
RouterB4# show ip route
Codes: C - connected, S - static, I - IGRP, R - RIP, M - mobile, B - BGP
   D - EIGRP, EX - EIGRP external, O - OSPF, IA - OSPF inter area
   N1 - OSPF NSSA external type 1, N2 - OSPF NSSA external type 2
   E1 - OSPF external type 1, E2 - OSPF external type 2, E - EGP
   i - IS_IS, L1 - IS_IS level-1, L2 - IS_IS level-2, * - candidate
      default
   U - per-user static route, o - ODR

Gateway of last resort is not set
   C 192.168.119.0/24 is directly connected, Ethernet0
   C 172.16.1.0/24 is directly connected, Ethernet1
   D 192.168.12.0/24 [9/1100] via 172.16.1.1, 00:00:33,
   D 192.168.125.0/24 [90/1100] via 172.16.1.1, 00:00:33,
   D 192.168.119.0/24 [90/1100] via 172.16.1.1, 00:00:33,
   D 192.168.97.0/24 [90/1100] via 172.16.1.3, 00:00:33
```

The routing table for B5 is shown in Listing 10.12.

LISTING 10.12   Routing Table of B5

```
RouterB5# show ip route
Codes: C - connected, S - static, I - IGRP, R - RIP, M - mobile, B - BGP
   D - EIGRP, EX - EIGRP external, O - OSPF, IA - OSPF inter area
   N1 - OSPF NSSA external type 1, N2 - OSPF NSSA external type 2
   E1 - OSPF external type 1, E2 - OSPF external type 2, E - EGP
   i - IS_IS, L1 - IS_IS level-1, L2 - IS_IS level-2, * - candidate
      default
   U - per-user static route, o - ODR

Gateway of last resort is not set
   C 192.168.97.0/24 is directly connected, Ethernet0
   C 172.16.1.0/24 is directly connected, Ethernet1
   D 192.168.12.0/24 [9/1100] via 172.16.1.1, 00:00:33,
   D 192.168.125.0/24 [90/1100] via 172.16.1.1, 00:00:33,
   D 192.168.119.0/24 [90/1100] via 172.16.1.1, 00:00:33,
   D 192.168.119.0/24 [90/1100] via 172.16.1.2, 00:00:33
```

Redistribution of Routes among Classless and Classful Routing Protocols

EIGRP being classful and IGRP being a classless routing protocol can lead to some problems while redistributing. In this section, you will look at these problems and ways to overcome them.

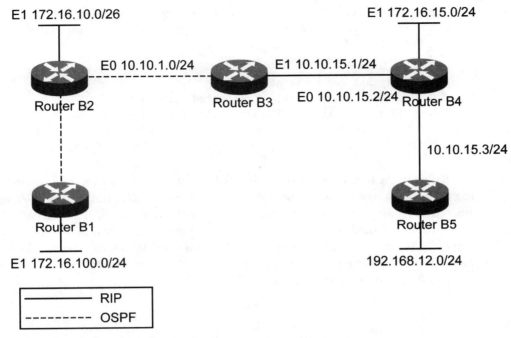

E1 172.16.10.0/26

E0 10.10.1.0/24

E1 10.10.15.1/24

E1 172.16.15.0/24

Router B2

Router B3

E0 10.10.15.2/24 Router B4

10.10.15.3/24

Router B1

Router B5

E1 172.16.100.0/24

192.168.12.0/24

RIP

OSPF

FIGURE 10.4 A connected network with RIP and OSPF domains.

Figure 10.4 shows five routers: B1, B2, B3, B4, and B5. B1 and B2 run OSPF exclusively, B4 and B5 run RIP exclusively, and B3 runs both RIP and OSPF. B3 is redistributing between RIP and OSPF so that all the networks have reachability with each other irrespective of whether they are connected to RIP or OSPF routers.

In Figure 10.4, B2 has a directly connected network 172.16.10.0/26 in the OSPF domain. B4 has a directly connected network 172.16.15.0/24 in the RIP domain. These two networks have different subnet masks within the same major network, and RIP would not recognize this. As a result, it would not be possible to route packets to 172.16.10.0/26, though it will be redistributed into RIP from OSPF in B3.

Listing 10.13 shows the routing table in B4.

LISTING 10.13 Routing Table of B4

```
RouterB4# show ip route
Codes: C - connected, S - static, I - IGRP, R - RIP, M - mobile, B - BGP
    D - EIGRP, EX - EIGRP external, O - OSPF, IA - OSPF inter area
    N1 - OSPF NSSA external type 1, N2 - OSPF NSSA external type 2
```

```
     E1 - OSPF external type 1, E2 - OSPF external type 2, E - EGP
     i - IS_IS, L1 - IS_IS level-1, L2 - IS_IS level-2, * - candidate
       default
     U - per-user static route, o - ODR

Gateway of last resort is not set
  C 172.16.15.0/24 is directly connected, Ethernet1
  C 10.10.15.0/24 is directly connected, Ethernet0
  R 172.16.10.0/24 [120/1] via 10.10.15.1, 00:00:16, Ethernet0
  R 172.16.100.0/24 [120/1] via 10.10.15.3, 00:00:16, Ethernet0
```

RIP is made to accept the route to 172.16.10.0/26 in its routing table by making it appear to have a subnet mask that is recognized by RIP. In the example considered in Figure 10.4, a static route is added to null0 to network 172.16.10.0/24 in B3 and redistributed to RIP. This route is accepted by RIP, and all traffic corresponding to network 172.16.10.0/26 will be forwarded to B3. As a result, B3 will have a more specific route to network 172.16.10.0/26, toward B2.

Configuration in Router B3 is:

```
!
router ospf 5
!
router rip
redistribute ospf 1
redistribute static
default-metric 1
!
ip route 172.16.10.0 255.255.255.0 null0
!
```

After this configuration, the routing table in Router B4 would appear as shown in Listing 10.14.

LISTING 10.14 Routing Table of B4

```
RouterB4# show ip route
Codes: C - connected, S - static, I - IGRP, R - RIP, M - mobile, B - BGP
  D - EIGRP, EX - EIGRP external, O - OSPF, IA - OSPF inter area
  N1 - OSPF NSSA external type 1, N2 - OSPF NSSA external type 2
  E1 - OSPF external type 1, E2 - OSPF external type 2, E - EGP
  i - IS_IS, L1 - IS_IS level-1, L2 - IS_IS level-2, * - candidate
    default
  U - per-user static route, o - ODR
```

```
Gateway of last resort is not set
  C 172.16.15.0/24 is directly connected, Ethernet1
  C 10.10.15.0/24 is directly connected, Ethernet0
  R 172.16.24.0/24 [120/1] via 10.10.15.1, 00:00:16, Ethernet0
  R 172.16.100.0/24 [120/1] via 10.10.15.1, 00:00:16, Ethernet0
  R 192.168.12.0/24 [120/1] via 10.10.15.3, 00:00:16, Ethernet0
```

There can be issues during redistribution, in case the other subnets in the 172.16.10.0/24 network do not belong to the OSPF domain.

CAUTION

Automatic redistribution is ensured among RIP v1 and RIP v2, because both of these protocols share the same database.

NOTE

Problems with Redistribution

Care should be taken to avoid routing loops and selection of non-optimal routes while configuring redistribution. Route maps and access-lists prevent these problems by controlling the networks being redistributed.

In Figure 10.5, Routers A1 and A2 are running RIP exclusively, and Routers A4 and A5 are running OSPF exclusively. A3 is running both RIP and OSPF and is redistributing between the two protocols.

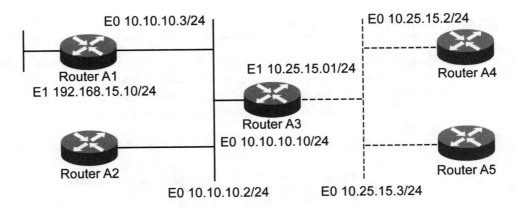

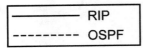

FIGURE 10.5 A connected network with RIP and OSPF domains.

In Figure 10.5, access-list 1 is used to avoid learning the RIP network 192.168.15.0/24 through OSPF after the route is redistributed into OSPF in A3. The configuration of Router A3 is shown in Listing 10.15.

LISTING 10.15 Configuration of Router A3

```
interface Ethernet 0
ip address 10.10.10.1 255.255.255.0
!
interface Ethernet 1
ip address 10.25.15.1 255.255.255.0
!
router rip
network 10.10.10.0
redistribute ospf metric 1
!
router ospf 1
network 10.25.15.0 0.0.0.255 area 0
redistribute rip metric 10 subnets
distribute-list 1 in Ethernet1
!
access-list 1 deny 192.168.15.0 0.0.0.255
access-list 1 permit any
```

ADMINISTRATIVE DISTANCES

Administrative Distance (AD) is a parameter that is used locally within a router to decide the preferred route of entries in any routing updates. It is the measure of trustworthiness of a routing protocol. Each routing protocol is assigned a pre-defined value administrative distance, which is listed in Table 10.2.

TABLE 10.2 Routing Protocols and Administrative Distance

Source	Administrative Distance
Connected interface	0
Static route	1
EIGRP summary route	5
EBGP	20
Internal EIGRP	90

TABLE 10.2   *(continued)*

Source	Administrative Distance
IGRP	100
OSPF	110
IS_IS	115
RIP	120
EGP	140
On Demand Routing	160
External EIGRP Route	170
IBGP	200
Unknown	255

Consider an example where there are three routers, A1, A2, and A3, and two routing domains running EIGRP and OSPF, respectively. A1 is running both OSPF and EIGRP, A2 is running EIGRP alone, and A3 is running OSPF alone. Both routing processes have run their respective algorithm to find the best path to the network 192.1.68.10.0/24. This example is illustrated in Figure 10.6.

In Figure 10.6, A1 is learning the network 192.168.10.0/24 from A2 via EIGRP, and from A3 via OSPF. The decision as to which path is to be installed in the

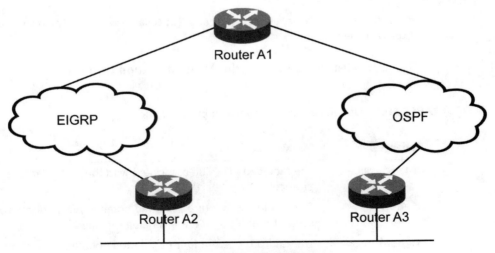

FIGURE 10.6   A connected network with EIGRP and OSPF domains.

routing table would be taken on the basis of AD. EIGRP has an AD of 90, and OSPF has an AD of 120. The path with the lower administrative distance is preferred. As a result, the next-hop for reaching network 192.1.68.10.0/24 from A1 would be A2.

Listing 10.16 shows the routing table of A1.

LISTING 10.16   Routing Table of A1

```
RouterA1# show ip route
Codes: C - connected, S - static, I - IGRP, R - RIP, M - mobile, B - BGP
   D - EIGRP, EX - EIGRP external, O - OSPF, IA - OSPF inter area
   N1 - OSPF NSSA external type 1, N2 - OSPF NSSA external type 2
   E1 - OSPF external type 1, E2 - OSPF external type 2, E - EGP
   i - IS_IS, L1 - IS_IS level-1, L2 - IS_IS level-2, * - candidate
      default
   U - per-user static route, o - ODR

Gateway of last resort is not set
   D 192.168.12.0/24 [90/1100] via 10.10.10.1, 00:00:33,
```

If A2 is unavailable, the path learned from OSPF will be installed in the routing table of A1, and traffic for the network 192.168.10.0/24 would be forwarded to A3.

Modification of Administrative Distance

The default values of ADs can be modified to optimize routing within a network. The syntax of the command to modify AD is:

```
distance distance-value {ip-address {wildcard-mask}} [ip-standard-list]
[ip-extended-list]
```

Table 10.3 lists the options in the AD modification command.

TABLE 10.3   Options in the AD Modification Command

Option	Meaning
distance-value	The modified AD value associated with the particular routing protocol.
ip-address	The IP address in four part dotted notation AD associated with routes from this IP address is being modified.
wildcard-mask	The wildcard mask in four part dotted notation.
ip-standard-list ip-extended-list	The standard or extended ip access list listing the routes whose AD is modified.

The modification of AD values needs to be done with utmost care, because it may lead to internal routing problems within an organization.

Consider an example where there are three routers, A1, A2, and A3, and two routing domains, running EIGRP and OSPF, respectively. A1 is running both OSPF and EIGRP, A2 is running EIGRP alone, and A3 is running OSPF alone. Both of the routing processes have run their respective algorithm to find the best path to the network 192.1.68.10.0/24. This network is shown in Figure 10.7.

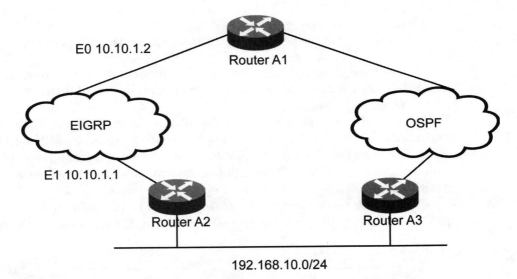

FIGURE 10.7 A connected network with EIGRP and OSPF domains.

By default, the path taken for traffic destined to network 192.168.10.0/24 from A1 will be via A2, as the AD of EIGRP is lower than that of OSPF. Consider a scenario where the AD is modified to make A3 the preferred path for 192.168.10.0/24. The AD of all OSPF learned routes in A1 is modified to 85 from the default value of 120. The configuration of Router A1 is:

```
!
router ospf 1
distance 85
!
```

As a result, all OSPF learned routes are preferred over the EIGRP learned routes in A1. The preferred path to network 192.168.10.0/24 is via A3.

The routing table at Router A1 is shown in Listing 10.17.

LISTING 10.17 Routing Table of A1

```
RouterA1# show ip route
Codes: C - connected, S - static, I - IGRP, R - RIP, M - mobile, B - BGP
  D - EIGRP, EX - EIGRP external, O - OSPF, IA - OSPF inter area
  N1 - OSPF NSSA external type 1, N2 - OSPF NSSA external type 2
  E1 - OSPF external type 1, E2 - OSPF external type 2, E - EGP
  i - IS_IS, L1 - IS_IS level-1, L2 - IS_IS level-2, * - candidate
    default
  U - per-user static route, o - ODR

Gateway of last resort is not set
  O 192.168.10.0/24 [85/25] via 10.10.10.1, 00:34:11, Ethernet0
```

Modification of the AD values associated with routing protocols is not recommended, because it may lead to less preferred routes, instead of better routes, entering the routing table.

Consider the example depicted in Figure 10.8. In the network, there are six routers, B1, B2, B3, B4, B5, and B6, and two routing protocols, RIP and OSPF. Some routers are running RIP or OSPF exclusively, while other routers are running both the protocols. B1, B2, B3, B4, B5, and B6 form the RIP routing domain, while B2, B3, B4, and B6 form the OSPF routing domain.

Consider the two networks, 192.168.10.0/24 and 10.10.1.0/24, shown in Figure 10.8. B2 learns both the networks via OSPF and RIP. OSPF has a lower AD value of

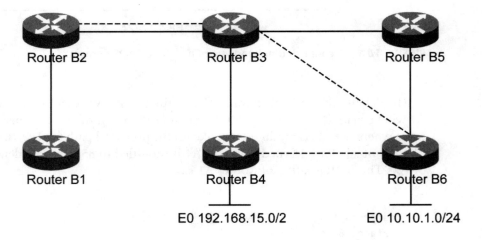

FIGURE 10.8 A connected network with RIP and OSPF domains.

110 as compared to 120 of RIP. As a result, the OSPF path is preferred and installed in B2's routing table. Three hops (B3-B6-B4) will be taken to reach the network 192.168.10.0/24 via OSPF from B2. The RIP path would take only two hops (B3-B4). In case of network 10.10.1.0/24, the OSPF path takes two hops (B3-B6), and the RIP path would have taken three hops (B3-B5-B6).

Listing 10.18 shows the routing table of B2.

LISTING 10.18 Routing Table of B2

```
RouterB2# show ip route
Codes: C - connected, S - static, I - IGRP, R - RIP, M - mobile, B - BGP
   D - EIGRP, EX - EIGRP external, O - OSPF, IA - OSPF inter area
   N1 - OSPF NSSA external type 1, N2 - OSPF NSSA external type 2
   E1 - OSPF external type 1, E2 - OSPF external type 2, E - EGP
   i - IS_IS, L1 - IS_IS level-1, L2 - IS_IS level-2, * - candidate
     default
   U - per-user static route, o - ODR

Gateway of last resort is not set
   O 192.168.15.0/24 [110/25] via 192.168.11.1, 00:34:11, Ethernet0
   O 10.10.1.0/24 [110/25] via 192.168.22.1, 00:34:11, Ethernet1
```

Let us modify the AD value to check and see whether we can have an optimum path to all the destinations. The AD associated with RIP is modified to 95 from the default 120. Now RIP has a lower AD than OSPF in B2. The configuration of B2 is:

```
!
router rip
distance 95
!
```

The routes for networks 192.168.10.0/24 and 10.10.1.0/24 from B1 will now be using the RIP path. Two hops (B3-B4) are required for reaching network 192.168.10.0/24 via RIP from B2, whereas the OSPF path would have taken three hops (B3-B6-B4). In case of network 10.10.1.0/24, the RIP path will take three hops (B3-B5-B6) and the OSPF path will take two hops (B3-B6). Modifying the AD value may not always lead to optimum route selection.

Listing 10.19 shows the routing table of B2.

LISTING 10.19 Routing Table of B2

```
RouterB2# show ip route
Codes: C - connected, S - static, I - IGRP, R - RIP, M - mobile, B - BGP
   D - EIGRP, EX - EIGRP external, O - OSPF, IA - OSPF inter area
```

```
N1 - OSPF NSSA external type 1, N2 - OSPF NSSA external type 2
E1 - OSPF external type 1, E2 - OSPF external type 2, E - EGP
i - IS_IS, L1 - IS_IS level-1, L2 - IS_IS level-2, * - candidate
   default
U - per-user static route, o - ODR

Gateway of last resort is not set
   R 192.168.15.0/24 [95/1] via 172.16.11.1, 00:34:11, Ethernet0
   R 10.10.1.0/24 [95/1] via 172.16.22.1, 00:34:11, Ethernet1
```

Consider the introduction of some selective configuration while modifying the AD values. Modification of AD of RIP for the route 192.168.10.0/24 ensures that the RIP path is preferred over the OSPF path. The configuration of Router B2 is:

```
!
router rip
distance 95 0.0.0.0 255.255.255.255 10
!
access-list 10 permit 192.168.10.0 0.0.0.255
```

In the configuration, an access-list 10 is defined to specify the network 192.168.10.0/24. This access-list is used to modify the AD value of RIP to 95, for updates received from any source for the network 192.168.10.0/24. Network 192.168.10.0/24 will take the RIP path, and network 10.10.1.0/24 will take the OSPF path.

Listing 10.20 shows the routing table of B2.

LISTING 10.20 Routing Table of B2

```
RouterB2# show ip route
Codes: C - connected, S - static, I - IGRP, R - RIP, M - mobile, B - BGP
   D - EIGRP, EX - EIGRP external, O - OSPF, IA - OSPF inter area
   N1 - OSPF NSSA external type 1, N2 - OSPF NSSA external type 2
   E1 - OSPF external type 1, E2 - OSPF external type 2, E - EGP
   i - IS_IS, L1 - IS_IS level-1, L2 - IS_IS level-2, * - candidate
      default
   U - per-user static route, o - ODR

Gateway of last resort is not set
   R 192.168.15.0/24 [95/1] via 172.16.11.1, 00:34:11, Ethernet0
   O 10.10.1.0/24 [110/25] via 192.168.22.1, 00:34:11, Ethernet1
```

As a result, optimum routing is achieved with the careful use of AD.

SUMMARY

In this chapter, we discussed redistribution process with respect to the different protocols. We also discussed the concept of Administrative Distance. The next chapter will discuss On Demand Routing and Default Routing.

POINTS TO REMEMBER

- Redistribution of routes is the sharing of routing information among different routing protocols.
- The syntax of the command for enabling redistribution is redistribute protocol [`process-id`] {level-1 | level-1-2 | level-2} [`as-number`] [metric metric-value] [metric-type type-value] [match {internal | external 1 | external 2}][tag tag-value] [route map map-tag] [subnets].
- The metric value can be assigned to a route using redistribute or default-metric commands.
- Route maps and access-lists prevent these problems by controlling the networks being redistributed.
- AD is a parameter, which is used locally within a router to decide the preferred route over other entries in any routing updates.
- The syntax of the command to modify AD is distance distance-value {ip-address {wildcard-mask}} [ip-standard-list] [ip-extended-list]

11

On Demand Routing and Default Routing

In the previous chapters, we looked at the various interior and exterior routing protocols such as Routing Information Protocol (RIP), Enhanced Interior Gateway Routing Protocol (EIGRP), Open Shortest Path First (OSPF), Intermediate System to Intermediate System (IS_IS), and Border Gateway Protocol (BGP). On Demand Routing (ODR) is not a conventional routing protocol, because it is used in a very limited domain of IP networks, such as stub networks. Most routing protocols are not specific to any type of network and are universal in nature (though the effectiveness of each IGP varies with respect to the type of network in which it is deployed). ODR is, however, classified with other IGPs because it facilitates routing of Internet Protocol (IP) traffic.

INTRODUCTION TO ODR

ODR is implemented for stub networks. A stub network is one that does not serve as a transit for data. Any traffic forwarded to a stub network is meant for that network only and is not forwarded to any other network. Stub networks have a hub and spoke topology. As a result, such networks have a central main office site with a high-end router, such as a Cisco 7500™ or 7200 series™ router, which serves as the hub. Multiple branch offices form the spokes and are connected to the hub via WAN links. The routers used in the spoke locations are low-end routers, such as 2500, 1600, and 1700. Figure 11.1 depicts a hub and spoke network, where Router B1 represents the hub, and the Routers B2, B3, B4, B5, B6, and B7 represent the spokes.

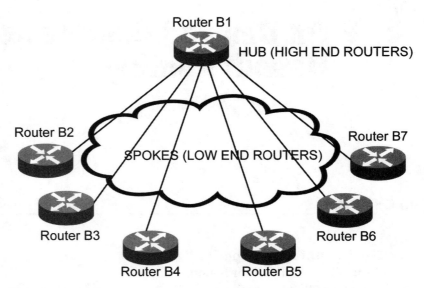

FIGURE 11.1   Hub and spoke network, where Router B1 is the hub and the Routers B2, B3, B4, B5, B6, and B7 are the spokes.

Today, this is one of the most common topologies used, with most organizations following this type of structure. All servers, such as Web and mail servers and those that run Enterprise Resource Planning (ERP) applications, are located at the central site or the hub. The branch offices connect to the central office to enable employees to use resources available at the hub. The growing popularity of hub and spoke topology has triggered the development and use of ODR as one of the best routing options for these types of networks.

The most important advantage of ODR over other dynamic routing protocols is the comparatively small amount of overhead. There are usually two types of overheads in IP networks: resource overhead, that is, high CPU and memory utilization of routers and network bandwidth, and maintenance overhead, which consists of configuration, maintenance, and monitoring requirements. The use of ODR helps reduce both of these types of overheads.

ODR is the best option for stub networks; the only disadvantage is that it is a Cisco-proprietary feature and works optimally only for Cisco networks. Any non-Cisco routers in the network can be used as spokes. In addition, a standard routing protocol, such as RIP or OSPF, can be used between the hub and spoke. The remaining spokes, which are Cisco routers, can run ODR.

Unlike other routing protocols, ODR does not function in Layer 3 of the OSI model. IP prefixes are advertised with Layer 2 Cisco Discovery Protocol (CDP) updates. ODR is an enhanced form of CDP. The updates contain an extra five bytes for each network, four of which are used for the network number or IP address; the fifth one is used for the IP prefix. The updates carry the prefix information, and as a result, ODR can understand and propagate Variable Length Subnet Mask (VLSM). Let us now look at CDP and its basic functionalities.

CDP is a Cisco-proprietary protocol that operates at Layer 2 of the OSI model. It is enabled by default on all interfaces of Cisco devices and collects information about the neighboring Cisco devices, such as routers and switches. The protocol running at the Network layer does not impact CDP operations, because it does not use Layer 3 information. CDP uses Layer 2 multicasts to discover all Cisco neighbors. CDP can be disabled or enabled on a per-interface basis. If CDP is disabled in a Cisco device, it will not be discovered as a CDP neighbor by any other Cisco device. If CDP is disabled only for a particular interface, the device will not be discovered as a neighbor by other Cisco devices on the same network segment as the interface.

NOTE

The term Cisco neighbors (in the context of this chapter) means all the Cisco devices that are on directly connected links with the subject device and are running CDP.

CDP packets are sent out every 60 seconds. Any CDP packet from a neighbor is held for 180 seconds, after which it is discarded. You can modify these timers according to your requirement. (This is discussed later in the chapter.)

All routing information is exchanged using CDP. The stub routers or the remote locations advertise their directly connected networks to the hub router, which advertises the default route to the stub routers. Updates exchanged are CDP updates at Layer 2 with minimal utilization of router resources. As a result, ODR handles the resource and configuration overheads for stub networks.

ODR is directly related to CDP updates. Thus, disabling CDP globally will disable ODR, and disabling CDP at a particular interface will disable ODR at that interface.

Figure 11.2 shows a stub network, where Routers A2, A3, and A4 are the stub routers. To understand how ODR works, let us first look at how CDP updates are exchanged among these routers.

You can use certain CDP commands, which demonstrate the type of data that is exchanged in CDP updates. Figure 11.2 also shows that CDP neighbor summary information is available on all the routers. Listing 11.1 shows the summary of CDP-enabled neighbors of Router A1, which has been collected by exchange of CDP updates.

LISTING 11.1 Summary of CDP-enabled Neighbors of Router A1

```
routerA1# show cdp neighbors
  Capability Codes: R - Router, T - Trans Bridge, B - Source Route Bridge
```

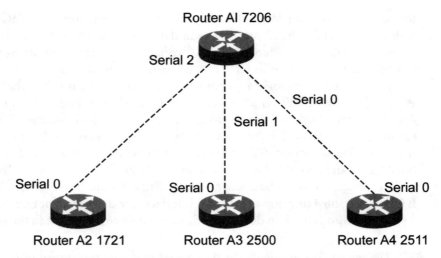

FIGURE 11.2 Stub network where Routers A2, A3, and A4 exchange routing information.

```
S - Switch, H - Host, I - IGMP, r - Repeater
Device ID    Local Intrfce    Holdtme    Capability    Platform    Port ID
RouterA2     Ser 2            151        R             1721        Ser 0
RouterA3     Ser 1            127        R             2500        Ser 0
RouterA4     Ser 0            158        R             2511        Ser 0
```

Listing 11.2 shows detailed information associated with the neighbor router A2, as available at Router A1. This information is collected through CDP.

LISTING 11.2 A2 Updates as Available at Router A1

```
RouterA1# show cdp entry RouterA2
-------------------
Device ID: RouterA2
Entry address(es):
IP address: 10.1.1.1
Platform: cisco 1721, Capabilities: Router
Interface: Serial2, Port ID (outgoing port): Serial0
Holdtime : 130 sec
Version :
Cisco Internetwork Operating System Software
IOS (tm) C1700 Software (C1700-Y-M), Version 12.2(4)YA2, EARLY
DEPLOYMENT RELEASE SOFTWARE (fc1)
```

```
Synched to technology version 12.2(5.4)T
TAC Support: http://www.cisco.com/tac
Copyright (c) 1986-2002 by cisco Systems, Inc.
Compiled Thu 11-Apr-02 21:54 by ealyon
advertisement version: 2
```

Listing 11.3 shows neighbor summary information, as available at Router A2. This information is collected through CDP.

LISTING 11.3   Neighbor Summary Information Available at Router A2

```
RouterA2# show cdp neighbors
Capability Codes: R - Router, T - Trans Bridge, B - Source Route Bridge
S - Switch, H - Host, I - IGMP, r - Repeater
Device ID    Local Intrfce    Holdtme    Capability    Platform    Port ID
RouterA1     Ser 0            144        R             7206        Ser 2
```

Listing 11.4 shows the neighbor summary information, as available at Router A3. This information is collected through CDP.

LISTING 11.4   Neighbor Summary Information Available at Router A3

```
RouterA3# show cdp neighbors
Capability Codes: R - Router, T - Trans Bridge, B - Source Route Bridge
S - Switch, H - Host, I - IGMP, r - Repeater
Device ID    Local Intrfce    Holdtme    Capability    Platform    Port ID
RouterA1     Ser 0            144        R             7206        Ser 1
```

Listing 11.5 shows the neighbor summary information, as available at Router A4. This information is collected through CDP.

LISTING 11.5   Neighbor Summary Information Available at Router A4

```
RouterA4# show cdp neighbors
Capability Codes: R - Router, T - Trans Bridge, B - Source Route Bridge
S - Switch, H - Host, I - IGMP, r - Repeater
Device ID    Local Intrfce    Holdtme    Capability    Platform    Port ID
RouterA1     Ser 0            144        R             7206        Ser 0
```

We will disable CDP in Serial0 Interface of Router A1 to understand the impact on neighbor information. Listing 11.6 shows the neighbor summary information, at Router A1.

LISTING 11.6   Neighbor Summary Information at Router A1

```
RouterA1# show cdp neighbors
Capability Codes: R - Router, T - Trans Bridge, B - Source Route Bridge
S - Switch, H - Host, I - IGMP, r - Repeater
Device ID     Local Intrfce   Holdtme   Capability   Platform   Port ID
RouterA2      Ser 2           151       R            2500       Ser 0
RouterA3      Ser 1           127       R            1721       Ser 0
```

After disabling CDP in interface Serial0, no neighbor information is learned via that particular interface.

ON DEMAND ROUTING CONFIGURATION

There are specific commands required to configure ODR in a stub network. In this section, we will look at configuration commands that are used to configure ODR for optimal network performance.

Basic ODR Configuration

There are some basic commands that are required to configure ODR in a network. Let us take the example of a stub network, as shown in Figure 11.3, and configure it using all available options.

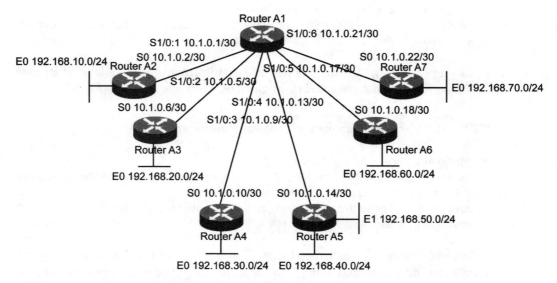

FIGURE 11.3   A stub network with hub and spoke topology.

In Figure 11.3, Router A1 represents the central site, while Routers A2, A3, A4, A5, A6, and A7 are the spokes, or branches, which are connected via WAN links. Each branch has one or more connected networks. Listing 11.7 depicts the configuration of ODR in the network represented shown in Figure 11.3.

LISTING 11.7   ODR Configuration

```
interface Serial1/0:1
ip address 10.1.0.1 255.255.255.252
!
interface Serial1/0:2
ip address 10.1.0.5 255.255.255.252
!
interface Serial1/0:3
ip address 10.1.0.9 255.255.255.252
!
interface Serial1/0:4
ip address 10.1.0.13 255.255.255.252
!
interface Serial1/0:5
ip address 10.1.0.17 255.255.255.252
!
interface Serial1/0:6
ip address 10.1.0.21 255.255.255.252
!
router odr
!
```

The only command required to configure ODR in the hub router is the router odr command. No additional command is required to enable ODR in the spoke routers. After ODR is enabled in the hub router, it starts installing stub networks in its routing table.

To enable ODR to run, do not configure any dynamic routing protocol on the stub routers.

CAUTION

Filtering ODR Information

ODR information can be filtered at the hub. However, it cannot be filtered at the spoke routers. The hub can decide which networks learned from the spoke routers should be included in its routing table. Taking the example of the network depicted in Figure 11.3, let us configure the ODR filter.

Listing 11.8 shows the configuration for filtering ODR at Router A1, the hub router.

LISTING 11.8   Configuration of ODR at Router A1

```
interface Serial1/0:1
ip address 10.1.0.1 255.255.255.252
!
interface Serial1/0:2
ip address 10.1.0.5 255.255.255.252
!
interface Serial1/0:3
ip address 10.1.0.9 255.255.255.252
!
interface Serial1/0:4
ip address 10.1.0.13 255.255.255.252
!
interface Serial1/0:5
ip address 10.1.0.17 255.255.255.252
!
interface Serial1/0:6
ip address 10.1.0.21 255.255.255.252
!
router odr
distribute-list 15 in
!
access-list 15 deny 192.168.50.0 0.0.0.255
access-list 15 permit any
!
```

The ODR routes received by A1 are filtered, using the access-list 15 in this case. As a result, A1 filters route 192.168.50.0/24 and accepts all other routes.

Redistributing ODR Information

The ODR routing information can also be redistributed into other routing protocols. In Figure 11.4, two new routers, B1 and B2, are added to the network shown in Figure 11.3.

In Figure 11.4, an OSPF domain has been added in the network. Routers A1, B1, and B2 are a part of the OSPF routing domain, where A1 has a stub network connected to it. ODR is running in the stub network, but Router A1 is also running OSPF. The ODR routing information needs to be available to the routers in the OSPF domain in order to forward packets destined to the stub network effectively. As a result, ODR routes need to be redistributed in OSPF.

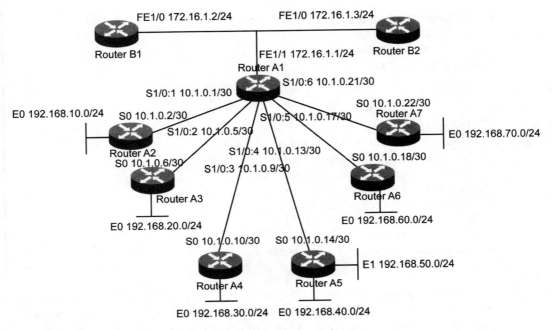

FIGURE 11.4 Routers B1 and B2 are added to the network.

Listing 11.9 shows the configuration of Router A1 for redistributing ODR routing information into OSPF.

LISTING 11.9 Configuration of A1 for Redistributing ODR Routing Information into OSPF

```
interface Serial1/0:1
ip address 10.1.0.1 255.255.255.252
!
interface Serial1/0:2
ip address 10.1.0.5 255.255.255.252
!
interface Serial1/0:3
ip address 10.1.0.9 255.255.255.252
!
interface Serial1/0:4
ip address 10.1.0.13 255.255.255.252
!
interface Serial1/0:5
ip address 10.1.0.17 255.255.255.252
!
interface Serial1/0:6
```

```
ip address 10.1.0.21 255.255.255.252
!
interface fastEthernet1/1
ip address 172.16.1.1 255.255.255.0
!
router odr
!
router ospf 1
network 172.16.1.0 0.0.0.255 area 0
passive-interface Serial1/0:1
passive-interface Serial1/0:2
passive-interface Serial1/0:3
passive-interface Serial1/0:4
passive-interface Serial1/0:5
passive-interface Serial1/0:6
redistribute odr metric 10 subnets
!
```

Here, ODR routes are redistributed into OSPF with a metric value of 10, with all available subnets. The serial interfaces that form the stub network have been configured as passive interfaces. This is to ensure that OSPF updates are not unnecessarily sent out through those interfaces.

Modifying ODR Timers

The default timers for ODR are the same as those of CDP: updates are sent every 60 seconds; routes become invalid if no updates are received every 180 seconds and are subsequently removed from the routing table after 240 seconds. In certain cases, however, these timers may have to be modified to allow for faster convergence.

Table 11.1 lists the commands for configuring CDP timers.

TABLE 11.1 Commands for Configuring CDP Timers

Command	Explanation
#cdp timer seconds	Modifies the CDP timer, that is, the interval at which updates are sent. The value varies between 5 to 900.
#cdp holdtime seconds	Modifies the CDP holdtime, that is, the interval after which updates are invalid. The value varies between 10 to 255.
#router odr	Configures the different timers of ODR.
#timers basic update invalid holddown flush [sleeptime]	Configures the different timers of ODR.

Configuring Spokes with Multiple Connections to a Hub

So far, we looked at stub networks with spokes having one connection to the hub. However, certain networks may also have spokes with multiple connections to the hub to ensure higher network availability. Usually, sensitive locations with huge volumes of data require multiple connections to the hub.

Figure 11.5 depicts a hub and spoke network, where Router A1 is the hub and Routers A2, A3, and A4 are the spokes. Router A3 has multiple connections to the hub via two WAN links, which can be used either in load balancing or backup mode.

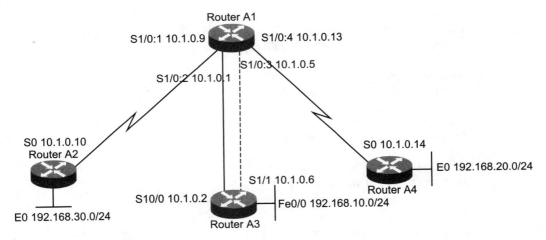

FIGURE 11.5   Hub and spoke network, where Router A1 is the hub and Routers A2, A3, and A4 are the spokes.

In ODR, load balancing among multiple routes is the norm. The hub load balances between two routes to the spoke and sends the default route through both the links to the spoke. As a result, outgoing traffic from the spoke is load-balanced.

However, load balancing may not be the desired option if there are links with unequal bandwidth. The link with higher bandwidth is usually used as the primary link, and the link with the lower bandwidth is used as the backup link. Listing 11.10 shows the configuration of Router A1, such that one of the links to Router A3 is the primary link and the other link is the backup link.

LISTING 11.10   Configuring Spokes with Multiple Connections to a Hub

```
interface Serial1/0:1
ip address 10.1.0.1 255.255.255.252
!
interface Serial1/0:2
ip address 10.1.0.5 255.255.255.252
```

```
!
interface Serial1/0:3
ip address 10.1.0.9 255.255.255.252
!
interface Serial1/0:4
ip address 10.1.0.13 255.255.255.252
!
router odr
distance 140 10.1.1.1 0.0.0.0
!
Router A3:
interface Serial1/0
ip address 10.1.0.2 255.255.255.252
!
interface Serial1/1
ip address 10.1.0.5 255.255.255.252
!
interface Ethernet0/0
ip address 192.168.10.1 255.255.255.0
!
ip route 0.0.0.0 0.0.0.0 10.1.0.1
ip route 0.0.0.0 0.0.0.0 10.1.0.5 100
!
```

In Listing 11.10, a preference for routes is set separately at the hub and the spoke. At the hub, one of the routes is preferred over the other by using the distance command for each CDP neighbor address. Here, the administrative distance for the router 10.1.1.1 is set to 140, which is lower than the default value of 170 for ODR.

At the spoke, this is achieved using static routes with different administrative distances. Here, the administrative distance of the route via 10.1.0.5 is set to 100, against the default value of 1 for static routes.

Configuring Spokes with Connection to Multiple Hubs

In the previous section, we looked at configuring ODR when a single spoke has multiple connections to a hub. Another serious concern may be the failure of the hub itself, which will lead to complete network disruption. To overcome this, certain stub networks are designed with multiple hubs, where each spoke is connected to more than one hub.

Figure 11.6 depicts a network with two hubs, C1 and B1, respectively, and three spokes, A2, A3, and A4 connected to both the hubs. If there are multiple connections between a hub and a spoke, load balancing is achieved by default.

To specify the preferences for one route over another, the static route with administrative distance variation needs to be configured.

DEFAULT ROUTES

The Internet is a huge IP network with millions of hosts. If each router had the routing information for each network of the Internet, the size of the routing table would have been huge. The maintenance of such a massive routing table would put tremendous pressure on the router resources, and the cost required to maintain high-end routers throughout the Internet would have resulted in a very restricted growth of the Internet.

Therefore, the concept of default routes was used for the development of the Internet. The presence of a default route ensures that each router is not required to have routing information about all networks in the Internet. It may have information about some networks and a default destination router where it can "dump" the traffic. A default route is the one that caters traffic destined for all destinations besides the ones that have an explicit entry in the routing table. If a route does not have a default route entry, all traffic destined for networks not having an explicit entry in a routing table will be dropped. The router specified as the next-hop in the default route is also referred to as the gateway of last resort.

Static Configuration of Default Routes

There are different ways to configure default routes statically. Let us use the example of the network of ABCD Inc. to understand the configuration options for default routes. Figure 11.7 depicts the network of ABCD Inc., with Routers A1 and A2; all traffic destined to the Internet from Router A2 has to pass through Router A1.

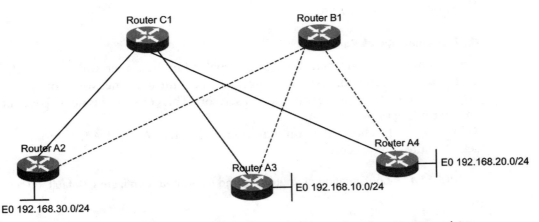

FIGURE 11.6 Load balancing between the hubs C1 and B1 and spokes A2, A3, and A4.

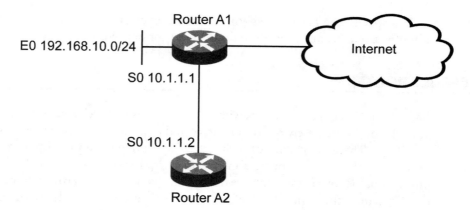

FIGURE 11.7 Network of ABCD Inc. with Routers A1 and A2.

Let us look at the different methods to configure a default route statically in a router.

The ip default-gateway Command

The ip default-gateway command is used when IP routing is disabled in a router. This is used only in special situations, like when the router is in boot mode and loading the image from a TFTP server. The configuration of the ip default-gateway command in Router A2, when it is in boot mode, is:

```
interface Serial0
ip address 10.1.1.2 255.255.255.252
!
ip default-gateway 10.1.1.1
```

The ip default-network Command

This command is used when IP routing is enabled in the router, unlike the ip default-gateway command. The router considers the route to the network, configured as the default network as the gateway of last resort. This command is not dependent on any routing protocols.

Listing 11.11 shows the configuration of Routers A1 and A2, without any default route configured.

LISTING 11.11 Configuration of Routers A1 and A2 Without Configuring Default Routes

```
interface Serial0
ip address 10.1.1.1 255.255.255.252
```

```
!
interface Ethernet0
ip address 192.168.10.1 255.255.255.0
!
Router A2:
interface Serial0
ip address 10.1.1.2 255.255.255.252
!
ip route 192.168.10.0 255.255.255.0 10.1.1.1
!
```

At this stage, a default network has not been configured in Router A2. The routing table of Router A2 is shown in Listing 11.12.

LISTING 11.12   Routing Table of Router A2

```
RouterA2# show ip route
Codes: C - connected, S - static, I - IGRP, R - RIP, M - mobile, B - BGP
   D - EIGRP, EX - EIGRP external, O - OSPF, IA - OSPF inter area
   N1 - OSPF NSSA external type 1, N2 - OSPF NSSA external type 2
   E1 - OSPF external type 1, E2 - OSPF external type 2, E - EGP
   i - IS_IS, L1 - IS_IS level-1, L2 - IS_IS level-2, * - candidate
     default
   U - per-user static route, o - ODR

Gateway of last resort is not set
   C 10.1.1.0/30 is directly connected, Serial0
   S 192.168.10.0 [1/0] via 10.1.1.1
```

The configuration for adding a default route in Router A2 is shown in Listing 11.13.

LISTING 11.13   Configuration for Adding a Default Route in Router A2

```
interface Serial0
ip address 10.1.1.2 255.255.255.252
!
ip route 192.168.10.0 255.255.255.0 10.1.1.1
!
ip default-network 192.168.10.0
!
```

The routing table of A2 is shown in Listing 11.14.

LISTING 11.14 Routing Table of A2

```
RouterA2# show ip route
Codes: C - connected, S - static, I - IGRP, R - RIP, M - mobile, B - BGP
   D - EIGRP, EX - EIGRP external, O - OSPF, IA - OSPF inter area
   N1 - OSPF NSSA external type 1, N2 - OSPF NSSA external type 2
   E1 - OSPF external type 1, E2 - OSPF external type 2, E - EGP
   i - IS_IS, L1 - IS_IS level-1, L2 - IS_IS level-2, * - candidate
     default
   U - per-user static route, o - ODR

Gateway of last resort is 10.1.1.1 to network 192.168.10.0
   C  10.1.1.0/30 is directly connected, serial0
   S* 192.168.10.0 [1/0] via 10.1.1.1
```

Multiple networks can be defined as the default network. The network with the lowest administrative distance is chosen as the default network. If all the defined networks have the same administrative distance, the network listed first in the routing table takes precedence.

The ip route 0.0.0.0 0.0.0.0 NXT HOP Command

This command can also be used to set the gateway of last resort in a router. Similar to the default-network command, this command is also not dependent on any routing protocol. The only prerequisite is that IP routing must be enabled in the router.

The configuration of the network depicted in Figure 11.7, using the ip route 0.0.0.0 0.0.0.0 NXT HOP command for configuring the default route, is shown in Listing 11.15 and Listing 11.16.

LISTING 11.15 Configuring Default Routes for Router A1

```
interface serial0
ip address 10.1.1.1 255.255.255.252
!
interface Ethernet0
ip address 192.168.10.1 255.255.255.0
!
```

LISTING 11.16 Configuring Default Routes for Router A2

```
interface serial0
ip address 10.1.1.2 255.255.255.252
!
ip route 0.0.0.0 0.0.0.0 10.1.1.1
!
```

The routing table of Router A2 is shown in Listing 11.17.

LISTING 11.17   Routing Table of A2

```
RouterA2# show ip route
Codes: C - connected, S - static, I - IGRP, R - RIP, M - mobile, B - BGP
   D - EIGRP, EX - EIGRP external, O - OSPF, IA - OSPF inter area
   N1 - OSPF NSSA external type 1, N2 - OSPF NSSA external type 2
   E1 - OSPF external type 1, E2 - OSPF external type 2, E - EGP
   i - IS_IS, L1 - IS_IS level-1, L2 - IS_IS level-2, * - candidate
      default
   U - per-user static route, o - ODR

Gateway of last resort is 10.1.1.1 to network 0.0.0.0
   C  10.1.1.0/30 is directly connected, Serial0
   S* 0.0.0.0/0 [1/0] via 10.1.1.1
```

If a router has both the ip route 0.0.0.0 0.0.0.0 and ip default-network commands configured, the default-network command will take precedence if a particular network is learned via static configuration. If the same is learned using a dynamic routing protocol, the static route to 0.0.0.0 takes precedence and is installed in the routing table as the gateway of last resort. If there are multiple ip route 0.0.0.0 0.0.0.0 statements, traffic is load balanced among the multiple default routes that have been defined.

Default Route Configurations of Routing Protocols

We looked at the different methods of configuring the gateway of last resort in a router. In the methods discussed so far, the default gateway information is configured individually for each router, regardless of whether they are members of any routing protocol domain. Individual configuration of default routes in each router may be avoided in a dynamic routing protocol domain, because a router can generate and advertise a default route. In this section, let us look at how different protocols generate and advertise default route information.

Configuring Default Routes in RIP

When a default route is configured in a RIP-enabled router, it does not propagate it by default if the route is not a RIP-learned route. Normally, in a routing domain, the default route would point out of the domain. RIP is an IGP, that is, a protocol that runs among the routers within an organization. The organization would be connected to the global Internet via single or multiple ISPs. The router at which the ISP WAN link would be terminated is typically the boundary of the RIP domain with static or BGP routing with the ISP. The default route would point toward the

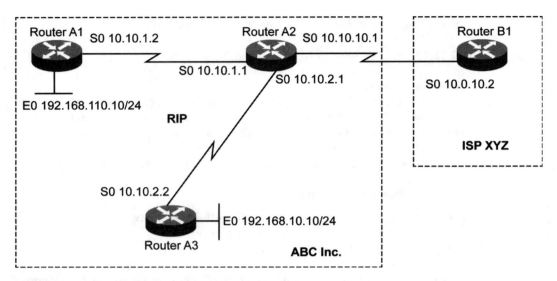

FIGURE 11.8 A complex network configured in RIP.

ISP WAN link, using the IP route 0.0.0.0 0.0.0.0 command or toward an ISP network, using the ip default-network command. In both scenarios, the default route
would not be a RIP-learned route; it would be either static or a BGP route.

There are two options to propagate the default route information: redistributing the static default route and using the default-information-originate at the
boundary router. Let us use the example of an RIP network to check these.

In Figure 11.8, ABCD Inc. has RIP running in its backbone with Routers A1,
A2, and A3. ABCD Inc. is connected to the Internet via ISP XYZ. ABCD Inc. and
ISP XYZ are connected via a WAN link terminating in Routers A2 and B1, respec-
tively, with static routing between them.

In Figure 11.8, ABCD Inc. and ISP XYZ are connected via a WAN link terminating in Routers A2 and B1, respectively.

Listing 11.18 shows the configuration of Router A2.

LISTING 11.18 Configuration of Router A2

```
!
interface Serial0
ip address 10.10.10.1 255.255.255.252
!
interface Serial1
ip address 10.10.1.1 255.255.255.252
!
interface Serial2
```

```
ip address 10.10.2.1 255.255.255.252
!
!
router rip
network 10.10.1.0
network 10.10.2.0
network 10.10.10.0
redistribute static metric 1
!
ip route 0.0.0.0 0.0.0.0 10.10.10.2
!
```

Listing 11.18 shows only that part of the configuration of Router A2 that is relevant to the current context.

In Listing 11.18, the default route is pointed statically toward Router B1 of ISP B. This default route is redistributed and advertised in the RIP domain. This becomes the default route for the entire RIP domain. Listing 11.19 shows a summarized version of the routing table of Router A3 in the RIP domain.

LISTING 11.19 Summarized Version of the Routing Table of A3 in the RIP Domain

```
RouterA3# show ip route
Codes: C - connected, S - static, I - IGRP, R - RIP, M - mobile, B - BGP
   D - EIGRP, EX - EIGRP external, O - OSPF, IA - OSPF inter area
   N1 - OSPF NSSA external type 1, N2 - OSPF NSSA external type 2
   E1 - OSPF external type 1, E2 - OSPF external type 2, E - EGP
   i - IS_IS, L1 - IS_IS level-1, L2 - IS_IS level-2, * - candidate
      default
   U - per-user static route, o - ODR

Gateway of last resort is 192.168.10.1 to network 0.0.0.0
   C  10.10.2.0/30 directly connected, Serial2
   C  10.10.10.0/30 directly connected, Serial0
   C  10.10.1.0/30 directly connected, Serial1
   R  192.168.11.0/24 [1/0] via 10.10.2.1
   R  192.168.10.0/24 [1/0] via 10.10.2.1
   R* 0.0.0.0/0 [1/0] via 192.168.10.1
```

There is another method of configuring and propagating a default route in a RIP domain, as shown in Listing 11.20. In this method, the default-information originate command is used to originate the default route in Router A2 and advertise it in the RIP domain.

LISTING 11.20   Originating the Default Route in A2 and Advertising it in the RIP Domain

```
!
interface Serial0
ip address 10.10.1.1 255.255.255.252
!
interface Ethernet0
ip address 192.168.10.1 255.255.255.0
!
router rip
network 10.10.1.0
network 192.168.10.0
default-information originate
!
ip route 0.0.0.0 0.0.0.0 10.10.1.2
!
```

Listing 11.21 shows the routing table of Router A3 in the RIP domain. The table shows the default route.

LISTING 11.21   Routing Table of A3 in the RIP Domain

```
RouterA3# show ip route
Codes: C - connected, S - static, I - IGRP, R - RIP, M - mobile, B - BGP
   D - EIGRP, EX - EIGRP external, O - OSPF, IA - OSPF inter area
   N1 - OSPF NSSA external type 1, N2 - OSPF NSSA external type 2
   E1 - OSPF external type 1, E2 - OSPF external type 2, E - EGP
   i - IS_IS, L1 - IS_IS level-1, L2 - IS_IS level-2, * - candidate
     default
   U - per-user static route, o - ODR

Gateway of last resort is 192.168.10.1 to network 0.0.0.0
   C  192.168.10.0/24 is directly connected, Ethernet0
   C  10.10.2.0/30 directly connected, Serial0
   R  192.168.11.0/24 [1/0] via 10.10.2.1
   R  192.168.10.0/24 [1/0] via 10.10.2.1
   R* 0.0.0.0/0 [1/0] via 192.168.10.1
```

Configuring Default Routes in IGRP/EIGRP

IGRP and EIGRP routing processes can be configured to accept or deny default routes and to send out updates containing the default route information. In Figure 11.9, Routers A1, A2, and A3 belong to ABCD Inc. and are running EIGRP among themselves. ABCD Inc. is connected to ISP XYZ via a static route.

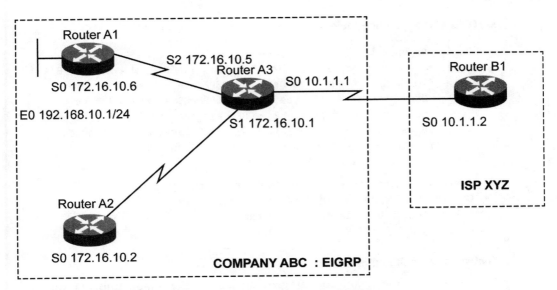

FIGURE 11.9   Routers A1, A2, and A3 of ABCD Inc. running EIGRP.

Router A3 generates a default route and propagates it in the EIGRP routing domain, as shown in Listing 11.22.

LISTING 11.22   Default Route Information Generated by Router A3

```
interface Serial0
ip address 10.1.1.1 255.255.255.252
!
interface Serial1
ip address 172.16.10.1 255.255.255.252
!
interface Serial2
ip address 172.16.10.5 255.255.255.252
!
interface Ethernet0
ip address 172.16.10.1 255.255.255.0
!
router eigrp 200
default-information out
!
ip route 0.0.0.0 0.0.0.0 10.10.1.2
!
```

Listing 11.23 shows the routing table of Router A1, which contains the default route information.

LISTING 11.23   Routing Table of A2

```
RouterA2# show ip route
Codes: C - connected, S - static, I - IGRP, R - RIP, M - mobile, B - BGP
   D - EIGRP, EX - EIGRP external, O - OSPF, IA - OSPF inter area
   N1 - OSPF NSSA external type 1, N2 - OSPF NSSA external type 2
   E1 - OSPF external type 1, E2 - OSPF external type 2, E - EGP
   i - IS_IS, L1 - IS_IS level-1, L2 - IS_IS level-2, * - candidate
     default
   U - per-user static route, o - ODR

Gateway of last resort is 172.16.10.1 to network 0.0.0.0
C   172.16.10.4/30 is directly connected, Serial0
C   192.168.10.0/24 is directly connected, Ethernet0
D*  0.0.0.0/0 [1/0] via 172.16.10.1
```

Configuring Default Routes in OSPF

In case of OSPF, the default route is generated, using the default-information orig-inate command. Figure 11.10 depicts the network of ABCD Inc., using OSPF with Routers A1, A2, and A3. ABCD Inc. is connected to ISP XYZ via a WAN link and configured with static default routing.

Let us configure the network represented in Figure 11.10 to configure the default route and inject into the OSPF routing domain. Listing 11.24 shows the configuration of Router A2.

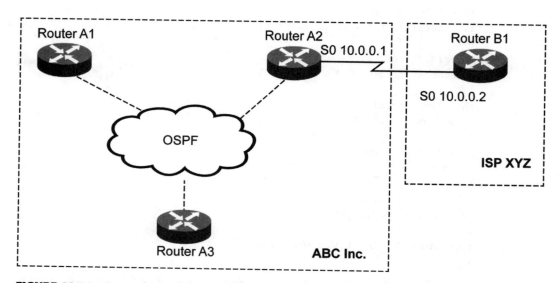

FIGURE 11.10   Network of ABCD Inc. with Routers A1, A2, and A3 connected to ISP XYZ via a WAN link.

LISTING 11.24   Configuration of Router A2

```
!
interface Serial0
ip address 10.10.1.1 255.255.255.252
!
router ospf 1
network 10.0.0.0 0.0.0.255 area 0
default-information originate metric 15
!
ip route 0.0.0.0 0.0.0. 10.10.1.2
!
```

In Router A2, there is a statically configured default route pointing to Router B1. The default-information originate command has been used to generate a default route in the OSPF domain and will be available to all routers in the OSPF domain.

A default route would be generated in this case only if a route to 0.0.0.0 is available in the routing table of Router A3. This default route may not always be a static route; it may be a dynamic default route learned from another routing domain. In such a situation, any flapping of the default route may render the OSPF domain unstable. To avoid this, an always keyword may be added to the default-information command; the command syntax is:

```
default-information originate always metric 15
```

Flapping refers to a state in which a route is intermittently available and unavailable within very short intervals of time.

In such a situation, the default route will always be advertised by Router A3 in the OSPF domain, regardless of whether the default route is available in the router at that time.

SUMMARY

In this chapter, you learned how ODR is used for stub networks. You also learned about the various configuration commands used to enable the ODR routing process. The next chapter will introduce the concepts pertaining to route maps and filtering.

POINTS TO REMEMBER

- On Demand Routing (ODR) is not a conventional routing protocol, because it is used only for limited types of IP networks, such as stub networks.

- Stub networks are those that have a hub and spoke topology. Any traffic forwarded to a stub network is meant for that network only and is not forwarded to any other network.
- ODR has very low resource and maintenance overheads as compared to other dynamic routing protocols.
- Unlike other routing protocols, ODR does not function in Layer 3 of the OSI model. IP prefixes are advertised with Layer 2 Cisco Discovery Protocol (CDP) updates.
- The default timers for ODR are the same as those of CDP; updates are sent every 60 seconds, and routes become invalid if no updates are received every 180 seconds.
- The # show cdp neighbors command is used to show the summary of CDP-enabled neighbors.
- The router odr command is used to configure ODR in the hub router.
- The # cdp timer seconds command modifies the CDP timer.
- The # router odr and # timers basic update invalid holddown flush [sleeptime] commands configure the different timers of ODR.
- The ip default-gateway command is used when IP routing is disabled in a router.
- The ip default-network command is used when IP routing is enabled in the router, unlike the ip default-gateway command.
- The ip route 0.0.0.0 0.0.0.0 NXT HOP command can be used to set the gateway of last resort in a router.

12 ⋮ Route Maps and Filtering

INTRODUCTION TO ROUTE MAPS

Route maps filter the routing updates of routing protocols and control redistribution among different routing protocols. They are also used to configure and control policy-based routing. Route maps add more complexities to access lists to perform multiple operations on selected routes. They contain sequential statements, each containing conditions and actions to perform.

CONFIGURING ROUTE MAPS

In this section, we will be looking at uses and configurations of route maps.

Policy-based Routing

All routing decisions are made on the basis of the destination IP address of the data packet. The routing table is the sum of all the static routes and the dynamically learned routes in a router. When a packet enters a router, the interface to which it has to be switched so that it reaches the destination is decided. The destination IP address is matched in the routing table. The most specific route matching is considered, and the packet is forwarded from the corresponding interface. If there is no match, the IP address is matched with the default route. In the absence of a default route, the packet is dropped.

There may be situations when specific policies need to be configured on traffic entering the router from a specific interface or having some specific source IP address. Cisco IOS provides the option called policy routing that allows data to be routed based on certain policies. These policies can be based on source IP address, protocol, port, or packet size. To implement policy-based routing, the policies need to be configured at the interface where the packet is received. The policy shows the route as available in the routing table.

Route maps are the primary tools for implementing policy routing.

Let us consider different policies that can be configured and the effect of these policies on the packets passing through the router. Policy routing can occur based on:

- Source IP address
- Other parameters

Policy Routing Based on Source IP Address

Consider the example depicted in Figure 12.1. The network company A consists of Routers A1, A2, A3, A4, A5, A6, and A7. Company A is connected to the Internet

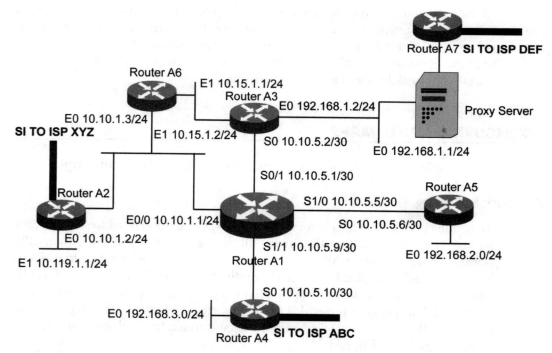

FIGURE 12.1 Policy routing based on source IP address.

via ISPs ABC, DEF, and XYZ. Router A1 is the central router where policy-based routing is configured to influence the routing decision. In Figure 12.1, all the routers are configured without policy routing. The configuration of Router A1 is shown in Listing 12.1.

LISTING 12.1   Configuration of Router A1

```
Interface Ethernet0
Ip address 10.10.1.1 255.255.255.0
!
Interface serial0/1
Ip address 10.10.5.1 255.255.255.252
!
Interface serial1/0
Ip address 10.10.5.5 255.255.255.252
!
Interface serial1/1
Ip address 10.10.5.9 255.255.255.252
!
ip route 192.168.1.0 255.255.255.0 10.10.5.2
ip route 192.168.2.0 255.255.255.0 10.10.5.6
ip route 192.168.3.0 255.255.255.0 10.10.5.10
ip route 10.15.1.0 255.255.255.0 10.10.1.3
ip route 0.0.0.0 0.0.0.0 10.10.1.2
ip route 0.0.0.0 0.0.0.0 10.10.5.10 100
!
```

The configuration of Router A2 is shown in Listing 12.2.

LISTING 12.2   Configuration of Router A2

```
interface Ethernet0
ip address 10.10.1.2 255.255.255.0
!
interface Ethernet1
ip address 10.119.1.1 255.255.255.0
!
Interface serial1
Ip address 172.16.2.1 255.255.255.252
!
ip route 0.0.0.0 0.0.0.0 Serial1
!
```

The configuration of Router A6 is shown in Listing 12.3.

LISTING 12.3 Configuration of Router A6

```
interface Ethernet0
ip address 10.10.1.3 255.255.255.0
!
interface Ethernet1
ip address 10.15.1.1 255.255.255.0
!
ip route 0.0.0.0 0.0.0.0 10.10.1.1
!
```

The configuration of Router A3 is shown in Listing 12.4.

LISTING 12.4 Configuration of Router A3

```
interface Ethernet0
ip address 192.168.1.1 255.255.255.0
!
interface Ethernet1
ip address 10.15.1.2 255.255.255.0
!
Interface serial0/1
Ip address 10.10.5.2 255.255.255.252
!
ip route 0.0.0.0 0.0.0.0 192.168.1.1
```

The configuration of Router A4 is shown in Listing 12.5.

LISTING 12.5 Configuration of Router A4

```
interface Ethernet0
ip address 192.168.2.1 255.255.255.0
!
Interface serial0
Ip address 10.10.5.6 255.255.255.252
!
Interface serial1
Ip address 172.16.1.1 255.255.255.252
!
ip route 0.0.0.0 0.0.0.0 serial1
!
```

The configuration of Router A5 is shown in Listing 12.6.

LISTING 12.6   Configuration of Router A5

```
interface Ethernet0
ip address 192.168.3.1 255.255.255.0
!
Interface serial0/1
Ip address 10.10.5.10 255.255.255.252
!
ip route 0.0.0.0 10.10.5.10 Serial1
```

The routing table at A1 will be as shown in Listing 12.7.

LISTING 12.7   Routing Table at A1

```
RouterA1# show ip route
Codes: C - connected, S - static, I - IGRP, R - RIP, M - mobile, B - BGP
  D - EIGRP, EX - EIGRP external, O - OSPF, IA - OSPF inter area
  N1 - OSPF NSSA external type 1, N2 - OSPF NSSA external type 2
  E1 - OSPF external type 1, E2 - OSPF external type 2, E - EGP
  i - IS_IS, L1 - IS_IS level-1, L2 - IS_IS level-2, * - candidate
    default
  U - per-user static route, o - ODR

Gateway of last resort is 10.1.1.2 to network 0.0.0.0
C   10.10.1.0/24 is directly connected, Ethernet1
C   10.10.5.0/30 is directly connected, Serial0/1
C   10.10.5.4/30 is directly connected, Serial1/0
C   10.10.5.8/30 is directly connected, Serial1/1
S   192.168.1.0 [1/0] via 10.10.5.2
S   192.168.2.0 [1/0] via 10.10.5.6
S   192.168.3.0 [1/0] via 10.10.5.10
S   10.15.1.0 [1/0] via 10.10.1.3
S*  0.0.0.0/0 [1/0] via 10.1.1.2
```

We configure two policies in A1 based on the source IP address of the packets arriving at different interfaces.

Policy 1

The configuration is performed such that all traffic from the network 192.168.2.0 moves to the Internet via ISP ABC, and the primary route to the Internet from A1 is via ISP XYZ. The configuration required for A1 is shown in Listing 12.8.

LISTING 12.8   Configuration of Router A1

```
interface serial1/0
ip address 10.10.5.1 255.255.255.252
```

```
ip policy route map change-def
!
route map change-def permit 10
match ip address 10
set ip next-hop 10.10.5.9
!
access-list 10 permit 192.168.2.0 0.0.0.255
!
```

The default route to the Internet for any packet entering A1 is via A2 to ISP XYZ. In the configuration in Listing 12.8, the default route of any packet from network 192.168.2.0/24 to ISP ABC has been changed to A4. If A4 is inaccessible, the packets from network 192.168.2.0/24 also take the default route via A2 to ISP XYZ.

You can also specify multiple options as the next-hop. If the default route of packets from network 192.168.2.0/24 is to be via ISP DEF, the configuration of A1 will be as shown in Listing 12.9.

LISTING 12.9   Configuration of Router A1

```
interface Serial1/0
ip address 10.10.5.1 255.255.255.252
ip policy route map change-def
!
route map change-def permit 10
match ip address 10
set ip next-hop 10.10.5.9 10.10.5.2
!
access-list 10 permit 192.168.2.0 0.0.0.255
!
```

Consider a scenario where both Routers A4 and A3 are down. In this case, only the default route via ISP XYZ will be considered.

Policy 2

The configuration is performed such that all traffic from network 10.15.1.0 accesses the Internet only via proxy server 192.168.1.1. The configuration is shown in Listing 12.10.

LISTING 12.10   Configuration of Router A1

```
interface Ethernet0/0
ip address 10.10.1.1 255.255.255.0
ip policy route map use-proxy
```

```
!
route map use-proxy permit 10
match ip address 20
set interface Serial0/1
!
access-list 20 permit 10.15.1.0 0.0.0.255
!
```

All traffic from network 10.15.1.0 will be sent to A1, because it is the configuration of the default route. The command route map use-proxy is used to change the egress interface for packets from network 10.15.1.0/24 to Serial0/1, which is connected to A3. This policy has been applied at the interface Ethernet0/0 from where the packets would enter A1.

If A3 is inaccessible, the packets will take the default route from A1 to A2, but per the policy description, the traffic from network 10.15.1.0/24 should not access the Internet from any other route except the proxy server 192.168.1.1. The configuration shown in Listing 12.11 enables the policy.

LISTING 12.11   Configuration for Traffic from 10.15.1.0/24 to Access the Internet via the 192.168.1.1 Proxy Server

```
interface Ethernet0/0
ip address 10.10.1.1 255.255.255.0
ip policy route map use-proxy
!
route map use-proxy permit 10
match ip address 20
set interface Serial0/1 Null0
!
access-list 20 permit 10.15.1.0 0.0.0.255
!
```

In this case, if A3 is inaccessible, all packets from network 10.15.1.0 will be moved via the Null 0 interface; that is, the packets will be dropped.

Policy Routing Based on Other Parameters

Selective policies can be applied based on the source port of a packet. This property is relevant in case of specific port attacks from a specific ingress interface. This type of configuration is used to protect the internal LAN of an organization from various virus and worm attacks. Consider the example illustrated in Figure 12.2. Routers B1-B2 and B1-B3 are connected via a serial link.

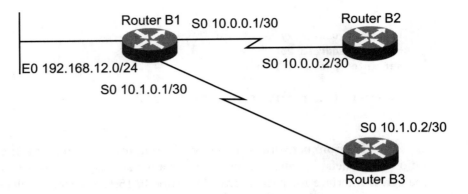

FIGURE 12.2 Policy routing based on packet length and port.

In this example, we are using policy routing based on the packet length and port to ensure that certain types of packets do not reach the network 192.168.12.0/24. The configuration of Router B1 is shown in Listing 12.12.

LISTING 12.12 Configuration of Router B1

```
interface Ethernet0
ip address 192.168.12.1 255.255.255.0
!
interface serial0
ip address 10.10.0.1 255.255.255.252
ip policy route map check-packet1
!
interface serial1
ip address 10.10.1.1 255.255.255.252
ip policy route map check-packet2
!
route map check-packet1 permit 20
match ip address 150
match length 92 96
set interface Null0
!
route map check-packet2 permit 20
match ip address 165
match length 42 42
set interface Null0
!
access-list 150 deny icmp any 192.168.12.0 0.0.0.255 echo
access-list 150 deny icmp any 192.168.12.0 0.0.0.255 echo-reply
```

```
access-list 150 permit icmp any any
!
access-list 165 deny icmp any 192.168.12.0 0.0.0.255 echo
access-list 165 permit icmp any any
!
```

In Listing 12.12, two policy routing instances have been defined corresponding to incoming packets at interfaces Serial 0 and Serial 1 of B1. Both the packet length and the port have been taken into consideration while making the policy routing decision.

The command route map check-packet1 applied at interface S0 matches the IP address and port of the packets using the extended IP access-list 150. The command also matches port of packets with the minimum and maximum packet lengths of 92 and 96, respectively. All packets arriving at B1 via S1 are matched against this route map and dropped by setting the interface as Null 0. Similarly, the packets entering via Interface S1 are matched against access-list 165 and packet length 42 and dropped in case of a match. Both the minimum and maximum packet length values have been set to the same value. All packets that match are prevented from reaching the network 192.168.12.0/24.

Route Tagging

In routing protocols, route tagging is implemented using route maps. Tag values are attached to routes, and redistribution of these routes is controlled using the corresponding tag value. Route tagging prevents routing loops and the non-optimal selection of routes. Tag-based redistribution does not work for RIP and IGRP.

Consider the example illustrated in Figure 12.3. In the network, Routers A1 and A2 run RIP v2, and Routers A4 and A5 run EIGRP. Routers A3 and A6 run both RIP v2 and EIGRP and redistribute between these two protocols.

In the example shown in Figure 12.3, we configure route-tagging using route maps. The configuration of Router A1 is shown in Listing 12.13.

LISTING 12.13 Configuration of Router A1

```
interface Ethernet0
ip address 192.168.15.2 255.255.255.0
!
interface Ethernet1
ip address 192.168.12.1  255.255.255.0
!
router rip
version 2
```

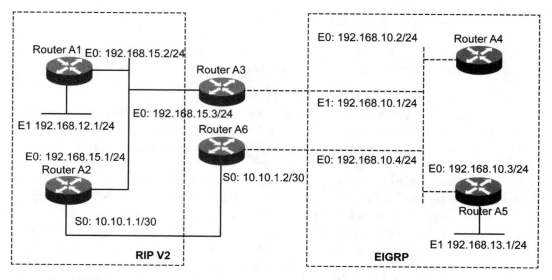

FIGURE 12.3 Route tagging is used to redistribute routes selectively in a network.

```
network 192.168.15.0
network 192.168.12.0
!
```

The configuration of Router A2 is shown in Listing 12.14.

LISTING 12.14 Configuration of Router A2

```
interface Ethernet0
ip address 192.168.15.1 255.255.255.0
!
interface serial0
ip address 10.10.1.1 255.255.255.252
!
router rip
version 2
network 192.168.15.0
!
```

The configuration of Router A3 is shown in Listing 12.15.

LISTING 12.15 Configuration of Router A3

```
interface Ethernet0
ip address 192.168.15.3 255.255.255.0
```

```
!
interface Ethernet1
ip address 192.168.10.1 255.255.255.0
!
router rip
version 2
network 192.168.10.0
network 192.168.15.0
passive-interface Ethernet1
redistribute eigrp 100 route map TEST1 metric 20000 100 250 5 1500
!
router eigrp 100
network 192.168.10.0
network 192.168.15.0
passive-interface Ethernet0
redistribute rip route map TEST2 metric 1
!
route map TEST1 deny 10
match tag 15
route map TEST1 permit 20
set tag 25
!
route map TEST2 deny 10
match tag 25
route map TEST2 permit 20
set tag 15
!
```

The configuration of Router A4 is shown in Listing 12.16.

LISTING 12.16 Configuration of Router A4

```
interface Ethernet0
ip address 192.168.10.2 255.255.255.0
!
router eigrp 100
network 192.168.10.0
!
```

The configuration of Router A5 is shown in Listing 12.17.

LISTING 12.17 Configuration of Router A5

```
interface Ethernet0
ip address 192.168.10.3.255.255.255.0
```

```
!
interface Ethernet1
ip address 192.168.13.1   255.255.255.0
!
router eigrp 100
network 192.168.10.0
network 192.168.13.0
!
```

The configuration of Router A6 is shown in Listing 12.18.

LISTING 12.18   Configuration of Router A6

```
interface Ethernet0
ip address 192.168.10.4 255.255.255.0
!
interface serial0
ip address 10.10.1.2 255.255.255.252
!
router rip
version 2
network 192.168.10.0
network 192.168.15.0
passive-interface Ethernet1
redistribute eigrp 100 route map TEST1 metric 20000 100 250 5 1500
!
router eigrp 100
network 192.168.10.0
network 192.168.15.0
passive-interface Ethernet0
redistribute rip route map TEST2 metric 1
!
route map TEST1 deny 10
match tag 15
route map TEST1 permit 20
set tag 25
!
route map TEST2 deny 10
match tag 25
route map TEST2 permit 20
set tag 15
!
```

In the example depicted in Figure 12.3, mutual redistribution occurs between EIGRP and RIP v2 in A3 and A6. Tags ensure that RIP v2 routes are not redistributed back into RIP v2 after being redistributed into EIGRP. A tag value of 15 is associated with routes redistributed into RIP v2 from EIGRP. The routes with this tag value are denied redistribution into EIGRP from RIP. Similarly, a tag value of 25 is associated with routes redistributed into EIGRP from RIP v2. The routes with this tag value are denied redistribution into RIP v2 from EIGRP.

The tag value associated with a route can be viewed using the show ip route command. The tag value associated with the routes 192.168.12.0/24 and 192.168.13.0/24 and available in A5 and A1, respectively, is as shown in Listing 12.19.

LISTING 12.19 Tag Values Associated with A5 and A1

```
Router A5# show ip route 192.168.12.0
Routing entry for 192.168.12.0/24
Known via "eigrp 100", distance 170, metric 2560512256
Tag 25, type external
Redistributing via eigrp 100
Last update from 192.168.10.4 on Ethernet0, 00:07:22 ago
Routing Descriptor Blocks:
* 192.168.10.4, from 192.168.10.4, 00:07:22 ago, via Serial0
Route metric is 2560512256, traffic share count is 1
Total delay is 20010 microseconds, minimum bandwidth is 1 Kbit
Reliability 1/255, minimum MTU 1 bytes
Loading 1/255, Hops 1
Router A1# show ip route 192.168.13.0
Routing entry for 192.168.13.0/24
Known via "rip", distance 120, metric 2
Tag 15
Redistributing via rip
Last update from 192.168.15.3 on Ethernet0, 00:00:15 ago
Routing Descriptor Blocks:
* 192.168.15.3, from 192.168.15.3, 00:00:15 ago, via Ethernet0
Route metric is 2, traffic share count is 1
```

Route Maps and Redistribution

Route maps are crucial in redistribution of connected and static routes and in the redistribution of routes among different routing protocols. Route maps enable selective redistribution of routes to ensure a routing policy per the requirement. This option also enables the prevention of routing loops and non-optimal routing.

Consider the example as depicted in Figure 12.4. The network has six routers, A1, A2, A3, A4, A5, and A6. Router A3 is the central router where redistribution is configured. A3 is also a part of the OSPF domain consisting of A1 and A2 and of the RIP domain consisting of A5 and A6. Router A3 is also connected to A4 via a WAN link, and static routing is configured for the same.

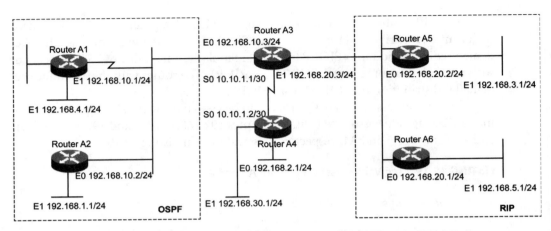

FIGURE 12.4 The network is configured to achieve selective redistribution into RIP.

In Figure 12.4, the network is configured to achieve selective redistribution into RIP. Listing 12.20 shows the configuration of A1.

LISTING 12.20 Configuration of Router A1

```
interface Ethernet0
ip address 192.168.10.1 255.255.255.0
!
interface Ethernet1
ip address 192.168.4.1 255.255.255.0
!
router ospf 1
network 192.168.10.0 0.0.0.255 area 0
network 192.168.4.0 0.0.0.255 area 0
!
```

The configuration of A2 is shown in Listing 12.21.

LISTING 12.21 Configuration of Router A2

```
interface Ethernet0
ip address 192.168.10.2 255.255.255.0
!
interface Ethernet1
ip address 192.168.1.1 255.255.255.0
!
router ospf 1
network 192.168.10.0 0.0.0.255 area 0
```

```
network 192.168.1.0 0.0.0.255 area 0
!
```

The configuration of A3 is shown in Listing 12.22.

LISTING 12.22   Configuration of Router A3

```
interface Ethernet0
ip address 192.168.10.3 255.255.255.0
!
interface Ethernet1
ip address 192.168.20.3 255.255.255.0
!
interface serial0
ip address 10.10.1.1 255.255.255.252
!
router ospf 1
network 192.168.10.0 0.0.0.255 area 0
network 192.168.4.0 0.0.0.255 area 0
network 10.10.1.0 0.0.0.255 area 0
passive-interface Ethernet1
passive-interface serial0
!
router rip
network 192.168.10.0 0.0.0.255
network 192.168.4.0 0.0.0.255
network 10.10.1.0
passive-interface Ethernet0
passive-interface serial0
default-metric 1
redistribute static
redistribute ospf 1
!
ip route 192.168.2.0 255.255.255.0 10.10.1.2
ip route 192.168.30.0 255.255.255.0 10.10.1.2
!
```

The configuration of A5 is shown in Listing 12.23.

LISTING 12.23   Configuration of Router A5

```
interface Ethernet0
ip address 192.168.20.2 255.255.255.0
!
interface Ethernet1
```

```
ip address 192.168.3.1 255.255.255.0
!
router rip
network 192.168.20.0
network 192.168.3.0
!
```

The configuration of A6 is shown in Listing 12.24.

LISTING 12.24 Configuration of Router A6

```
interface Ethernet0
ip address 192.168.20.1 255.255.255.0
!
interface Ethernet1
ip address 192.168.5.1 255.255.255.0
!
router rip
network 192.168.20.0
network 192.168.5.0
!
```

In the configuration shown in Listing 12.24, all the static and OSPF routes are redistributed into RIP at A3. The routing table of A6 in the RIP domain would look like Listing 12.25.

LISTING 12.25 Routing Table for Router A6

```
RouterA6# show ip route
Codes: C - connected, S - static, I - IGRP, R - RIP, M - mobile, B - BGP
    D - EIGRP, EX - EIGRP external, O - OSPF, IA - OSPF inter area
    N1 - OSPF NSSA external type 1, N2 - OSPF NSSA external type 2
    E1 - OSPF external type 1, E2 - OSPF external type 2, E - EGP
    i - IS_IS, L1 - IS_IS level-1, L2 - IS_IS level-2, * - candidate
      default
    U - per-user static route, o - ODR

Gateway of last resort is not set
    C  192.168.5.0/24 is directly connected, Ethernet1
    C  192.168.20.0/24 is directly connected, Ethernet0
    R  192.168.3.0/24 [120/1] via 192.168.20.2, 00:00:12, Ethernet0
    R  192.168.1.0/24 [120/1] via 192.168.20.1, 00:00:12, Ethernet0
    R  192.168.2.0/24 [120/1] via 192.168.20.1, 00:00:15, Ethernet0
    R  10.10.1.0/30 [120/1] via 192.168.20.1, 00:00:12, Ethernet0
    R  192.168.10.0/24 [120/1] via 192.168.20.1, 00:00:13, Ethernet0
```

All routes are available in the routing table of A6. Route maps can be used to perform selective redistribution into RIP. The new configuration of A3 is shown in Listing 12.26.

LISTING 12.26 New Configuration of Router A3

```
interface Ethernet0
ip address 192.168.10.3 255.255.255.0
!
interface Ethernet1
ip address 192.168.20.3 255.255.255.0
!
interface serial0
ip address 10.10.1.1 255.255.255.252
!
router ospf 1
network 192.168.10.0 0.0.0.255 area 0
network 192.168.4.0 0.0.0.255 area 0
network 10.10.1.0 0.0.0.255 area 0
passive-interface Ethernet1
passive-interface Serial0
!
router rip
network 192.168.10.0 0.0.0.255
network 192.168.4.0 0.0.0.255
network 10.10.1.0
passive-interface Ethernet0
passive-interface Serial0
redistribute static route map redis-stat
redistribute ospf 1 route map redis-ospf
!
ip route 192.168.2.0 255.255.255.0 10.10.1.2
ip route 192.168.30.0 255.255.255.0 10.10.1.2
!
route map redis-ospf permit 10
match ip address 10
set metric 2
!
route map redis-stat permit 10
match ip address 10
set metric 1
!
access-list 10 deny 192.168.4.0 0.0.0.255
access-list 10 permit any
```

```
access-list 20 permit 192.168.2.0 0.0.0.255
!
```

The configuration of A3 has been modified to demonstrate the use of route maps in selective redistribution. Route maps redis-ospf and redis-stat are used to control redistribution from static and OSPF.

In case of static routing, we only want the route 192.168.2.0/24 to be redistributed into RIP with a metric of 2. In case of OSPF, all networks other than 192.168.4.0/24 are to be redistributed with a metric of 1. Look at the routing table of A6 shown in Listing 12.27.

LISTING 12.27   Routing Table for Router A6

```
RouterA6# show ip route
Codes: C - connected, S - static, I - IGRP, R - RIP, M - mobile, B - BGP
   D - EIGRP, EX - EIGRP external, O - OSPF, IA - OSPF inter area
   N1 - OSPF NSSA external type 1, N2 - OSPF NSSA external type 2
   E1 - OSPF external type 1, E2 - OSPF external type 2, E - EGP
   i - IS_IS, L1 - IS_IS level-1, L2 - IS_IS level-2, * - candidate
     default
   U - per-user static route, o - ODR

Gateway of last resort is not set
   C   192.168.5.0/24 is directly connected, Ethernet1
   C   192.168.20.0/24 is directly connected, Ethernet0
   R   192.168.3.0/24 [120/1] via 192.168.20.2, 00:00:12, Ethernet0
   R   192.168.4.0/24 [120/1] via 192.168.20.1, 00:00:12, Ethernet0
   R   192.168.1.0/24 [120/1] via 192.168.20.1, 00:00:12, Ethernet0
   R   192.168.2.0/24 [120/1] via 192.168.20.1, 00:00:15, Ethernet0
   R   192.168.30.0/24 [120/1] via 192.168.20.1, 00:00:16, Ethernet0
   R   10.10.1.0/30 [120/1] via 192.168.20.1, 00:00:12, Ethernet0
   R   192.168.10.0/24 [120/1] via 192.168.20.1, 00:00:13, Ethernet0
```

Route maps can be used in redistribution into IGRP or EIGRP routing protocols. Consider the example depicted in Figure 12.5. In the figure, Routers B1, B2, and B3 form the EIGRP routing domain and B2, B4, and B5 form the RIP routing domain.

In Figure 12.5, redistribution of RIP routes into EIGRP occurs in B2, using route maps to control the routes being redistributed. The configuration of router B1 is shown in Listing 12.28.

LISTING 12.28   Configuration of Router B1

```
interface Ethernet0
ip address 192.168.1.1 255.255.255.0
```

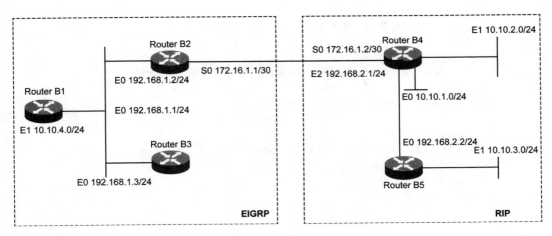

FIGURE 12.5   Redistribution of RIP routes into EIGRP protocols.

```
!
interface Ethernet1
ip address 10.10.4.1 255.255.255.0
!
router eigrp 100
network 10.10.4.0
network 192.168.1.0
!
```

The configuration of Router B2 is shown in Listing 12.29.

LISTING 12.29   Configuration of Router B2

```
interface Ethernet0
ip address 192.168.1.2 255.255.255.0
!
interface serial0
ip address 172.16.1.1 255.255.255.252
!
router eigrp 100
network 172.16.1.0
network 192.168.1.0
passive-interface Serial 0
redistribute rip metric 10000 20 250 5 1500
!
router rip
network 172.16.1.0
network 192.168.1.0
```

```
passive-interface Ethernet0
!
```

The configuration of router B3 is shown in Listing 12.30.

LISTING 12.30   Configuration of Router B3

```
interface Ethernet0
ip address 192.168.1.2 255.255.255.0
!
router eigrp 100
network 192.168.1.0
!
```

The configuration of Router B4 is shown in Listing 12.31.

LISTING 12.31   Configuration of Router B4

```
interface Ethernet0
ip address 10.10.1.1 255.255.255.0
!
interface Ethernet1
ip address 10.10.2.1 255.255.255.0
!
interface Ethernet2
ip address 192.168.2.1 255.255.255.0
!
interface serial0
ip address 172.16.1.2 255.255.255.252
!
router rip
network 10.10.1.0
network 10.10.2.0
network 172.16.1.0
network 192.168.2.0
!
```

The configuration of Router B5 is shown in Listing 12.32.

LISTING 12.32   Configuration of Router B5

```
interface Ethernet0
ip address 192.168.2.2 255.255.255.0
!
```

```
interface Ethernet1
ip address 10.10.3.1 255.255.255.0
!
router rip
network 10.10.3.0
network 192.168.2.0
!
```

The routes learned from RIP are redistributed into EIGRP with the metric values 10000, 20, 250, 5, and 1500. The routing table of Router B3 is shown in Listing 12.33

LISTING 12.33   Routing Table for Router B3

```
RouterB3# show ip route
Codes: C - connected, S - static, I - IGRP, R - RIP, M - mobile, B - BGP
   D - EIGRP, EX - EIGRP external, O - OSPF, IA - OSPF inter area
   N1 - OSPF NSSA external type 1, N2 - OSPF NSSA external type 2
   E1 - OSPF external type 1, E2 - OSPF external type 2, E - EGP
   i - IS_IS, L1 - IS_IS level-1, L2 - IS_IS level-2, * - candidate
     default
   U - per-user static route, o - ODR

Gateway of last resort is not set
   C  192.168.1.0/24 is directly connected, Ethernet0
   D  10.10.4.0/24 [90/1100] via 192.168.1.1, 00:00:30
   D  172.16.1.0/24 [90/1100] via 192.168.1.2, 00:00:32
   D  10.10.1.0/24 [90/3000] via 192.168.1.2, 00:00:32
   D  10.10.2.0/24 [90/3000] via 192.168.1.2, 00:00:31
   D  10.10.3.0/24 [90/3000] via 192.168.1.2, 00:00:33
   D  192.168.2.0/24 [90/3000] via 192.168.1.2, 00:00:32
```

Route maps are used selectively to allow routes for redistribution into EIGRP from RIP. The configuration of Router B2 is shown in Listing 12.34.

LISTING 12.34   Configuration of Router B2

```
interface Ethernet0
ip address 192.168.1.2 255.255.255.0
!
interface serial0
ip address 172.16.1.1 255.255.255.252
!
router eigrp 100
```

```
network 172.16.1.0
network 192.168.1.0
passive-interface serial0
redistribute rip route map redis-rip
!
router rip
network 172.16.1.0
network 192.168.1.0
passive-interface Ethernet0
!
route map redis-rip permit 20
match ip address 15
set metric 10000 20 250 5 1500
!
access-list 15 deny 10.10.2.0 0.0.0.255
access-list 15 permit any
!
```

In Listing 12.34, route map redis-rip is used to redistribute RIP routes selectively into EIGRP. All RIP routes except 10.10.2.0/24 are allowed to be redistributed. Listing 12.35 shows the routing table of B3.

LISTING 12.35 Routing Table of Router B3

```
RouterB3# show ip route
Codes: C - connected, S - static, I - IGRP, R - RIP, M - mobile, B - BGP
   D - EIGRP, EX - EIGRP external, O - OSPF, IA - OSPF inter area
   N1 - OSPF NSSA external type 1, N2 - OSPF NSSA external type 2
   E1 - OSPF external type 1, E2 - OSPF external type 2, E - EGP
   i - IS_IS, L1 - IS_IS level-1, L2 - IS_IS level-2, * - candidate
      default
   U - per-user static route, o - ODR

Gateway of last resort is not set
   C  192.168.1.0/24 is directly connected, Ethernet0
   D  10.10.4.0/24 [90/1100] via 192.168.1.1, 00:00:30
   D  172.16.1.0/24 [90/1100] via 192.168.1.2, 00:00:32
   D  10.10.1.0/24 [90/3000] via 192.168.1.2, 00:00:32
   D  10.10.3.0/24 [90/3000] via 192.168.1.2, 00:00:33
   D  192.168.2.0/24 [90/3000] via 192.168.1.2, 00:00:32
```

Now you will look at the use of route maps to redistribute routes into the OSPF routing protocol. Consider the example depicted in Figure 12.6. Routers B3, B4, and B5 are running OSPF. Routers B1 and B2 are configured using static routing protocols.

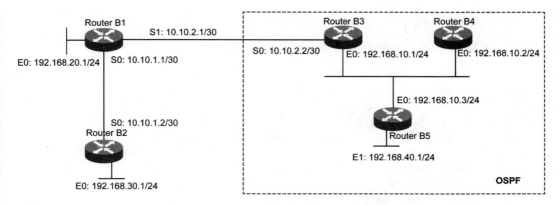

FIGURE 12.6 Use of route maps to redistribute routes into the OSPF routing protocol.

The configuration of Router B1 is shown in Listing 12.36.

LISTING 12.36 Configuration of Router B1

```
interface serial0
ip address 10.10.1.1 255.255.255.252
!
interface serial1
ip address 10.10.2.1 255.255.255.252
!
interface Ethernet0
ip address 192.168.20.1 255.255.255.0
!
ip route 0.0.0.0 0.0.0.0 10.10.2.2
ip route 192.168.30.0 255.255.255.0 10.10.1.2
!
```

The configuration of Router B2 is shown in Listing 12.37.

LISTING 12.37 Configuration of Router B2

```
interface serial0
ip address 10.10.1.2 255.255.255.252
!
interface Ethernet0
ip address 192.168.30.1 255.255.255.0
!
ip route 0.0.0.0 0.0.0.0 10.10.1.1
!
```

The configuration of Router B3 is shown in Listing 12.38.

LISTING 12.38 Configuration of Router B3

```
interface serial0
ip address 10.10.2.2 255.255.255.252
!
interface Ethernet0
ip address 192.168.10.1 255.255.255.0
!
router ospf 1
network 10.10.2.0 0.0.0.255 area 0
network 192.168.10.0 0.0.0.255 area 0
passive-interface serial0
redistribute static metric 10
!
ip route 192.168.20.0 255.255.255.0 10.10.2.1
ip route 192.168.20.0 255.255.255.0 10.10.2.1
ip route 10.10.1.0 255.255.255.252 10.10.2.1
!
```

The configuration of Router B4 is shown in Listing 12.39.

LISTING 12.39 Configuration of Router B4

```
interface Ethernet0
ip address 192.168.10.2 255.255.255.0
!
router ospf 1
network 192.168.10.0 0.0.0.255 area 0
!
```

The configuration of Router B5 is shown in Listing 12.40.

LISTING 12.40 Configuration of Router B5

```
interface Ethernet0
ip address 192.168.10.3 255.255.255.0
!
interface Ethernet1
ip address 192.168.40.1 255.255.255.0
!
router ospf 1
network 192.168.10.0 0.0.0.255 area 0
network 192.168.40.0 0.0.0.255 area 0
!
```

The routing table of Router B4 will be as shown in Listing 12.41.

LISTING 12.41 Routing Table of Router B4

```
RouterB4# show ip route
Codes: C - connected, S - static, I - IGRP, R - RIP, M - mobile, B - BGP
  D - EIGRP, EX - EIGRP external, O - OSPF, IA - OSPF inter area
  N1 - OSPF NSSA external type 1, N2 - OSPF NSSA external type 2
  E1 - OSPF external type 1, E2 - OSPF external type 2, E - EGP
  i - IS_IS, L1 - IS_IS level-1, L2 - IS_IS level-2, * - candidate
    default
  U - per-user static route, o - ODR

Gateway of last resort is not set
  C  192.168.10.0/24 is directly connected, Ethernet0
  O  192.168.30.0/24 [110/25] via 192.168.10.3, 00:34:11, Ethernet0
  O  10.10.2.0/24 [110/25] via 192.168.10.1, 00:32:11, Ethernet0
  O  192.168.20.0/24 [110/25] via 192.168.10.1, 00:34:11, Ethernet0
  O  192.168.40.0/24 [110/25] via 192.168.10.1, 00:34:11, Ethernet0
  O  10.10.1.0/24 [110/25] via 192.168.10.1, 00:34:11, Ethernet0
```

Routes are selectively redistributed into OSPF by using route maps. The new configuration of Router B3 is shown in Listing 12.42.

LISTING 12.42 Configuration of Router B3

```
interface serial0
ip address 10.10.2.2 255.255.255.252
!
interface Ethernet0
ip address 192.168.10.1 255.255.255.0
!
router ospf 1
network 10.10.2.0 0.0.0.255 area 0
network 192.168.10.0 0.0.0.255 area 0
passive-interface serial0
redistribute static route map redis-stat
!
ip route 192.168.20.0 255.255.255.0 10.10.2.1
ip route 192.168.30.0 255.255.255.0 10.10.2.1
ip route 10.10.1.0 255.255.255.252 10.10.2.1
!
route map redis-stat permit 20
match ip address 10
set metric 10
```

```
!
access-list 10 permit 192.168.30.0 0.0.0.255
access-list 10 permit 10.10.1.0 0.0.0.3
!
```

The statically learned networks 192.168.30.0 /24 and 10.10.1.0/30 are redistributed into OSPF, but the network 192.168.20.0/24 is denied access. The routing table at Router B4 is shown in Listing 12.43.

LISTING 12.43   Routing Table for Router B4

```
RouterB4# show ip route
Codes: C - connected, S - static, I - IGRP, R - RIP, M - mobile, B - BGP
  D - EIGRP, EX - EIGRP external, O - OSPF, IA - OSPF inter area
  N1 - OSPF NSSA external type 1, N2 - OSPF NSSA external type 2
  E1 - OSPF external type 1, E2 - OSPF external type 2, E - EGP
  i - IS_IS, L1 - IS_IS level-1, L2 - IS_IS level-2, * - candidate
    default
  U - per-user static route, o - ODR

Gateway of last resort is not set
  C  192.168.10.0/24 is directly connected, Ethernet0
  O  192.168.30.0/24 [110/25] via 192.168.10.3, 00:34:11, Ethernet0
  O  10.10.2.0/24 [110/25] via 192.168.10.1, 00:32:11, Ethernet0
  O  192.168.40.0/24 [110/25] via 192.168.10.1, 00:34:11, Ethernet0
  O  10.10.1.0/24 [110/25] via 192.168.10.1, 00:34:11, Ethernet0
```

ROUTE FILTERING

Route filtering enables you to decide the routes that are required to be a part of the routing table. A router running a distance vector routing protocol sends routes to the neighbors based on the entries in the routing table. These entries form a part of the routing update sent out to the neighbors. Route filtering is used to regulate those routes that are to be sent out in updates and accepted to form a part of the routing table.

In the case of link state routing protocols, the routing table is determined based on entries in the link state database. The updates sent out are not actual route information; they are link states. As a result, route filtering is not relevant in link state routing protocols. However, route filtering can be implemented in case of link state routing protocols. There may not be any control exercised on the Link State Acknowledgments (LSAs) sent and received through the various interfaces, but the routes to be installed in the routing table can be controlled. As a result, there can be inbound route filtering in case of link state routing protocols, whereas distance vector routing protocols provide the flexibility of both inbound and outbound route filtering.

Configuring Route Filtering

In this section, we will discuss the configuration of route filtering in the case of both distance vector and link state routing protocols.

Route Filtering in Distance Vector Routing Protocols

Consider the example shown in Figure 12.7. The figure depicts a RIP routing domain with three routers, B1, B2, and B3.

The configuration of Router B1 without implementing route filtering is shown in Listing 12.44.

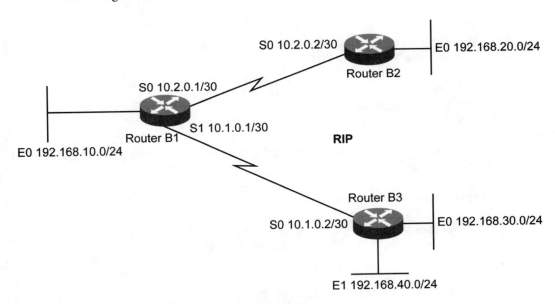

FIGURE 12.7 RIP routing domain with three routers.

LISTING 12.44 Configuration of Router B1

```
interface Ethernet0
ip address 192.168.10.1 255.255.255.0
!
interface serial0
```

```
ip address 10.2.0.1 255.255.255.252
!
interface serial1
ip address 10.1.0.1 255.255.255.252
!
router rip
network 192.168.10.0
network 10.2.0.0
network 10.1.0.0
```

The configuration of Router B2 without implementing route filtering is shown in Listing 12.45.

LISTING 12.45 Configuration of Router B2

```
interface Ethernet0
ip address 192.168.20.1 255.255.255.0
!
interface serial0
ip address 10.2.0.1 255.255.255.252
!
router rip
network 192.168.20.0
network 10.2.0.0
```

The configuration of Router B3 without implementing route filtering is shown in Listing 12.46.

LISTING 12.46 Configuration of Router B3

```
interface Ethernet0
ip address 192.168.30.1 255.255.255.0
!
interface Ethernet1
ip address 192.168.40.1 255.255.255.0
!
interface serial1
ip address 10.1.0.2 255.255.255.252
!
router rip
network 192.168.30.0
network 192.168.40.0
network 10.1.0.0
!
```

The route updates of all routes are sent from all the interfaces when route filtering is not implemented. Let us consider scenarios where inbound and outbound route filtering occur.

Inbound Route Filtering

Consider implementing inbound route filtering at Router B1. We do not want the route 192.168.30.0/24 to be a part of the routing table of B1. The configuration of Router B1 is shown in Listing 12.47.

LISTING 12.47   Configuration of Router B1

```
interface Ethernet0
ip address 192.168.10.1 255.255.255.0
!
interface serial0
ip address 10.2.0.1 255.255.255.252
!
interface serial1
ip address 10.1.0.1 255.255.255.252
!
router rip
network 192.168.10.0
network 10.2.0.0
network 10.1.0.0
distribute-list 2 in Serial1
!
access-list 2 deny 192.168.30.0 0.0.0.255
access-list 2 permit any
!
```

In Listing 12.47, the access-list 2 is used to identify the route that is to be filtered. Access-list 2 is used to filter updates received at interface Serial 1 because the updates of the route to be filtered are received from the interface Serial 1 of B1.

Outbound Route Filtering

Consider implementing inbound route filtering at B1. We do not want the route 192.168.10.0/24 to be advertised to B2 specifically. Listing 12.48 shows the configuration of Router B1.

LISTING 12.48   Configuration of Router B1

```
interface Ethernet 0
ip address 192.168.10.1 255.255.255.0
```

```
!
interface serial0
ip address 10.2.0.1 255.255.255.252
!
interface serial1
ip address 10.1.0.1 255.255.255.252
!
router rip
network 192.168.10.0
network 10.2.0.0
network 10.1.0.0
distribute-list 3 out serial0
!
access-list 3 deny 192.168.10.0 0.0.0.255
access-list 3 permit any
!
```

The access-list 3 identifies the route to be filtered. Access-list 3 is used to filter updates out of interface Serial 0 of B1, which is connected Router B2. This prevents sending updates of network 192.168.10.0/24 to B2.

There can be a maximum of one inbound and one outbound access-list per interface per routing protocol.

Route Filtering in Link State Routing Protocols

In OSPF, route filtering is performed using route maps to prevent certain OSPF-learned routes from being installed in the routing table. The route maps can match any attribute of the OSPF route. Only effective inbound route filtering occurs in the case of an OSPF-routing protocol.

Look at the example shown in Figure 12.8. In this network, there are four routers, A1, A2, A3, and A4, running a link state routing protocol, OSPF.

The configuration of Router A1, without implementing route filtering, is shown in Listing 12.49.

LISTING 12.49 Configuration of Router A1

```
interface Ethernet0
ip address 192.168.11.1 255.255.255.0
!
interface serial0
ip address 10.0.0.1 255.255.255.252
!
```

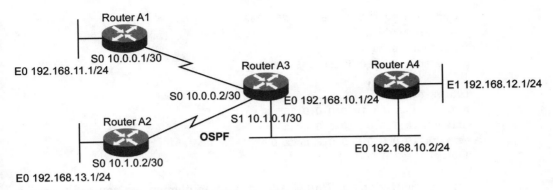

FIGURE 12.8 Route filtering in link state routing protocols.

```
router ospf 1
network 192.168.11.0 0.0.0.255 area 0
network 10.0.0.0 0.0.0.3 area 0
!
```

The configuration of Router A2, without implementing route filtering, is shown in Listing 12.50.

LISTING 12.50 Configuration of Router A2

```
interface Ethernet0
ip address 192.168.13.1 255.255.255.0
!
interface serial0
ip address 10.1.0.2 255.255.255.252
!
router ospf 1
network 192.168.13.0 0.0.0.255 area 0
network 10.1.0.0 0.0.0.3 area 0
!
```

The configuration of Router A3, without implementing route filtering, is shown in Listing 12.51.

LISTING 12.51 Configuration of Router A3

```
interface Ethernet0
ip address 192.168.10.1 255.255.255.0
!
interface serial0
```

```
ip address 10.0.0.2 255.255.255.252
!
interface serial1
ip address 10.1.0.1 255.255.255.252
!
router ospf 1
network 192.168.10.0 0.0.0.255 area 0
network 10.1.0.0 0.0.0.3 area 0
network 10.0.0.0 0.0.0.3 area 0
!
```

The configuration of Router A4, without implementing route filtering, is shown in Listing 12.52.

LISTING 12.52 Configuration of Router A4

```
interface Ethernet0
ip address 192.168.10.2 255.255.255.0
!
interface Ethernet1
ip address 192.168.12.1 255.255.255.0
!
router ospf 1
network 192.168.10.0 0.0.0.255 area 0
network 192.168.12.0 0.0.0.255 area 0
!
```

All the routes derived from the link state database of a router are installed in its routing table because no route filtering has been configured in the routers. In Figure 12.8, consider a scenario where the route 192.168.11.0/24 is made unavailable in the routing table of A1. The configuration of Router A1 is shown in Listing 12.53.

LISTING 12.53 Configuration of Router A1

```
interface Ethernet0
ip address 192.168.11.1 255.255.255.0
!
interface serial0
ip address 10.0.0.1 255.255.255.252
!
router ospf 1
network 192.168.11.0 0.0.0.255 area 0
network 10.0.0.0 0.0.0.3 area 0
distribute-list 9 in
!
```

```
access-list 9 deny 192.168.12.0 0.0.0.255
access-list permit any
!
```

In Listing 12.53, the route 192.168.12.0/24 is prevented from being installed in the routing table of A1, using access-list 9.

In Figure 12.8, consider a scenario where only routes matching the supernet 192.168.0.0/16 are available in the routing table of Router A4. Listing 12.54 shows the configuration of Router A4.

LISTING 12.54   Configuration of Router A4

```
interface Ethernet0
ip address 192.168.10.2 255.255.255.0
!
interface Ethernet1
ip address 192.168.12.1 255.255.255.0
!
router ospf 1
network 192.168.10.0 0.0.0.255 area 0
network 192.168.12.0 0.0.0.255 area 0
distribute-list 15 in
!
access-list 15 permit 192.168.0.0 0.0.255.255
!
```

All routes shown in Figure 12.8, in addition to 10.0.0.0/30 and 10.1.0.0/30, would now be available in the routing table of A4.

Configuring Redistribution and Route Filtering

Redistributed routes can also be filtered specifically. This section discusses the correlation between redistribution and route filtering. Consider the network shown in Figure 12.9. It depicts a redistribution scenario where route filtering is configured. The network is running on both RIP and EIGRP. Router B2 runs solely on RIP,

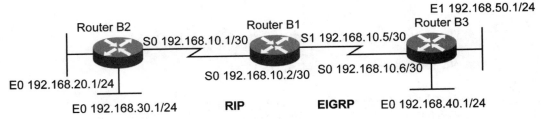

FIGURE 12.9   Configuring redistribution and route filtering.

Router B3 runs only on EIGRP, and Router B1 runs on both RIP and EIGRP and
redistributes between the two protocols.

The configuration of Router B2 is shown in Listing 12.55.

LISTING 12.55 Configuration of Router B2

```
interface Ethernet0
ip address 192.168.20.1 255.255.255.0
!
interface Ethernet1
ip address 192.168.30.1 255.255.255.0
!
interface serial 0
ip address 192.168.10.1 255.255.255.252
!
router rip
network 192.168.10.0
network 192.168.20.0
network 192.168.30.0
!
```

The configuration of Router B3 is shown in Listing 12.56.

LISTING 12.56 Configuration of Router B3

```
interface Ethernet0
ip address 192.168.40.1 255.255.255.0
!
interface Ethernet1
ip address 192.168.50.1 255.255.255.0
!
interface serial0
ip address 192.168.10.6 255.255.255.252
!
router eigrp 100
network 192.168.10.0
network 192.168.40.0
!
```

The configuration of Router B1 is shown in Listing 12.57.

LISTING 12.57 Configuration of Router B1

```
interface serial0
ip address 192.168.10.2 255.255.255.252
```

```
!
interface Serial1
ip address 192.168.10.5 255.255.255.252
!
router rip
network 192.168.10.0
passive-interface Serial1
redistribute eigrp 100 metric 1
redistribute-list 15 out eigrp 100
!
router eigrp 100
network 192.168.10.0
passive-interface Serial0
!
access-list 15 deny 192.168.50.0 0.0.0.255
```

Route filtering has been configured in B1 for redistribution into RIP using access-list 15. Any outgoing update of RIP, sourced from EIGRP originally, is checked against access-list 15. The update is sent in case of a match. In Figure 12.9, the sending of updates, corresponding to the EIGRP-learned route 192.168.50.0/24 through RIP interfaces, is restricted. The RIP interface is Serial 0 in this case.

SUMMARY

In this chapter, you learned about the concept of route filtering with the use of route maps. You also learned about the various configuration commands to enable the route filtering process.

POINTS TO REMEMBER

- Route maps filter the routing updates of routing protocols, control redistribution among different routing protocols, and configure and control policy-based routing.
- Cisco IOS provides the option called policy routing that allows data to be routed based on certain policies.
- Route maps are the primary tools for implementing policy routing.
- Policy routing can occur based on the source IP address and other parameters.
- Selective policies can be applied based on the source port of a packet.
- In routing protocols, route tagging is implemented using route maps.
- Route tagging prevents routing loops and the non-optimal selection of routes.

- Tag-based redistribution does not work for RIP and IGRP.
- The tag value associated with a route can be viewed using the show ip route command.
- Route maps enable selective redistribution of routes to ensure a routing policy and prevention of routing loops and non-optimal routing.
- Route maps redis-ospf and redis-stat are used to control redistribution from OSPF and static protocols.
- Route maps are used selectively to allow routes for redistribution into EIGRP from RIP.
- Route filtering is used to regulate the routes to be sent out in updates and the routes to be accepted to form a part of the routing table.
- In the case of link state routing protocols, the routing table is determined based on entries in the link state database.
- There can be a maximum of one inbound and one outbound access-list per interface per routing protocol.
- In OSPF, route filtering is performed using route maps to prevent certain OSPF-learned routes from being installed in the routing table.

Part

II

Case Studies

13 | BGP Case Study

et us look at a real-life scenario implementing BGP routing policies. A fairly big network is depicted in Figure 13.1, showing three ISPs and two companies running BGP with their respective ISPs. The organizations are Company A (AS 5555), Company B (AS 8888), ISP X(AS 9999), ISP Y(AS 6666), and ISP Z(AS 7777).

Company A has subscribed to WAN bandwidth from ISP Y, whereas Company B is multi-homed to ISP X and ISP Y. ISP X is peering only with ISP Z at a single point of interface, a WAN link between B3 and E1. ISP Z is connected to ISP Y via multiple points, which are B3-A3 and B1-A1. There is no direct connectivity

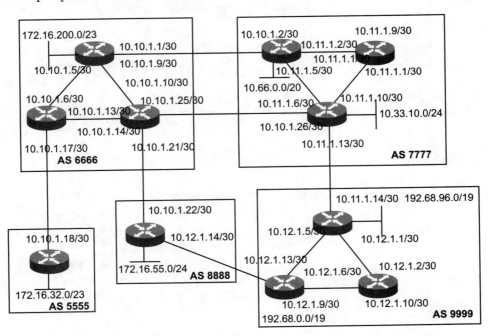

FIGURE 13.1 A case study of BGP.

between ISP X and ISP Y. ISP Z acts as the transit AS for traffic moving from ISP X to ISP Y.

The configuration of all the routers used in this setup are listed in Table 13.1.

TABLE 13.1 Configuration of Routers in BGP Network

ASN	Router	Configuration of Router
6666	A1	router bgp 6666 neighbor 10.10.1.2 remote-as 7777 neighbor 10.10.1.2 route map med1 out neighbor 10.10.1.6 remote-as 6666 neighbor 10.10.1.10 remote-as 6666 network 172.16.200.0 mask 255.255.254.0 ! route map med1 permit 10 set metric 100 !
6666	A2	router bgp 6666 neighbor 10.10.1.18 remote-as 5555 neighbor 10.10.1.14 remote-as 6666 neighbor 10.10.1.10 remote-as 6666
6666	A3	router bgp 6666 neighbor 10.10.1.22 remote-as 8888 neighbor 10.10.1.26 remote-as 7777 neighbor 10.10.1.26 route map med2 out neighbor 10.10.1.13 remote-as 6666 neighbor 10.10.1.9 remote-as 6666 ! route map med2 permit 10 set metric 200 !
5555	C1	router bgp 5555 neighbor 10.10.1.17 remote-as 6666 network 172.16.32.0 mask 255.255.254.0
7777	B1	router bgp 7777 neighbor 10.10.1.1 remote-as 6666 neighbor 10.11.1.1 remote-as 7777 neighbor 10.11.1.6 remote-as 7777 network 10.66.0.0 mask 255.255.240.0

(continued)

TABLE 13.1 *(continued)*

ASN	Router	Configuration of Router
7777	B2	router bgp 7777 neighbor 10.11.1.2 remote-as 7777 neighbor 10.11.1.10 remote-as 7777
7777	B3	router bgp 7777 neighbor 10.11.1.14 remote-as 9999 neighbor 10.11.1.14 route map weight1 in neighbor 10.10.1.25 remote-as 6666 neighbor 10.10.1.25 route map weight2 in neighbor 10.11.1.25 route map as1 out neighbor 10.11.1.1 remote-as 7777 neighbor 10.11.1.9 remote-as 7777 network 10.33.10.0 mask 255.255.255.0 ! route map weight1 permit 10 match ip address 1 set weight 100 ! route map weight1 permit 20 match ip address 2 ! route map weight2 permit 10 match ip address 1 set weight 200 ! route map weight1 permit 20 match ip address 2 ! route map as1 out permit 10 match ip address 3 set as-path-prepend 7777 7777 7777 ! access-list 1 permit 172.16.55.0 0.0.0.255 access-list 2 permit any access-list 3 permit 10.33.10.0 0.0.0.255 !
8888	D1	router bgp 8888 neighbor 10.10.1.22 remote-as 6666

(continued)

TABLE 13.1 *(continued)*

ASN	Router	Configuration of Router
		neighbor 10.12.1.13 remote-as 9999 neighbor 10.12.1.13 distribute-list 2 in network 172.16.55.0 mask 255.255.255.0 ! access-list 2 deny 0.0.0.0 255.255.255.255 access-list 2 permit any !
9999	E1	router bgp 9999 neighbor 10.11.1.13 remote-as 7777 neighbor 10.12.1.6 remote-as 9999 neighbor 10.12.1.2 remote-as 9999 network 192.168.96.0 mask 255.255.2224.0
9999	E2	router bgp 9999 neighbor 10.12.1.14 remote-as 8888 neighbor 10.12.1.5 remote-as 9999 neighbor 10.12.1.10 remote-as 9999 network 192.168.0.0 mask 255.255.2224.0
9999	E3	router bgp 9999 neighbor 10.12.1.1 remote-as 9999 neighbor 10.12.1.9 remote-as 9999

The routing policies that are implemented in the network depicted in Figure 13.1 are:

1. A3 is the preferred entry point to AS 6666 from AS 7777, as compared to A1. This is decided by the MED values that are configured in outgoing updates from A1 and A3.
2. AS 6666 is the preferred route to network 172.16.55.0/24 from AS 7777, as compared to AS 9999. This is decided by the weight values set in the incoming route map in B3.
3. AS 9999 is the preferred path to network 10.33.10.0/24 from AS 8888. The path via AS 6666 is made less preferable because the AS path is longer. The outgoing updates have been modified in B3.
4. A3 in AS 6666 is the preferred exit point to the Internet from AS 8888. The default route is denied updates received from ISP X. There is no such restriction in updates received from ISP Y.

14 | EIGRP Case Study

odham College has standardized on Cisco routers. They use networks having classless addresses that need to be interconnected. Taking advantage of the full Cisco environment, they have identified EIGRP as the routing protocol for the entire campus.

Figure 14.1 shows the way in which the Mathematics, Science, and Arts departments are interconnected. The serial links connecting the three departments use a /30 (i.e., 255.255.255.252) subnet to eliminate the waste of address space.

Rodham College uses an arbitrary process ID of 1 for all its routers. The configuration for the Mathematics department router is shown in Listing 14.1.

LISTING 14.1 Configuration for Mathematics Department Router

```
Mathematics(config)# router eigrp 1
Mathematics(config-router)# network 172.16.0.0
Mathematics(config-router)# network 200.200.1.0
```

The configuration for the Science department router is shown in Listing 14.2

LISTING 14.2 Configuration for Science Department Router

```
Science(config)# router eigrp 1
Science(config-router)# network 172.16.0.0
Science(config-router)# network 200.200.2.0
```

The configuration for the Arts department router is shown in the Listing 14.3.

LISTING 14.3 Configuration for Arts Department Router

```
Arts(config)# router eigrp 1
Arts(config-router)# network 172.16.0.0
Arts(config-router)# network 200.200.3.0
```

Different subnets of 172.16.0.0 would show up in the routing table of the routers as variably subnetted.

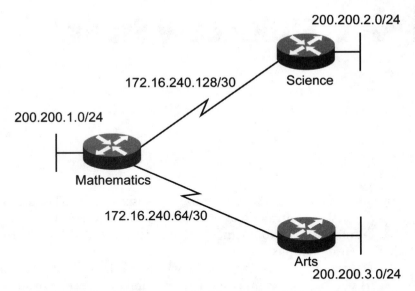

FIGURE 14.1   Interconnection of Mathematics, Science, and Arts departments at Rodham College using EIGRP.

MUTUAL REDISTRIBUTION BETWEEN IGRP AND EIGRP

Rodham College recently decided to connect the Commerce department to its mainstream network. The Commerce department, belonging to an earlier setup, is using the IGRP routing protocol. This posed integration challenges, because IGRP cannot understand different classless subnets.

It was observed that IGRP and EIGRP, both being Cisco proprietary routing protocols, redistribute routing information among them as long as they have the same routing process ID. Figure 14.2 shows how the Commerce department is connected to the college network.

The configuration of the Mathematics router is shown in Listing 14.4. Both EIGRP and IGRP are using the process ID of 1, which would lead to an automatic redistribution of the routing information. The use of passive-interface prevents advertisements of EIGRP out of Serial 2 and IGRP out of Serial 0 and 1.

LISTING 14.4   Mathematics Router Configuration

```
Mathematics(config)# router eigrp 1
Mathematics(config-router)# passive-interface serial 2
Mathematics(config-router)# network 172.16.0.0
Mathematics(config-router)# network 200.200.1.0
```

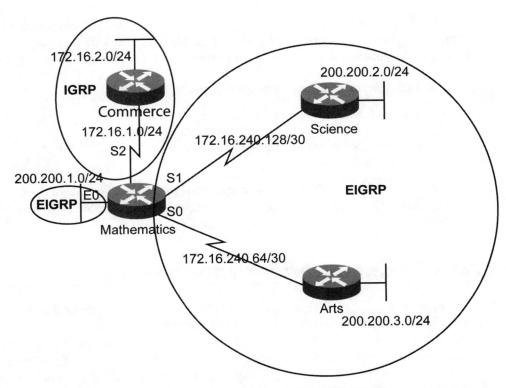

FIGURE 14.2 The Commerce department, which is using IGRP, has been connected to the main network at the Mathematics department.

```
Mathematics(config)# router igrp 1
Mathematics(config-router)# passive-interface serial0
Mathematics(config-router)# passive-interface serial1
Mathematics(config-router)# network 172.16.0.0
```

Similarly, the Commerce router is configured for IGRP. The configuration for the Commerce router is shown in Listing 14.5.

LISTING 14.5 Commerce Router Configuration

```
Commerce(config)# router igrp 1
Commerce(config-router)# network 172.16.0.0
```

While the routing information gets automatically redistributed due to the use of the same process IDs, IGRP does not have the capability to understand the variably subnetted routes to 172.16.240.192/30 and 172.16.240.64/30. The Mathematics router generates a summary address of 172.16.240.0/24 at the Serial 2 interface for ease of understanding the Commerce router, as shown in Listing 14.6.

LISTING 14.6 Summary Address Generation by the Mathematics Router

```
Mathematics(config)# interface serial2
Mathematics(config-if)# ip address 172.16.1.1 255.255.255.0
Mathematics(config-if)# ip summary-address eigrp 1 172.16.240.0
   255.255.255.0
```

This generated summary address would show in the routing table of the Commerce router. The LAN segment of 172.16.2.0/24 would be marked as an external EIGRP route in the routing table of Science and Arts routers.

EIGRP MD5 AUTHENTICATION FOR ROUTING UPDATES

The management of Rodham College received credible information that several students have been trying to hack into the network by introducing their own router within the campus-wide EIGRP network. To prevent such attempts, the management decided to introduce MD5 authentication in the EIGRP routing updates such that no unauthorized router can participate in the campus EIGRP network. However, before rolling out the authentication across all routers in the campus, they want to be sure by enabling it only on the link between Mathematics and Science routers. Figure 14.3 depicts authentication between Mathematics and Science routers but not between Mathematics and Arts routers.

The configuration steps for enabling MD5 authentication on a router are:

```
Router(config)# interface type number : To go to the interface
   configuration mode
Router(config-if)# ip authentication mode eigrp autonomous-system md5:
   Enables the md5 authentication mode for EIGRP updates.
Router(config-if)# ip authentication key-chain eigrp autonomous-system
   key-chain: Links the authentication with its corresponding key-chain
Router(config)# key-chain name-of-chain: Defines the key-chain
Router(config-key-chain)# key number: Defines the key within the
   key-chain
Router(config-key-chain-key)# key-string text: Defines a key string
   within the key
```

The configuration of the Mathematics router is given in Listing 14.7 as having a key-chain named Rodhamchain, a key number of 1, and a key-string rodham.

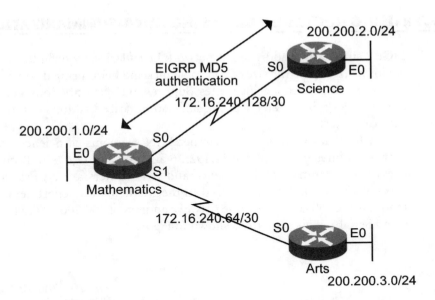

FIGURE 14.3   Mathematics and Science routers use EIGRP MD5 authentication for routing updates with no authentication between Mathematics and Arts.

LISTING 14.7   Configuration of Mathematics Router

```
Mathematics(config)# interface serial0
Mathematics(config-if)# ip authentication mode eigrp 1 md5
Mathematics(config-if)# ip authentication key-chain eigrp 1 Rodhamchain
Mathematics(config)# key-chain Rodhamchain
Mathematics(config-key-chain)# key 1
Mathematics(config-key-chain-key)# key-string rodham
```

Similarly, the configuration of the Science router is given in Listing 14.8.

LISTING 14.8   Configuration of Science Router

```
Science(config)# interface serial0
Science(config-if)# ip authentication mode eigrp 1 md5
Science(config-if)# ip authentication key-chain eigrp 1 Rodhamchain
Science(config)# key-chain Rodhamchain
Science(config-key-chain)# key 1
Science(config-key-chain-key)# key-string rodham
```

There is no authentication between Mathematics and Arts routers.

ADVERTISING ROUTES BY DISABLING AUTO-SUMMARIZATION

EIGRP summarizes addresses at network boundaries, by default. This default behavior helps reduce the size of routing tables and helps bring down network traffic by suppressing more specific routes. Suppression of specific routes can also lead to a routing loop if a router sees the same summarized address coming from more than one interface.

At Rodham College, the Ethernet LAN segments of Science and Arts have network addresses of 200.200.12.192/26 and 200.200.12.64/26. Both of these addresses are summarized by Science and Arts routers at network boundaries of 200.200.12.0/24 and get advertised to the Mathematics department. The Mathematics router observes the same network address of 200.200.12.0/24 on both Serial 0 and Serial 1. This scenario is shown in Figure 14.4.

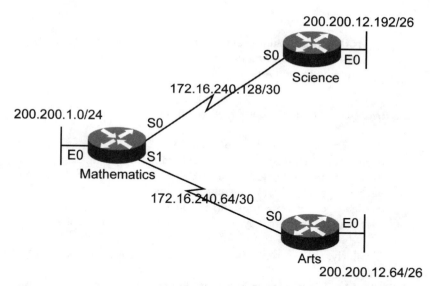

FIGURE 14.4   The Mathematics router forms a loop because of the default auto-summarization by Science and Arts.

The default behavior of auto-summarization in EIGRP should be switched off, using the no auto-summary command, as shown in the configuration in Listing 14.9.

LISTING 14.9   Output of Auto-summarization Command

```
Science(config)# router eigrp 1
Science(config-router)# network 172.16.0.0
Science(config-router)# network 200.200.12.0
Science(config-router)# no auto-summary

Arts(config)# router eigrp 1
Arts(config-router)# network 172.16.0.0
Arts(config-router)# network 200.200.12.0
Arts(config-router)# no auto-summary
```

The no auto-summary command causes the Science and Arts routers to advertise the routes 200.200.12.192/26 and 200.200.12.64/26 to the Mathematics router.

15 IGRP Case Study

The organization Larsen Steel has opened a new plant of small capacity and size. The organization wants to connect the three routers, Anderson, Lee, and Jeff, using IGRP. It has no future plans of scaling to a huge network. The management wants to use a simple routing protocol and save on the cost of highly skilled network administrators.

Larsen Steel has the liberty to use any address of their choice, because the address is not going to be connected to any external network. The organization's consultant has advised them to use three classes of addresses, for security purposes. Figure 15.1 shows Anderson, Lee, and Jeff routers. These routers use different address classes.

The commands used to start the IGRP routing process and send routing updates of specified interfaces are:

```
Router(config)# router igrp as-number
Router(config-router)# network network-number
```

These commands are started in the global configuration mode. The Router(config)# router igrp as-number command enables the IGRP routing process for the particular Autonomous System Number (ASN). The command Router(config-router)# network network-number defines the network in the routing process.

Each of the three routers uses 1 as the process ID or ASN. Anderson advertises the three classful networks of 200.200.1.0, 200.200.2.0, and 200.200.3.0 as shown in Listing 15.1.

LISTING 15.1 Advertising Classful Networks

```
Anderson(config)# router igrp 1
Anderson(config-router)# network 200.200.1.0
Anderson(config-router)# network 200.200.2.0
Anderson(config-router)# network 200.200.3.0
```

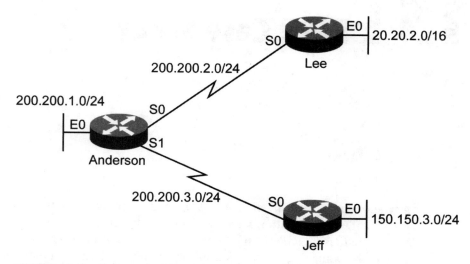

FIGURE 15.1 Classes of addresses are summarized at the network boundaries.

The configurations of Lee and Jeff are shown in Listing 15.2 and Listing 15.3. Specify the classful network address while using the network command.

LISTING 15.2 Configuration of Lee Router

```
Lee(config)# router igrp 1
Lee(config-router)# network 200.200.2.0
Lee(config-router)# network 20.0.0.0
```

LISTING 15.3 Configuration of Jeff Router

```
Jeff(config)# router igrp 1
Jeff(config-router)# network 200.200.3.0
Jeff(config-router)# network 150.150.0.0
```

IGRP does not carry any subnet mask information. As a result, IGRP summarizes the subnets at major network boundaries. Anderson and Jeff see an entry of 20.0.0.0 and not of 20.20.0.0 in their routing tables.

LOAD BALANCING OVER PATHS WITH UNEQUAL COSTS

Larsen Steel observed that the load between Anderson and Jeff has increased and has crossed the design estimates. The budget constraint prevents the management from adding another 1.544-Mbps link connecting two routers. In contrast, they improved the bandwidth by adding another 64-Kbps link and using IGRP for balancing the load across the links.

To understand load balancing across both links, recall the calculation for the IGRP metric.

Metric = [K1*BWigrp(min) + K2*BWigrp(min)/(256-load) + K3*DLYigrp (min)]*[K5/(reliability + K4)]

The default values of the constants are K1 = K3 = 1 and K2 = K4 = K5 = 0. Substituting these values in the equation,

Metric = BWigrp(min) + DLYigrp(min)

Table 15.1 lists the bandwidth and delay values for different media.

TABLE 15.1 Bandwidth and Delay Values for Different Media

Media	BWigrp	DLYigrp
Ethernet	1000	100
1544 Kbps	6476	2000
64 Kbps	156250	2000

The IGRP metric between Anderson and Jeff over the 1.544-Mbps link is BWigrp(min) + DLYigrp(min) = 6476 + [2000 + 100] = 8576.

The IGRP metric between Anderson and Jeff over the 64-Kbps link is BWigrp(min) + DLYigrp(min) = 156250 + [2000 + 100] = 158350.

Figure 15.2 depicts how IGRP balances load between paths having unequal costs. The variance command enables load balancing between different alternate paths having different metrics. The default value of variance is 1, which means that IGRP will load balance only between paths having equal costs.

The metric across the 64-Kbps link is 158350, while that across the 1544-Kbps link is 8576. The metric on the 64-Kbps link is 18.4 times the other. To perform load balancing over these unequal cost paths using IGRP, the value of variance should be 19. The configuration of Anderson and Jeff routers is shown in Listing 15.4 and Listing 15.5.

LISTING 15.4 Configuration of Anderson Router

```
Anderson(config)# router igrp 1
Anderson(config-router)# network 200.200.1.0
Anderson(config-router)# network 200.200.2.0
Anderson(config-router)# network 200.200.3.0
```

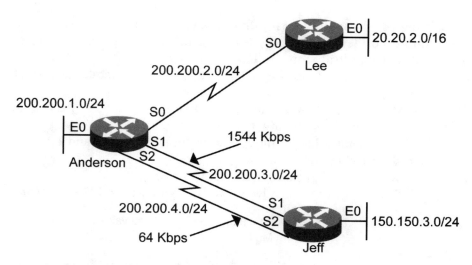

FIGURE 15.2 Larsen Steel uses IGRP to balance load between paths having unequal costs.

```
Anderson(config-router)# network 200.200.4.0
Anderson(config-router)# variance 19
```

LISTING 15.5 Configuration of Jeff Router

```
Jeff(config)# router igrp 1
Jeff(config-router)# network 150.150.0.0
Jeff(config-router)# network 200.200.3.0
Jeff(config-router)# network 200.200.4.0
Jeff(config-router)# variance 19
```

LOAD BALANCING OVER TWO PATHS WITH THIRD AS FALLBACK

Lack of funds forced Larsen Steel to balance the load between the Anderson and Jeff routers across the 1544-Kbps and 64-Kbps links. The metric of the 1544-Kbps link is 19 times more than the other. Therefore, the results obtained after load balancing were not satisfactory. Now, Larsen Steel is financially better and wants to add a 512-Kbps link between the Anderson and Jeff links. It wants to balance the load between 1544-Kbps and 512-Kbps links. Instead of discarding the 64-Kbps link, it wants to keep that configured as a fallback to these two primary links.

The value of the IGRP metric between Anderson and Jeff over the 512-Kbps link is BWigrp(min) + DLYigrp(min) = 19531 + [2000 + 100] = 21631.

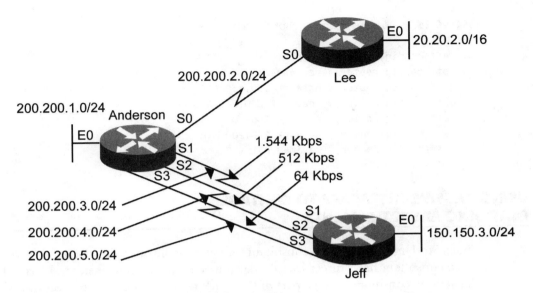

FIGURE 15.3 Larsen Steel uses the third path as the fallback.

The newly added link of 512-Kbps has a metric that is 2.5 times that of the 1544-Kbps link. This would certainly improve the quality of load balancing across the Anderson and Jeff routers. The value of variance does not change, because the newly introduced link has a metric value between the earlier two.

By default, IGRP has maximum paths of four defined for balancing load across multiple paths. If Larsen Steel were to go by this default setting, then it would be balancing across all of the three links instead of keeping the 64-Kbps link as the fallback. As a result, the maximum-paths setting needs to be modified to two. Figure 15.3 shows the use of the third path as fallback.

The configuration of the Anderson and Jeff routers are shown in Listing 15.6 and Listing 15.7.

LISTING 15.6 Modified Configuration for Anderson Router

```
Anderson(config)# router igrp 1
Anderson(config-router)# variance 19
Anderson(config-router)# network 200.200.1.0
Anderson(config-router)# network 200.200.2.0
Anderson(config-router)# network 200.200.3.0
Anderson(config-router)# network 200.200.4.0
Anderson(config-router)# network 200.200.5.0
Anderson(config-router)# maximum-paths 2
```

LISTING 15.7   Modified Configuration for Jeff Router

```
Jeff(config)# router igrp 1
Jeff(config-router)# variance 19
Jeff(config-router)# network 150.150.0.0
Jeff(config-router)#  network 200.200.3.0
Jeff(config-router)# network 200.200.4.0
Jeff(config-router)# network 200.200.5.0
Jeff(config-router)# maximum-paths 2
```

USING PASSIVE-INTERFACE TO CONTROL OUTGOING ADVERTISEMENTS

Larsen Steel hired an external consultant for auditing the operations of Jeff. The external consultant has a direct-leased line connected with Jeff. Larsen Steel has not allowed the consultant to be a part of the IGRP routing process to protect the internal sensitive network. The consultant is using a static route to reach the LAN segment of Jeff. Figure 15.4 shows Jeff using a passive-interface to stop sending the IGRP advertisement.

The network address of the path connecting the external consultant with Jeff is a subnet belonging to 150.150.0.0. Jeff needs to add a passive interface on Serial 1 to ensure that it does not send out IGRP advertisements to the consultant. The configuration of Jeff is shown in Listing 15.8.

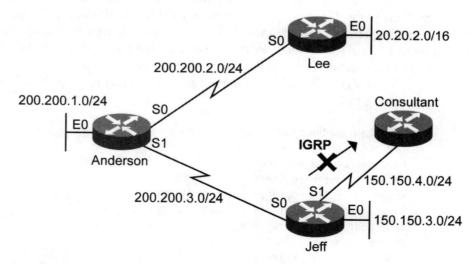

FIGURE 15.4   Jeff uses a passive interface to stop sending the IGRP advertisement.

LISTING 15.8   Configuration of Jeff Router

```
Jeff(config)# router igrp 1
Jeff(config-router)# passive-interface serial 1
Jeff(config-router)# network 150.150.0.0
Jeff(config-router)# network 200.200.3.0
```

CUSTOMIZING TIMERS TO SUIT LINK CONDITIONS

Larsen Steel has temporarily connected a remote site. Being a temporary site, the resources are not adequate. The link speed connecting to the Anderson router is slow, and the router at the remote site is also of low capacity. Considering the constraints, Larsen Steel decided to modify the IGRP timers. Figure 15.5 shows a remote router with a timer to suit the capacity of the slow-link and low-end routers.

The default update value is 90 seconds, invalid value is 270 seconds, hold-down value is 360 seconds, and the flush value is 630 seconds. The remote router has modified the values as shown in Listing 15.9.

LISTING 15.9   Modified Values of Remote Router

```
Remote(config-router)# timers basic 180 540 720 1260
```

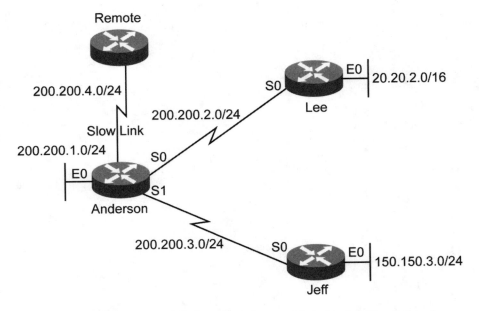

FIGURE 15.5   Remote router showing the connectivity with the slow-link and low-end router.

16 IS_IS Case Study

This case study will provide an understanding of the real configuration of the IS_IS protocol in a Cisco router. The scenario depicted in Figure 16.1 is discussed in detail in the section "Scenario Description."

SCENARIO DESCRIPTION

There are three intermediate systems/routers (Router 1, Router 2, and Router 3) in the IS_IS network depicted in Figure 16.1.

> IP address of Router 1: 172.32.1.1
>
> IP address of Router 2: 172.32.2.1
>
> IP address of Router 3: 172.32.3.1

Router 1 is connected to Router 2 using the Fast Ethernet FE1. Router 2 is connected to Router 3 using the interface pos3. The IP addresses of the individual interfaces are given in Figure 16.1.

CONFIGURATION SECTION

The configuration section deals with the configuration of the individual intermediate systems in the IS_IS network for Figure 16.1. Listing 16.1 shows the configuration for Router1.

LISTING 16.1   Router 1 Configuration

```
interface create ip fe1/1 address-netmask 14.2.2.8/24 port et.1.1
interface create ip fe2/1 address-netmask 172.32.1.1/24 port gi.3.1
!
```

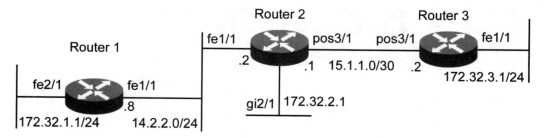

FIGURE 16.1 IS_IS case study scenario.

```
isis add area 49.0001
isis add interface fe1/1
isis add interface fe2/1
isis set system-id 0000.0000.0001
isis start
!
ip-router policy redistribute from-proto direct to-proto isis-level-1
!
system set name Router1
```

The configuration of Router 1 starts with the creation of individual interfaces. For Router 1, FE1 and FE2 represent the Fast Ethernet interfaces that connect Router 1 to Router 2 or to any other router in the network. The area in which Router 1 is available is added after creating interfaces and assigning IP addresses. After creating interfaces in the area, interfaces are added to the area just added. In the case of Router 1, interfaces FE1 and FE2 are added to the area with the area ID 49.001. After adding interfaces to the area, the system ID of Router 1 is configured as 0000.0000.0001. After performing these steps, the IS_IS routing protocol is enabled in this intermediate system (Router 1) using the command isis start. The routing policy for Router 1 is set for redistribution using the command ip-router policy redistribute. In the end, the system is assigned the name Router 1, which completes the configuration of Router 1.

Router 2 and Router 3 use similar configuration steps as discussed for Router 1 for their configuration in the IS_IS network. Listing 16.2 shows the configuration for Router 2.

LISTING 16.2 Router 2 Configuration

```
interface create ip pos3/1 address-netmask 15.1.1.1/30 port so.3.1
interface create ip fe1/1 address-netmask 14.2.2.2/24 port et.1.1
```

```
interface create ip gi2/1 address-netmask 172.32.2.1/24 port gi.2.1
!
isis add area 49.0002
isis add interface pos3/1
isis add interface fe1/1
isis add interface gi2/1
isis set system-id 0000.0000.0002
isis start
!
ip-router policy redistribute from-proto direct to-proto isis-level-1
!
system set name Router2
```

The Router 2 configuration starts creating interfaces and assigning IP addresses to them. An area with area ID 49.0002 is added to the IS_IS network, and the interfaces of Router 2 (pos3, FE1, gi2) are added to this area. After this step, the system ID for Router 2 is set to 0000.0000.0002, and the IS_IS routing protocol is enabled in Router 2, using the command isis start. In the end, the redistribution policy and system name of Router 2 are set, using commands ip-router policy redistribute and system set name.

Listing 16.3 shows the configuration for Router 3.

LISTING 16.3   Router 3 Configuration

```
interface create ip pos3/1 address-netmask 15.1.1.2/30 port so.3.1
interface create ip fe1/1 address-netmask 172.32.3.1/24 port et.1.1
!
isis add area 49.0003
isis add interface pos3/1
isis set system-id 0000.0000.0003
isis start
!
ip-router policy redistribute from-proto direct to-proto isis-level-1
!
system set name Router3
```

Configuration of Router 3 begins with the creation of two-interface pos3 and FE1 and the addition of area with the area ID 49.0003 to the IS_IS network. The interface pos3/1 is added to this area with the ID 49.0003. The system ID of Router 3 is set to 0000.0000.0003, and the IS_IS routing protocol is enabled on Router 3. In the end, the redistribution policy and system name of Router 3 are set, using commands ip-router policy redistribute and system set name.

VERIFICATION SECTION

The verification section deals with the various IS_IS commands that are used for verifying the configuration for the scenario depicted in Figure 16.1. Various commands that can be used for verification of IS_IS configurations are shown in Listing 16.4.

LISTING 16.4   Verification Commands for IS_IS Configurations

```
Router2# isis show adjacencies
Adjacencies
Circuit name: pos3/1
  SystemID: 0000.0000.0200
  State: up Usage: 10 Type: isHold: 30
  3-way: Neighbor Circuit: 00000005
  Areas: 49.0003
  Supported protocols: inet4
  Ifaddr: 15.1.1.2
Circuit name: fe1/1
  Level 1 adjacencies: 1
  SystemID: 0000.0000.0300Snpa: 802.2 0:e0:63:18:f6:61Pri: 64
  State: up Type: l1-is Hold: 9
  Areas: 49.0001
  Supported protocols: inet4
  Ifaddr: 14.2.2.8
  No level 2 adjacencies
```

The command in Listing 16.4 shows all the adjacency information for Router 2. These give the interface ID as Ifaddr. For instance, the Ifaddr of pos3 is 15.1.1.2. In addition, this command specifies the name of the interface (circuit name) and the level of adjacencies supported, as well as supported protocols for these interfaces.

Listing 16.5 shows link state database for Router 2.

LISTING 16.5   Link State Database for Router 2

```
Router2# isis show lsp-database
Task ISIS: Link State Database
IS-IS Level-1 Link State Database (* - originated local)
LSPID LSP Seq Num LSP Checksum LSP Holdtime ATT/P/OL
0000.0000.0200.0000 0X00000005 0X4334 502 0/0/0
0000.0000.0100.0300 0X00000004 0XEBE8 775 0/0/0
0000.0000.0100.0000* 0X00000000 0X1C9A 822 1/0/0
IS-IS Level-2 Link State Database (* - originated local)
```

```
LSPID LSP Seq Num LSP Checksum LSP Holdtime ATT/P/OL
0000.0000.0200.0000* 0X00000028 0XC876 651 0/0/0
0000.0000.0100.0000 0X00000021 0XA050 888 0/0/0
```

The command in Listing 16.5 shows the details of the IS_IS LSP database for Router2. It provides details about the LSPID, LSP Sequence Number, Checksum, and the Hold Time for Level 1 and Level 2 link state database of Router 2.

Listing 16.6 shows the routes for Router 1 using the command ip show routes.

LISTING 16.6   Router 1 Routes

```
Router1# ip show routes
Destination           Gateway              Owner      Netif
-----------           -------              -----      -----
default               14.2.2.2             ISIS_L1    fe1/
15.1.1.0/30           14.2.2.2             ISIS_L1    fe1/1
14.2.2.0/24           directly connected   -          fe1/1
172.32.1.0/24         directly connected   -          fe2/1
172.32.2.0/24         14.2.2.2             ISIS_L1    fe1/1
```

This command shows routes for Router 1. It shows details of the route followed by an LSP generated by Router 1.

17 ODR Case Study

The network of ABCD Inc. is connected to ISP XYZ using static routes. The network of ABCD Inc. is partial stub, where Router A1 is the hub and Routers A2, A3, A4, and A5 are the spokes. ODR is used in this part of the network. This is depicted in Figure 17.1.

The rest of the network, which consists of Routers A1, B1, and B2, is running OSPF. Router B2 is the OSPF Autonomous System Boundary Router (ASBR), which is connected to Router C1 of ISP XYZ via static routes.

ODR is configured in the stub part of the network. This information is partially redistributed into OSPF in Router A1 by filtering ODR updates. The updates of the network 192.168.50.0/24 have been filtered out, while updates of all other networks have been included. Further, the default route is added in Router B2 and propagated to all routers in the OSPF domain.

Listing 17.1 shows the configuration of Router A1 in the network depicted in Figure 17.1.

LISTING 17.1 Configuration of Router A1

```
interface Serial1/0:1
ip address 10.1.0.1 255.255.255.252
!
interface Serial1/0:2
ip address 10.1.0.5 255.255.255.252
!
interface Serial1/0:3
ip address 10.1.0.9 255.255.255.252
!
interface Serial1/0:4
ip address 10.1.0.13 255.255.255.252
!
interface FastEthernet1/1
ip address 172.16.1.1 255.255.255.0
```

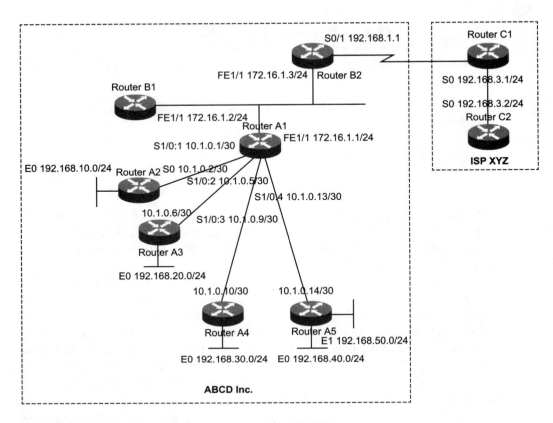

FIGURE 17.1 Network of ABCD Inc. connected to ISP XYZ.

```
!
router odr
distribute-list 10 in
!
router ospf 1
network 172.16.1.0 0.0.0.255 area 0
passive-interface Serial1/0:1
passive-interface Serial1/0:2
passive-interface Serial1/0:3
passive-interface Serial1/0:4
redistribute odr metric 10 subnets
!
access-list 10 deny 192.168.50.0 0.0.0.255
access-list 10 permit any
!
```

Listing 17. 2 shows the configuration of Router A2.

LISTING 17.2   Configuration of Router A2

```
!
interface Serial0
ip address 10.1.0.2 255.255.255.252
!
interface Ethernet0
ip address 192.168.10.1 255.255.255.0
!
```

Listing 17.3 shows the configuration of Router A3.

LISTING 17.3   Configuration of Router A3

```
interface Serial0
ip address 10.1.0.6 255.255.255.252
!
interface Ethernet0
ip address 192.168.20.1 255.255.255.0
!
```

Listing 17.4 shows the configuration of Router A4.

LISTING 17.4   Configuration of Router A4

```
interface Serial0
ip address 10.1.0.10 255.255.255.252
!
interface Ethernet0
ip address 192.168.30.1 255.255.255.0
!
```

Listing 17.5 shows the configuration of Router A5.

LISTING 17.5   Configuration of Router A5

```
interface Serial0
ip address 10.1.0.14 255.255.255.252
!
interface Ethernet0
ip address 192.168.40.1 255.255.255.0
```

```
!
interface Ethernet1
ip address 192.168.50.1 255.255.255.0
!
```

Listing 17.6 shows the configuration of Router B2.

LISTING 17.6 Configuration of Router B2

```
interface Serial0/1
ip address 192.168.1.1 255.255.255.0
!
interface FastEthernet1/1
ip address 172.16.1.3 255.255.255.0
!
router ospf 1
network 172.16.1.0 0.0.0.255 area 0
network 192.168.1.0 0.0.0.3 area 0
default-information originate metric 15
!
ip route 0.0.0.0 0.0.0.0 192.168.1.2
```

In Router A1, ODR has been configured, with some filtering. Access-list 10 is being used to filter out network 192.168.50.0/24; all other networks are being allowed. The ODR information is redistributed into OSPF in Router A1. At Router B3, the default route is injected into the OSPF domain. Listing 17.7 shows the routing table of a few routers in ABCD Inc.

LISTING 17.7 Routing Tables of Routers in Network of ABCD Inc.

```
RouterA4#show ip route
Codes: C - connected, S - static, I - IGRP, R - RIP, M - mobile, B - BGP
   D - EIGRP, EX - EIGRP external, O - OSPF, IA - OSPF inter area
   N1 - OSPF NSSA external type 1, N2 - OSPF NSSA external type 2
   E1 - OSPF external type 1, E2 - OSPF external type 2, E - EGP
   i - IS-IS, L1 - IS-IS level-1, L2 - IS-IS level-2, * - candidate
     default
   U - per-user static route, o - ODR

Gateway of last resort is 10.1.0.9 to network 0.0.0.0
   C  10.1.0.0/24 is directly connected, Serial0
   C  192.168.30.0/24 is directly connected, Ethernet0
   o* 0.0.0.0/0 [100/1] via 3.3.4.1, 00:00:39, Serial0
```

```
RouterB1# show ip route
Codes: C - connected, S - static, I - IGRP, R - RIP, M - mobile, B - BGP
   D - EIGRP, EX - EIGRP external, O - OSPF, IA - OSPF inter area
   N1 - OSPF NSSA external type 1, N2 - OSPF NSSA external type 2
   E1 - OSPF external type 1, E2 - OSPF external type 2, E - EGP
   i - IS-IS, L1 - IS-IS level-1, L2 - IS-IS level-2, * - candidate
     default
   U - per-user static route, o - ODR

Gateway of last resort is 172.16.1.1 to network 0.0.0.0
   C  172.16.1.0/24 is directly connected, Ethernet0
   O  E1 192.168.10.0/24 [110/12] via 172.16.1.1, 00:34:00 Ethernet0
   O  E1 192.168.20.0/24 [110/12] via 172.16.1.1, 00:33:01, Ethernet0
   O  E1 192.168.30.0/24 [110/12] via 172.16.1.1, 00:33:00, Ethernet0
   O  E1 192.168.40.0/24 [110/12] via 172.16.1.1, 00:34:01, Ethernet0
   O* 0.0.0.0/0 [110/15] via 172.16.1.1, 00:32:11, Ethernet0
```

18 OSPF Case Study

The objective of this case study is to make you familiar with the real scenario configuration of OSPF in multiple areas. There are no default routes configured for Autonomous System Boundary Routers (ASBRs), as shown in Figure 18.1.

The equipment used in the OSPF case study depicted in Figure 18.1 is:

1. ABR-1—Cisco Router 2621
2. ABR-2—Cisco Router 1720
3. ASBR—Cisco Router 2622
4. Internal Router (IR)—Cisco Router 1720
5. ABR-1 connects Area 31 to Area 0
6. ABR-2 connects Area 2 to Area 0

In Figure 18.1, ASBR connects Area 0 with another Autonomous System1, which has routers with the router IDs 11.0.0.0, 12.0.0.0, and 13.0.0.0 through the Ethernet interface 10.1.0.0.

The configurations of ABR-1, ABR-2, and IR all follow the normal sequence of configuration steps as listed:

1. Configure host name using hostname command.
2. Assign IP addresses to Ethernet and Serial interfaces.
3. Enable OSPF routing protocol on the router, using the router ospf command, and identify network interface that is a part of OSPF network, using the network command.
4. Set priority of the router. This is an optional step.

Listing 18.1 shows the general ASBR configuration.

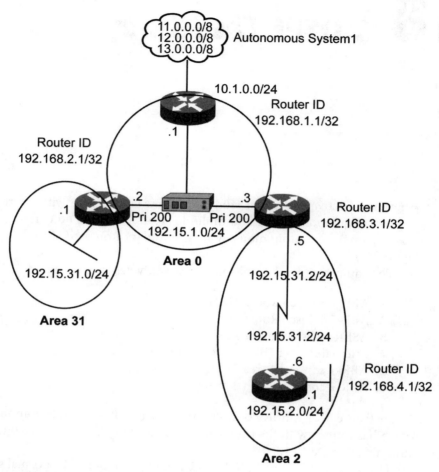

FIGURE 18.1 An OSPF Configuration.

LISTING 18.1 General ASBR Configuration

```
!
hostname ASBR
!
interface Loopback0
ip address 192.168.1.1 255.255.255.255
!
interface FastEthernet0/0
ip address 192.15.1.1 255.255.255.0
!
```

```
interface FastEthernet0/1
ip address 10.1.0.0 255.255.255.0
!
router ospf 1
redistribute static
network 192.15.1.0 0.0.0.255 area0
!
ip classless
ip route 11.0.0.0 255.0.0.0 Null0
ip route 12.0.0..0 255.0.0.0 Null0
ip route 13.0.0.0 255.0.0.0 Null0
!
```

Listing 18.2 shows the ABR-1 configuration.

LISTING 18.2   ABR-1 Configuration

```
!
hostname ABR-1
!
interface Loopback0
ip address 192.168.2.1 255.255.255.255
!
interface Loopback1
ip address 192.15.31.1 255.255.255.0
!
interface FastEthernet0/0
ip address 192.15.1.2 255.255.255.0
ip ospf priority 400
!
router ospf 1
network 192.15.1.0 0.0.0.255 area 0
network 192.15.31.0 0.0.0.255 area 31
!
ip classless
!
```

Listing 18.3 shows the ABR-2 configuration.

LISTING 18.3   ABR-2 Configuration

```
!
hostname ABR-2
!
interface Loopback1
ip address 192.168.3.1 255.255.255.255
!
```

```
interface Serial0
ip address 192.15.1.5 255.255.255.252
clockrate 64000
!
interface FastEthernet0
ip address 192.15.2.1 255.255.255.0
!
router ospf 1
network 192.15.1.0 0.0.0.255 area 0
network 192.15.1.2 0.0.0.3 area 2
!
ip classless
!
```

Listing 18.4 shows the IR configuration.

LISTING 18.4   IR Configuration

```
!
hostname internal-router
!
interface Loopback0
ip address 192.168.4.1 255.255.255.255
!
interface Serial0
ip address 192.15.1.6 255.255.255.252
clockrate 64000
!
interface FastEthernet0
ip address 192.15.2.1 255.255.255.0
!
router ospf 1
network 192.15.2.0 0.0.255.255 area 2
!
ip classless
!
```

The output for Router ASBR# show ip route command is shown in Listing 18.5.

LISTING 18.5   Output for Router ASBR# show ip route Command

```
Router ASBR# show ip route
Codes: C - connected, S - static, I - IGRP, R - RIP, M - mobile, B - BGP
   D - EIGRP, EX - EIGRP external, O - OSPF, IA - OSPF inter area
   N1 - OSPF NSSA external type 1, N2 - OSPF NSSA external type 2
   E1 - OSPF external type 1, E2 - OSPF external type 2, E - EGP
```

```
   i - IS-IS, L1 - IS-IS level-1, L2 - IS-IS level-2, ia - IS-IS inter area
   * - candidate default, U - per-user static route, o - ODR
   P - periodic downloaded static route

Gateway of last resort is not set
       192.15.0.0/16 is variably subnetted, 4 subnets, 3 masks
  O IA 192.15.31.1/32 [110/2] via 192.15.1.2, 00:02:54, FastEthernet0/0
  O IA 192.15.2.0/24 [110/783] via 192.15.1.3, 00:02:54,
     FastEthernet0/0
  O IA 192.15.1.2/30 [110/782] via 192.15.1.3, 00:02:54,
     FastEthernet0/0
  C 192.15.1.0/24 is directly connected, FastEthernet0/0
       10.0.0.0/24 is subnetted, 1 subnets
  C  10.1.0.0 is directly connected, FastEthernet0/1
  S  11.0.0.0/8 is directly connected, Null0
  S  12.0.0.0/8 is directly connected, Null0
       192.168.1.0/32 is subnetted, 1 subnets
  C 192.168.1.1 is directly connected, Loopback0
  S  13.0.0.0/8 is directly connected, Null0
```

The show ip route command on the ASBR shows the OSPF inter-area routes represented by IA. The IP addresses of the inter-areas are interconnected using a specific interface. For example, 172.16.51.1 is connected via 192.15.1.2 using the Fast Ethernet interface. The output for Router ASBR# show ip ospf neighbor command is shown in Listing 18.6.

LISTING 18.6   Output for Router ASBR# show ip ospf neighbor Command

```
Router ASBR# show ip ospf neighbor
Neighbor ID Pri State Dead Time Address Interface
192.168.3.1 100 FULL/BDR 00:00:37 192.15.1.3 FastEthernet0/0
192.168.2.1 200 FULL/DR 00:00:33 192.15.1.2 FastEthernet0/0
```

The router with the highest ip-ospf priority is the DR, and BDR is the router with the next-highest priority value. Listing 18.7 shows the output for the Router ASBR# show ip ospf command.

LISTING 18.7   Output for Router ASBR# show ip ospf Command

```
Router ASBR# show ip ospf
Routing Process "ospf 1" with ID 192.168.1.1
Supports only single TOS(TOS0) routes
It is an autonomous system boundary router
Redistributing External Routes from,
```

```
static
SPF schedule delay 15 secs, Hold time between two SPFs 20 secs
Minimum LSA interval 5 secs. Minimum LSA arrival 1 secs
Number of external LSA 3. Checksum Sum 0x97E3
Number of DCbitless external LSA 0
Number of DoNotAge external LSA 0
Number of areas in this router is 1. 1 normal 0 stub 0 nssa
External flood list length 0
Area BACKBONE(0)
Number of interfaces in this area is 1
Area has no authentication
SPF algorithm executed 36 times
Area ranges are
Number of LSA 8. Checksum Sum 0x507DB
Number of DCbitless LSA 0
Number of indication LSA 0
Number of DoNotAge LSA 0
Flood list length 0
```

The show ip ospf command on ASBR shows that the router is located at the Backbone area. Other details about ASBR are also available in the output. Listing 18.8 shows the output for the Router ABR-1#s how ip ospf command.

LISTING 18.8   Output for Router ABR-1# show ip ospf Command

```
Router ABR-1# show ip ospf
Routing Process "ospf 1" with ID 192.168.2.1
Supports only single TOS(TOS0) routes
It is an area border router
SPF schedule delay 5 secs, Hold time between two SPFs 10 secs
Minimum LSA interval 5 secs. Minimum LSA arrival 1 secs
Number of external LSA 3. Checksum Sum 0x97E3
Number of DCbitless external LSA 0
Number of DoNotAge external LSA 0
Number of areas in this router is 2. 2 normal 0 stub 0 nssa
External flood list length 0
Area BACKBONE(0)
Number of interfaces in this area is 1
Area has no authentication
SPF algorithm executed 38 times
Area ranges are
Number of LSA 11. Checksum Sum 0x752B2
Number of DCbitless LSA 0
Number of indication LSA 0
```

```
Number of DoNotAge LSA 0
Flood list length 0
Area 31
Number of interfaces in this area is 1
Area has no authentication
SPF algorithm executed 17 times
Area ranges are
Number of LSA 5. Checksum Sum 0x32392
Number of DCbitless LSA 0
Number of indication LSA 0
Number of DoNotAge LSA 0
Flood list length 0
```

The show ip ospf command on router ABR-1 shows that it connects the backbone area and Area 31. Listing 18.9 shows the output for the Router ABR-2# show ip ospf command.

LISTING 18.9   Output for Router ABR-2# show ip ospf Command

```
Router ABR-2# show ip ospf
Routing Process "ospf 1" with ID 192.168.3.1
Supports only single TOS(TOS0) routes
It is an area border router
SPF schedule delay 5 secs, Hold time between two SPFs 10 secs
Minimum LSA interval 5 secs. Minimum LSA arrival 1 secs
Number of external LSA 3. Checksum Sum 0x91E6
Number of DCbitless external LSA 0
Number of DoNotAge external LSA 0
Number of areas in this router is 2. 1 normal 1 stub 0 nssa
External flood list length 0
Area BACKBONE(0)
Number of interfaces in this area is 1
Area has no authentication
SPF algorithm executed 10 times
Area ranges are
Number of LSA 8. Checksum Sum 0x4FFDF
Number of DCbitless LSA 0
Number of indication LSA 0
Number of DoNotAge LSA 0
Flood list length 0
Area 2
Number of interfaces in this area is 1
It is a stub area, no summary LSA in this area
generates stub default route with cost 1
Area has no authentication
```

```
SPF algorithm executed 10 times
Area ranges are
Number of LSA 3. Checksum Sum 0x182B0
Number of DCbitless LSA 0
Number of indication LSA 0
Number of DoNotAge LSA 0
Flood list length 0
```

The show ip ospf command on the Router ABR-2 shows that it connects the backbone area and Area 2. Other details about ABR-2 are also available in the output. Listing 18.10 shows the output for the Router Internal#show ip ospf command.

LISTING 18.10 Output for Router Internal# show ip ospf Command

```
Router Internal# show ip ospf
Routing Process "ospf 1" with ID 192.168.4.1
Supports only single TOS(TOS0) routes
SPF schedule delay 5 secs, Hold time between two SPFs 10 secs
Minimum LSA interval 5 secs. Minimum LSA arrival 1 secs
Number of external LSA 0. Checksum Sum 0x0
Number of DCbitless external LSA 0
Number of DoNotAge external LSA 0
Number of areas in this router is 1. 0 normal 1 stub 0 nssa
External flood list length 0
Area 2
Number of interfaces in this area is 2
It is a stub area
Area has no authentication
SPF algorithm executed 8 times
Area ranges are
Number of LSA 3. Checksum Sum 0x182B0
Number of DCbitless LSA 0
Number of indication LSA 0
Number of DoNotAge LSA 0
Flood list length 0
```

The show ip ospf command on the Router IR shows that the router is located at Area 2. Other details about the IR are also available in the output.

19 Redistribution Case Study

Consider a company, XYZ Inc., which has a mixed network of legacy routers running RIP and new advanced routers running OSPF. XYZ Inc. is also running BGP with its Internet Service Provider, ISP A. This example is depicted in Figure 19.1.

ASN 6666 represents company XYZ, and ISP A is represented by ASN 7777. Within AS 6666, OSPF-enabled routers are X1, X3, X4, X5, and X6. The RIP-enabled routers are X1, X6, X7, and X8. Both X1 and X6 are part of both RIP and

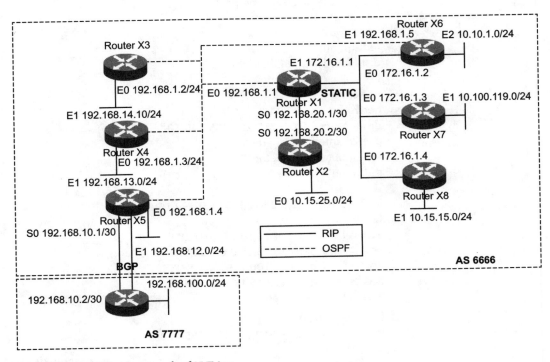

FIGURE 19.1 Mixed network of XYZ Inc.

OSPF routing domains, but redistribution between the two routing protocols is configured in X1. BGP is running between X5 of AS 6666 and A1 of AS 7777.

Table 19.1 shows the configuration for the different routers used in the network.

TABLE 19.1   Routers and Configurations

Router	Configuration
X1	interface Serial0
	ip address 192.168.20.1 255.255.255.252
	!
	interface Ethernet0
	ip address 192.168.1.1 255.255.255.0
	!
	interface Fast Ethernet1
	ip address 172.16.1.1 255.255.255.0
	!
	router ospf 1
	network 192.168.1.0 0.0.0.255 area 5
	passive-interface FastEthernet1
	passive-interface Serial0
	redistribute static metric 20
	redistribute rip metric 20
	distribute-list 2 in Ethernet0
	!
	router rip
	network 172.16.1.0
	distance 95 0.0.0.0 255.255.255.255 10
	passive-interface Ethernet0
	redistribute static metric 2
	redistribute ospf metric 2
	!
	ip route 10.15.25.0 255.255.255.0 192.168.20.2
	!
	access-list 10 permit 10.10.1.0 0.0.0.255
	access-list 2 deny 10.15.15.0 0.0.0.255
	access-list 2 deny 10.100.119.0 0.0.0.255
	access-list 2 deny 172.16.1.0 0.0.0.255
	access-list 2 permit any
X2	interface Serial0
	ip address 192.168.20.2 255.255.255.252
	!

(continued)

TABLE 19.1 *(continued)*

Router	Configuration
	interface Ethernet0 ip address 10.15.25.1 255.255.255.0 ! ip route 0.0.0.0 0.0.0.0 192.168.20.1 !
X3	interface Ethernet0 ip address 192.168.1.2 255.255.255.0 ! interface Ethernet1 ip address 192.168.14.1 255.255.255.0 ! router ospf 1 network 192.168.1.0 0.0.0.255 area 5 network 192.168.14.0 0.0.0.255 area 5 !
X4	interface Ethernet0 ip address 192.168.1.3 255.255.255.0 ! interface Ethernet1 ip address 192.168.13.1 255.255.255.0 ! router ospf 1 network 192.168.1.0 0.0.0.255 area 5 network 192.168.13.0 0.0.0.255 area 5 !
X5	interface Ethernet0 ip address 192.168.1.4 255.255.255.0 ! interface Ethernet1 ip address 192.168.12.1 255.255.255.0 ! interface Serial0 ip address 192.168.10.1 255.255.255.252 ! router ospf 1 network 192.168.1.0 0.0.0.255 area 5 network 192.168.12.0 0.0.0.255 area 5

(continued)

TABLE 19.1 *(continued)*

Router	Configuration
	redistribute bgp 6666 ! router bgp 6666 neighbor 192.168.10.2 remote-as 7777 redistribute ospf route-map select-routes ! route-map select-routes permit 10 match ip address 50 ! access-list 50 permit 10.100.119.0 0.0.0.255 access-list 50 permit 192.168.14.0 0.0.0.255
X6	interface Ethernet0 ip address 192.168.1.5 255.255.255.0 ! interface FastEthernet1 ip address 172.16.1.2 255.255.255.0 ! interface Ethernet2 ip address 10.10.1.1 255.255.255.0 ! router ospf 1 network 192.168.1.0 0.0.0.255 area 5 network 10.10.1.0 0.0.0.255 area 5 ! router rip network 172.16.1.0 network 10.10.1.0 passive-interface Ethernet0 !
X7	interface Ethernet0 ip address 172.16.1.3 255.255.255.0 ! interface Ethernet1 ip address 10.100.119.1 255.255.255.0 ! router rip network 172.16.1.0 network 10.100.119.0 !

(continued)

TABLE 19.1   *(continued)*

Router	Configuration
X8	interface Ethernet0 ip address 172.16.1.4 255.255.255.0 ! interface Ethernet1 ip address 10.15.15.1 255.255.255.0 ! router rip network 172.16.1.0 network 10.15.15.0 ! A1 ! interface Serial0 ip adress 192.168.10.2 255.255.255.252 ! interface Ethernet0 ip adress 192.168.100.2 255.255.255.0 ! router bgp 7777 reighbor 192.168.10.1 remote-as 6666 network 192.168.100.0 !

X1 receives two routes corresponding to network 10.10.1.0/24—a RIP route via 172.16.1.2 and an OSPF route via 192.168.1.5. The OSPF route is seleceted by default, because of a lower AD value. However, the RIP path is a higher bandwidth (Fast Ethernet) path as compared to the OSPF path (Ethernet). The AD value of the route 10.10.1.0/24 in X1 is modified to 95 via access-list 10. As a result, the RIP route is selected.

Access-list 2 is used in X1 to filter OSPF updates received on Ethernet 0, to avoid routing loops. X1 and X2 are connected by a WAN link, and a static route for network 10.15.25.0/24 is pointed to X2 from X1. This network, in turn, is redistributed into RIP with a metric of 2 and OSPF with a metric of 20. Also in X1, RIP and OSPF routes are being redistributed to each other.

X5, which belongs to AS 6666, is running BGP with A1 in AS 7777. In X5, only certain OSPF routes are being redistributed into BGP. The selection of routes to be redistributed into BGP is being done using the route-map select-routes and access-list 50. Also in X5, BGP learned routes are being redistributed into OSPF. Listing 19.1 shows the routing table for X8.

LISTING 19.1   Routing Table of X8

```
RouterX8# show ip route
Codes: C - connected, S - static, I - IGRP, R - RIP, M - mobile, B - BGP
   D - EIGRP, EX - EIGRP external, O - OSPF, IA - OSPF inter area
   N1 - OSPF NSSA external type 1, N2 - OSPF NSSA external type 2
   E1 - OSPF external type 1, E2 - OSPF external type 2, E - EGP
   i - IS-IS, L1 - IS-IS level-1, L2 - IS-IS level-2, * - candidate
      default
   U - per-user static route, o - ODR

Gateway of last resort is not set
C   10.15.15.0/24 is directly connected, Ethernet1
C   172.16.1.0/24 is directly connected, Ethernet0
R   10.100.119.0/24 [120/1] via 172.16.1.3, 00:00:16, Ethernet0
R   10.10.1.0/24 [120/1] via 172.16.1.2, 00:00:16, Ethernet0
R   10.15.25.0/24 [120/2] via 172.16.1.1, 00:00:16, Ethernet0
R   192.168.14.0/24 [120/2] via 172.16.1.1, 00:00:16, Ethernet0
R   192.168.13.0/24 [120/2] via 172.16.1.1, 00:00:16, Ethernet0
R   192.168.12.0/24 [120/2] via 172.16.1.1, 00:00:16, Ethernet0
R   192.168.100.0/24 [120/2] via 172.16.1.1, 00:00:16, Ethernet0
R   192.168.1.0/24 [120/2] via 172.16.1.1, 00:00:16, Ethernet0
R   192.168.10.0/30 [120/2] via 172.16.1.1, 00:00:16, Ethernet0
```

Listing 19.2 shows the routing table for X5.

LISTING 19.2   Routing Table of X5

```
RouterX5# show ip route
Codes: C - connected, S - static, I - IGRP, R - RIP, M - mobile, B - BGP
   D - EIGRP, EX - EIGRP external, O - OSPF, IA - OSPF inter area
   N1 - OSPF NSSA external type 1, N2 - OSPF NSSA external type 2
   E1 - OSPF external type 1, E2 - OSPF external type 2, E - EGP
   i - IS-IS, L1 - IS-IS level-1, L2 - IS-IS level-2, * - candidate
      default
   U - per-user static route, o - ODR

Gateway of last resort is not set
C   192.168.12.0/24 is directly connected, Ethernet1
C   192.168.1.0/24 is directly connected, Ethernet0
C   192.168.10.0/30 is directly connected, Serial0
O   192.168.13.0/24 [110/10] via 192.168.1.1, 00:33:19, Ethernet0
```

```
O   192.168.14.0/24 [110/10] via 192.168.1.2, 00:33:19, Ethernet0
O   10.15.25.0/24 [110/20] via 192.168.1.1, 00:33:19, Ethernet0
O   10.10.1.0/24 [110/20] via 192.168.1.1, 00:33:19, Ethernet0
O   10.100.119.0/24 [110/20] via 192.168.1.1, 00:33:19, Ethernet0
O   10.15.15.0/24 [110/20] via 192.168.1.1, 00:33:19, Ethernet0
O   172.16.1.0/24 [110/20] via 192.168.1.1, 00:33:19, Ethernet0
B   192.168.100.0/24 [200/0] via 192.168.10.2, 07:56:03
```

Listing 19.3 shows the routing table for A1.

LISTING 19.3   Routing Table of A1

```
RouterA1# show ip route
Codes: C - connected, S - static, I - IGRP, R - RIP, M - mobile, B - BGP
   D - EIGRP, EX - EIGRP external, O - OSPF, IA - OSPF inter area
   N1 - OSPF NSSA external type 1, N2 - OSPF NSSA external type 2
   E1 - OSPF external type 1, E2 - OSPF external type 2, E - EGP
   i - IS-IS, L1 - IS-IS level-1, L2 - IS-IS level-2, * - candidate
      default
   U - per-user static route, o - ODR

Gateway of last resort is not set
C   192.168.100.0/24 is directly connected, Ethernet0
C   192.168.10.0/30 is directly connected, Serial0
B   10.100.119.0/24 [200/0] via 192.168.10.1, 07:56:03
B   192.168.14.0/24 [200/0] via 192.168.10.1, 04:56:03
```

20 RIP Case Study

Rigmore Plaster, an organization that manufactures different types of plaster, has three departments, Management, Operations, and Materials. These are interconnected using RIP. While the rest of the organization is on various subnets of 172.16.0.0/16, the Management router uses a 10.0.0.0/21 subnet on its LAN for security reasons. RIP, being a classful routing protocol, does not carry any masking information. Figure 20.1 shows the configuration of Rigmore Plaster.

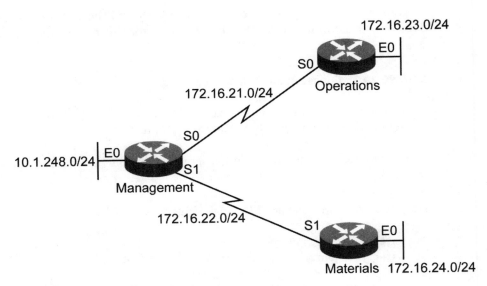

FIGURE 20.1 RIP configuration using different classes of networks.

RIP is configured in each of the three routers using the commands:

Router(config)# router rip
Router(config-router)# network ip-address

The network command should specify classful addresses as its argument, because RIP does not carry any masking information. Instead of 172.16.21.0, use the classful network address of 172.16.0.0 as an argument to the network command.

The Management router has a LAN configured on 10.1.248.0/21, and the serial interfaces are configured on 172.16.21.0/24 and 172.16.22.0/24. These networks are represented by 10.0.0.0 and 172.16.0.0, respectively, because the network command of RIP should include the classful address. This is shown in Listing 20.1.

LISTING 20.1   Configuration of Management Router

```
Management(config)# router rip
Management(config-router)# network 172.16.0.0
Management(config-router)# network 10.0.0.0
```

Similarly, the Operations router has its LAN and the WAN interfaces on the 172.16.0.0 classful network. The configuration for this router is shown in Listing 20.2.

LISTING 20.2   Configuration of Operations Router

```
Operations(config)# router rip
Operations(config-router)# network 172.16.0.0
```

The Materials router has a similar setup, having the configuration given in Listing 20.3.

LISTING 20.3   Configuration of Materials Router

```
Materials(config)# router rip
Materials(config-router)# network 172.16.0.0
```

The routing table of the Management router has separate entries for reaching subnets 172.16.23.0/24 and 172.16.24.0/24, instead of a single classful entry of 172.16.0.0/16. This is because the router Management has interfaces attached to the network 172.16.0.0. Operations and Materials do not have any interface attached to the 10.0.0.0 network. As a result, they receive the classful summarized address of 10.0.0.0 from the Management router.

Distance vector protocol follows the split-horizon rule, whereby it does not send out routing updates of networks from the same interface where the protocol has learned the routing updates. This behavior in RIP, a distance vector protocol, can also be observed in this scenario by executing the debug ip rip command on either the Operations or Materials router. The advertisement sent out from the Operations router's Ethernet interface should not contain any entry for 172.16.23.0/24 because it learns about the subnet 172.16.23.0/24 from its Ethernet interface. This phenomenon of split-horizon in RIP can be observed by following the debugger output.

CONTROLLING RIP ADVERTISEMENTS

There has been a security violation in the Operations department of Rigmore Plaster. As a result, the management is taking safety measures. This is done by ensuring that different networks within the organization are not seen and accessible by LAN users of the Operations department until the security issue is resolved.

The RIP advertisement going out of an interface is controlled using the passive-interface command. There is no mechanism to include a serial interface for a RIP advertisement while excluding the other, because the Management router has both the serial interfaces on the 172.16.0.0 subnet. Therefore, the best option is to execute the passive-interface command on the required interface to listen only to incoming advertisements but not send out any outgoing advertisements. This scenario is depicted in Figure 20.2.

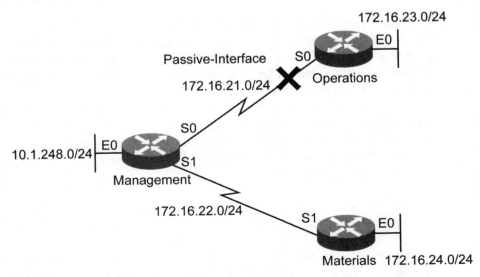

FIGURE 20.2   No RIP routing updates are sent between Operations and Management routers.

The configuration done at the Management router stops any outgoing RIP advertisement out of the Serial interface 0 connected to the Operations router. This is shown in Listing 20.4.

LISTING 20.4 Configuration to Stop Outgoing RIP Advertisements

```
Management(config)# router rip
Management(config-router)# passive-interface Serial0
```

The passive-interface command is also enabled at the Operations router to prevent the Management router from getting any routing updates for the network 172.16.23.0/24. This is shown in Listing 20.5.

LISTING 20.5 Enabling the passive-interface Command

```
Operations(config)# router rip
Operations(config-router)# passive-interface Serial0
Operations(config-router)# network 172.16.0.0
```

Rigmore Plaster uses the passive-interface command to control the outgoing RIP advertisements from any router interface.

CONTROLLING RIP TRAFFIC

Rigmore Plaster has hired an external consultant for three months to review its various management practices. During this time, it is estimated that there will be a huge flow of RIP traffic that may adversely affect the performance of the network. The network administrator has suggested the use of unicast traffic for periodic RIP updates to improve the network traffic.

By default, RIP works in the broadcast mode. To stop RIP from sending out a broadcast of a particular interface, the passive-interface command is used. The neighbor command is used to invoke unicast communication with a particular neighbor. This scenario is depicted in Figure 20.3. The syntax for neighbor and passive-interface commands are:

```
Router(config-router)# neighbor ip-address
Router(config-router)# passive-interface interface
```

The Management router has stopped all broadcasts over its Ethernet interface connecting to "Consultants" by the use of the passive-interface command.

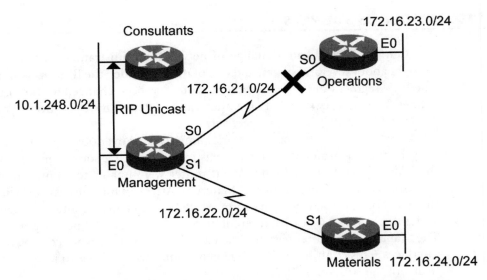

FIGURE 20.3   Management and consultants use the RIP unicast to bring down the network traffic.

Next, it only sends unicasts directed to the Ethernet interface address of the "Consultant" router. This reduces the network traffic generated by the RIP - advertisements. The configuration of the Management router is shown in Listing 20.6.

LISTING 20.6   Configuration of the Management Router

```
Management(config)# router rip
Management(config-router)# passive-interface Ethernet0
Management(config-router)# network 172.16.0.0
Management(config-router)# network 10.0.0.0
Management(config-router)# neighbor 10.1.248.2
```

Similarly, the Consultant router used the neighbor command to ensure unicast RIP update traffic with the Management router. This is shown in Listing 20.7.

LISTING 20.7   Unicast RIP Update Traffic with the Management Router

```
Consultant(config)# router rip
Consultant(config-router)# passive-interface Ethernet0
Consultant(config-router)# network 10.0.0.0
Consultant(config-router)# neighbor 10.1.248.1
```

DISCONTIGUOUS SUBNETS

Rigmore Plaster had a change of network administrators. The new person brought about several changes, and one of them included the IP address allocation. The LAN segment of the Management router has been changed to the 20.1.1.0/24 subnet. The LAN segments of Operations and Materials have also been changed to 10.1.1.0/24 and 10.2.2.0/24, respectively.

During changes, the new network administrator overlooked the fact that RIP v1 is the routing protocol used in the existing network. He soon found that users in the LAN segment of the Management router were unable to communicate with LAN users of both Operations and Materials. After careful analysis, he realized that there had been a major network loop created due to two discontiguous subnets of 10.1.1.0 and 10.2.2.0 at two different ends with a 172.16.0.0 network in between. The Management router sees the 10.0.0.0 major classful network coming from two interfaces. This confuses the Management router when it tries to forward a packet to any of the 10.1.1.0 or 10.2.2.0 subnets.

The network administrator at Rigmore Plaster had no way of reverting back the IP addresses. The only option left was to find a solution to this problem. The best solution was to use a routing protocol that carried the subnet mask information. RIP v2 is the optimal option.

Another solution would be to mislead the Management router into believing that it too has interfaces in the 10.0.0.0 network (when in reality it does not), and then it would have separate entries of 10.1.1.0 and 10.2.2.0 in its routing table. After it has separate entries for 10.1.1.0 and 10.2.2.0 subnets, it would not have to deal with the single entry of 10.0.0.0 coming from two different sources leading to a routing loop. In order to trick the Management router into having interfaces in the 10.0.0.0 network, the network administrator should configure secondary IP addresses. This scenario is shown in Figure 20.4.

The configuration of the secondary address and the subsequent advertisement of this newly created interface are shown in Listing 20.8.

LISTING 20.8   Configuration of the Secondary Address

```
Management(config)# interface serial0
Management(config-if)# ip address 10.3.3.3 255.255.255.0 secondary
Management(config)# interface serial1
Management(config-if)# ip address 10.4.4.4 255.255.255.0 secondary
Management(config)# router rip
Management(config-router)# network 172.16.0.0
Management(config-router)# network 10.0.0.0
```

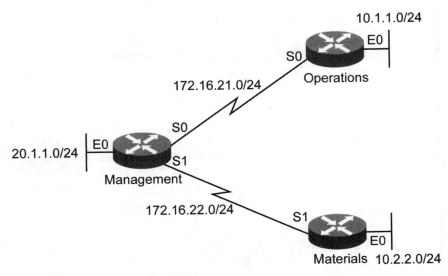

FIGURE 20.4   The LAN segments of Operations and Materials appear as
10.0.0.0 network to the Management router causing confusion.

A secondary interface needs to be created at the other end of the link at the
Operations and Materials router. This is done as shown in Listing 20.9.

LISTING 20.9   Creation of Secondary Interface

```
Operations(config)# interface serial0
Operations(config-if)# ip address 10.3.3.5 255.255.255.0 secondary
Operations(config)# router rip
Operations(config-router)# network 172.16.0.0
Operations(config-router)# network 10.0.0.0
Materials(config)# interface serial1
Materials(config-if)# ip address 10.4.4.5 255.255.255.0 secondary
Materials(config)# router rip
Materials(config-router)# network 172.16.0.0
Materials(config-router)# network 10.0.0.0
```

MD5 AUTHENTICATION

Rigmore Plaster observed that there were several attempts by unauthorized routers
to enter the RIP network. The network administrator did not take any chances and
introduced a secure MD5 authentication for RIP updates. This was to prevent any

unauthorized introduction of a router into the Rigmore Plaster network. Here, only the authorized person knows the password "key," and this would check for any new router trying to participate in the network. Figure 20.5 shows this scenario.

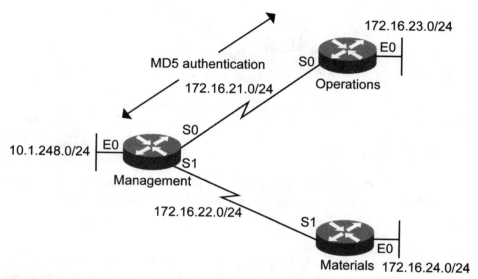

FIGURE 20.5 Management and Operations routers use RIP authentication for RIP updates.

MD5 is configured as shown in Listing 20.10

LISTING 20.10 Configuration of MD5

```
Router(config-if)# ip rip authentication key-chain name-of-key-chain
Router(config-if)# ip rip authentication mode {text | md5}
Router(config)# key-chain name-of-the-key-chain
Router(config-key-chain)# key number
```

MD5 authentication is not supported in RIP v1. However, RIP v2 supports this authentication and must be configured on the interface for RIP updates. Listing 20.11 shows the configuration of the Management router.

LISTING 20.11   Configuration of the Management Router

```
Management(config)# interface serial 0
Management(config-if)# ip rip authentication key-chain newchain
Management(config-if)# ip rip authentication mode md5
Management(config-if)# ip rip send version 2
Management(config-if)# ip rip receive version 2
Management(config)# key-chain newchain
Management(config-key-chain)# key cisco
```

Listing 20.12 shows the configuration of the Operations router.

LISTING 20.12   Configuration of the Operations Router

```
Operations(config)# interface serial 0
Operations(config-if)# ip rip authentication key-chain newchain
Operations(config-if)# ip rip authentication mode md5
Operations(config-if)# ip rip send version 2
Operations(config-if)# ip rip receive version 2
Operations(config)# key-chain newchain
Operations(config-key-chain)# key cisco
```

The RIP updates with the Management router are not authenticated because the Materials department never had an incident of unauthorized entries.

PATH SELECTION BY MODIFYING METRICS

Rigmore Plaster has a direct serial link between the Operations and Materials departments. The purpose of this link is to act as a backup to the primary link via the Management router.

The default behavior of RIP is to calculate metrics based on the hop-count. This would mean that Operations would take the direct link to reach Materials as opposed to the link via the Management router, because the hop-count is low. The metrics need to be modified to influence the path taken by RIP because this is against the network policy of Rigmore Plaster. This scenario is shown in Figure 20.6.

The command for achieving the same is Router(config-router)# offset-list {access-list-number | access-list-name} {in | out} offset [interface-type interface-name].

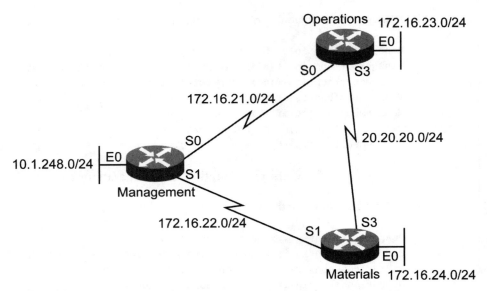

FIGURE 20.6 Operations and Materials have a serial link, which should act as a backup to the primary link.

The offset-list given in Listing 20.13 introduces an additional five hops for all incoming RIP updates that match the access list. All incoming routing updates having the network address 20.20.20.0 to Serial 3 of the Operations router will have their hop-counts incremented by 5. This increase in the hop-count makes the direct link to Materials appear longer than via the Management router. This makes the direct link from Operations to Materials act as a backup link to the link via the Management router.

LISTING 20.13 Offset-list Command at Operations Router

```
Operations(config)# access-list 1 permit 20.20.20.0 0.0.0.255
Operations(config)# router rip
Operations(config-router)# network 172.16.0.0
Operations(config-router)# network 20.0.0.0
Operations(config-router)# offset-list 1 in 5 Serial3
```

A similar configuration of the offset-list needs to be made at the Materials router. This is shown in Listing 20.14.

LISTING 20.14   Offset-list Command at Materials Router

```
Materials(config)# access-list 1 permit 20.20.20.0 0.0.0.255
Materials(config)# router rip
Materials(config-router)# network 172.16.0.0
Materials(config-router)# network 20.0.0.0
Materials(config-router)# offset-list 1 in 5 serial 3
```

21 Route Maps Case Study

Consider a mixed network running RIP and EIGRP routing protocols. Figure 21.1 depicts Routers A1 and A3 running RIP only. Routers A4, A5, and A6 are running EIGRP only. Router A2 is running both RIP and EIGRP and is redistributing between the two protocols.

In the network shown in Figure 21.1, route maps are configured for implementing policy routing and redistribution among the two routing protocols. Route filtering is also implemented in this network. The configuration for Router A1 is shown in Listing 21.1.

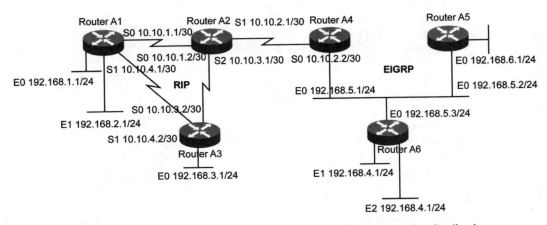

FIGURE 21.1 Configuring route maps for implementing policy routing and redistribution.

LISTING 21.1 Configuration of Router A1

```
interface Ethernet0
ip address 192.168.1.1 255.255.255.0
!
```

```
interface Ethernet1
ip address 192.168.2.1 255.255.255.0
!
interface serial0
ip address 10.10.1.1 255.255.255.252
!
interface serial1
ip address 10.10.4.1 255.255.255.252
!
router rip
network 192.168.1.0
network 192.168.2.0
network 10.10.1.0
network 10.10.2.0
!
```

The configuration of Router A2 is shown in Listing 21.2.

LISTING 21.2 Configuration of Router A2

```
interface serial0
ip address 10.10.1.2 255.255.255.252
!
interface serial1
ip address 10.10.2.1 255.255.255.252
!
interface serial2
ip address 10.10.3.1 255.255.255.252
!
router rip
network 10.10.1.00
network 10.10.2.0
network 10.10.3.0
passive-interface serial1
!
router eigrp 100
network 10.10.1.0
network 10.10.2.0
network 10.10.3.0
passive-interface serial0
passive-interface serial2
redistribute rip route map ALLOW2
!
route map ALLOW2 permit 10
match ip address 15
set metric 10000 200 250 2 1500
```

```
!
access-list 15 deny 192.168.1.0 0.0.0.255
access-list 15 permit any
!
```

The configuration of Router A3 is shown in Listing 21.3.

LISTING 21.3 Configuration of Router A3

```
interface Ethernet0
ip address 192.168.3.1 255.255.255.0
!
interface serial0
ip address 10.10.3.2 255.255.255.252
!
interface serial1
ip address 10.10.4.2 255.255.255.252
ip policy route map ALLOW1
!
router rip
network 192.168.3.0
network 10.10.3.0
network 10.10.4.0
!
route map ALLOW1 permit 10
match ip address 10
set ip next-hop Null 0
!
access-list 10 permit 192.168.2.0 0.0.0.255
```

The configuration of Router A4 is shown in Listing 21.4.

LISTING 21.4 Configuration of Router A4

```
interface Ethernet0
ip address 192.168.5.1 255.255.255.0
!
interface serial0
ip address 10.10.2.2 255.255.255.252
!
router eigrp 100
network 192.168.5.0
network 10.10.2.0
!
```

The configuration of Router A5 is shown in Listing 21.5.

LISTING 21.5 Configuration of Router A5

```
interface Ethernet0
ip address 192.168.5.2 255.255.255.0
!
interface Ethernet0
ip address 192.168.6.1 255.255.255.0
!
router eigrp 100
network 192.168.5.0
network 192.168.6.0
!
```

The configuration of Router A6 is shown in Listing 21.6.

LISTING 21.6 Configuration of Router A6

```
interface Ethernet0
ip address 192.168.5.3 255.255.255.0
!
interface Ethernet1
ip address 192.168.4.1 255.255.255.0
!
interface Ethernet2
ip address 192.168.10.1 255.255.255.0
!
router eigrp 100
network 192.168.5.0
network 192.168.4.0
network 192.168.10.0
distribute-list 1 in Ethernet0
distribute-list 2 out Ethernet0
!
access-list 1 deny 192.168.3.0 0.0.0.255
access-list 1 permit any
!
access-list 2 permit  192.168.4.0 0.0.0.255
!
```

Policy routing is configured in A3 using the ALLOW1 route map. Packets belonging to network 192.168.2.0/24 arrive via Serial 1 interface of A3. These packets are dropped without being forwarded. Route map ALLOW2 has been configured in A2 to allow selective redistribution into EIGRP from RIP. The RIP route

192.168.1.0/24 is denied redistribution into EIGRP, whereas all other routes are re-distributed.

Route filtering has been configured at A6 in both inbound and outbound directions, corresponding to interface Ethernet 0. The outbound update only contains the route 192.168.4.0/24. The other available route at Router A6, 192.168.10.0/24, is not a part of the advertisements. In the inbound direction, the route 192.168.3.0/24 is not installed on the routing table, even if it is received in routing updates.

Appendix ■ **About the CD-ROM**

The CD-ROM included with *Enabling IP Routing with Cisco Routers* includes all the code listings, tables, and images from the various examples found in the book. In addition, this CD-ROM includes an exhaustive Question Bank for the readers to test their knowledge as well as some demo software.

CD-ROM FOLDERS

Listings: Contains all the code listings from examples covered in each chapter.

Figures: Contains all the images from each chapter.

Tables: Contains all the tables from each chapter.

Question Bank: Contains multiple choice questions to test the learner's knowledge on Cisco Designing concepts.

Demo Software: Contains the Boson Router Simulator and WildPackets Network Calculator 3.2.1.

OVERALL SYSTEM REQUIREMENTS: RECOMMENDED

- Windows® 98, Windows NT, Windows 2000, or Windows XP Professional
- Pentium III processor or greater
- CD-ROM drive
- Hard drive
- 128 MB of RAM, 256 MB recommended
- 20 MB of hard drive space for the code examples

Instructions for use of the QuestionBank

1. Insert the CD-ROM in the CD-ROM drive
2. Double-click the QuestionBank folder
3. Double-click the Enabling-Setup.zip file
4. Extract the setup files in a folder (example, create a folder called EnablingQB in your C, D, or any other hard disk drive).
5. Double-click the EnablingQB folder
6. Double-click the Enabling-Setup folder
7. Double-click the setup.exe file
8. Follow the setup instructions
9. Once the setup is over, you can access the Question Bank from the Start‡ Programs Enabling IP Routing with Cisco Routers

OVERVIEW AND INSTALLATION FOR THE DEMO SOFTWARE

WildPackets Network Calculator 3.2.1

The Network Calculator™ is a multi-function Windows 95/98/NT/ME/2000 utility for the network analysis professional. The Network Calculator includes the Hexpert Calculator (binary-decimal-hexadecimal base conversion), the IP Subnet Calculator (subnet/supernet address computation), and the Latency Calculator (latency calculation across multiple LANs and WANs) in one, convenient program. To install double Click on the Network Calc 3.2.1 application file and follow the on screen instructions. In order to unlock the application, you will need to use the password of "nothing.but.net" without the quotes.

System Requirements for the Network Calculator

- Windows 95/98/ME/NT/2000 or XP is required
- Pentium 90 or better processor is recommended.

Boson NetSim 5.27

The **Boson NetSim™**, which includes the Boson® Router Simulator®, was designed to help you develop the skills necessary to pass the new CCNA® ICND (640-811) and CCNA® Intro (641-821) exam. The combination of Cisco® IOS® and Catalyst® simulated command line interface (CLI), establishes a foundation of knowledge that is integral to your success, both in taking the exam and in the I.T. profession. The **Boson NetSim™** includes a drag-and-drop Network Designer that supports up to 200 devices using 47 different device models. You can even have up to 200 active telnet sessions, configuring devices inside the simulator. There are no other aftermarket router simulators that offer this much versatility and support. The **Boson NetSim**™ also includes a comprehensive lab menu that contains 60 lessons and labs covering Routing Protocols, Cisco® Devices, Switching, Topological Design, and Much More. To install double click on the Boson NetSim 5.27 application file and follow the on screen instructions.

System Requirements for the Boson NetSim

Supported Platforms:

■ Windows™ 98/ME/NT/2000/XP

Installation Requirements for the Boson NetSim

■ 20MB Free Space
■ 64MB RAM
■ 800 × 600 Screen Resolution
■ Active Internet connection

Index